Mining Impacts on the Environment in the Central African Copperbelt of Zambia and the Democratic Republic of Congo

Ambayeba Muimba-Kankolongo •
Célestin Banza Lubaba Nkulu •
Jacob Mwitwa • Florence Kampemba
Mujinga • Misery Mulele Nabuyanda

Mining Impacts on the Environment in the Central African Copperbelt of Zambia and the Democratic Republic of Congo

Springer

Ambayeba Muimba-Kankolongo
Ottawa, ON, Canada

Jacob Mwitwa
Kapasa Makasa University
Chinsali, Zambia

Misery Mulele Nabuyanda
School of Mines and Mineral Sciences
Copperbelt University
Kitwe, Zambia

Célestin Banza Lubaba Nkulu
Faculty of Medicine, School of Public Health, Toxicology and Environment Unit
University of Lubumbashi
Lubumbashi, Democratic Republic of the Congo

Florence Kampemba Mujinga
Faculty of Agricultural Sciences, Department of Animal Sciences
University of Lubumbashi
Lubumbashi, Democratic Republic of the Congo

ISBN 978-3-031-96785-6 ISBN 978-3-031-96786-3 (eBook)
https://doi.org/10.1007/978-3-031-96786-3

This Springer imprint is published by the registered company Springer Nature Switzerland AG
The registered company address is: Gewerbestrasse 11, 6330 Cham, Switzerland

We dedicate this book to the resilient souls dwelling in both rural and urban communities around the mines and metallurgic industries in the central African Copperbelt of Zambia and the Democratic Republic of Congo. In solidarity and with boundless admiration and heartfelt gratitude, this book is a tribute to your unwavering spirit and undying commitment to your livelihoods. It is with deep empathy and utmost admiration that this book on the pollution to your environment and ecosystems by the mining and metallurgic industries is dedicated to you. Within this book, we shed light on the profound sacrifices and hardships you endure from the adversity of mining and metallurgic industries' salient impacts in the pursuit of your households' life. As the relentless wheels of national economic progress, it is often you who bear the burden of the environmental consequences. Year after year you face injustices from your daily cries which have not been heard by government officials. The toll on your health, the loss of your lands, and the erosion of your cultural heritage are not to be taken lightly. You, the guardians of nature's beauty and integrity, deserve recognition for the challenges you face daily. Your lives, once harmoniously intertwined with nature, now bear the weight of disruption, pollution, and environmental degradation. The lands, once original and abundant, now bear witness to the scars left by industrialization. The air, once pure and refreshing, now hangs heavy with pollutants, and your water, once crystal clear, now flows tainted with the remnants of progress. But through it all, you continue persisting. Your unwavering determination to provide for your families and communities shines brightly amidst the darkest of skies. You are the embodiment of resilience, fighting for a better future despite the adversities you face. This book dedicated to you serves as an ideal for hope and a catalyst for change. It is our fervent wish that through the understanding and awareness about the catastrophic environmental degradation, all stakeholders can work together toward finding sustainable solutions to the challenges you endure.

Foreword by Paul Chisale

It is well recognized that mining and metallurgic industries have a history of polluting the environment and causing harmful effects on communities. Their activities have various negative impacts on the environment, such as water and soil contamination, deforestation, and habitat destruction. These activities also lead to air pollution, noise pollution, and the release of toxic substances into the environment, affecting resident's health and the overall well-being. As their operations often involve the use of chemicals, heavy machinery, and the excavation of large areas of land, they further disrupt ecosystems and harm local wildlife. Additionally, the runoff from mining and metallurgy sites contaminate nearby water sources, affecting both aquatic life and communities that rely on these sources for drinking water and agriculture.

As the Chancellor of the Copperbelt University (CBU), I am honored to write this foreword for the book *Mining Impacts on the Environment in the Central African Copperbelt of Zambia and the DR Congo*, a ground-breaking work that shines a much-needed light on one of the most pressing global issues of our time. In an era of unbridled industrialization and resource extraction, the human and environmental toll of mining operations have far too often been overlooked or downplayed. A team of experts from CBU in Kitwe, Zambia, and UNILU in Lubumbashi, in the DR Congo, had conducted a meticulous research and wrote a powerful storytelling that has revealed the profound and lasting damage that mining and metallurgic industries′ pollution inflicts on vulnerable communities around the central African Copperbelt region.

Through in-depth case studies and rigorous data analysis, this book lays bare the devastating effects of contaminated water resources, ambient air, cultivated food crops, and the local land. We witness firsthand how these environmental harms translate into the destruction of traditional livelihoods, food insecurity, and severe public health crises. Perhaps most importantly, the authors have provided a clear call for good industrial governance and accountability, highlighting the urgent need for the extractive companies and both national and provincial governments to prioritize the well-being of local populations over short-term margins than financial benefits.

As an institution of higher learning committed to advancing knowledge and catalyzing positive social change, CBU is proud to add its voice in support of this essential work. "Trace Metals Silent Threats in the Central African Copperbelt of Zambia and the DR Congo: DelPHE Project Catalogs Salient Impacts" is a courageous and vital contribution to the global dialogue on environmental justice, human rights, and sustainable development. I am confident that it will inspire readers in Zambia and the DR Congo, and around the world, to demand more ethical and responsible mining practices, and to stand in solidarity with the communities bearing the brunt of this crisis. Moreover, I believe that this book will serve as a crucial resource for policymakers, environmental organizations, and the global community in their efforts to address the challenges posed by trace metal pollution in the Central African Copperbelt. By shedding light on this critical issue, we can work together to safeguard the health and well-being of the people who call this region home and ensure a more sustainable future for all.

CBU Vice Chancellor Paul Chisale
Riverside, Kitwe, Zambia

Foreword by Professor Gilbert Kishiba Fitula

Mining has been a fundamental pillar of human civilization, a source of valuable resources that propelled societies forward. From the Bronze Age to the Information Age, the extraction of minerals has been the backbone of technological advancements and economic growth. However, as we gaze upon this illustrious history, we must not overlook the darker side of mining—the contamination and pollution that often accompany this industrial sector. In the heart of African continent lies the Copperbelt, a region spanning across the borders of Zambia and the Democratic Republic of the Congo (DR Congo). For over a century, these rich lands endowed with copper and cobalt deposits have attracted numerous mining companies from around the globe. Yet, while these minerals have brought prosperity to the two countries, they have also left behind a bittersweet legacy in the form of contamination and pollution.

It is with great concern and determination that we delve into this issue through this book, shining light on the environmental and socio-economic impacts of mining in the African Copperbelt region. Through meticulous research and personal testimonies, we aim to illustrate the gravity of the situation and advocate for sustainable solutions that will safeguard the livelihoods of those living in this region for generations to come. The chapters of this book will guide you through a journey beginning with the history of mining in the Copperbelt, an exploration that traces the highs and lows of the industry's development. We will then explore into the nature of contamination and pollution caused by mines, discussing their devastating consequences on the environment, ecosystems, human health, and the local communities.

With expert analysis and research-driven insights, we will uncover the intricate systems of mining governance, and regulations that have struggled to address these issues effectively. Furthermore, we will draw attention to the need for a robust framework to ensure responsible mining practices that mitigate contamination and pollution from the source. While this book may expose the grim reality of mining in the African Copperbelt region, it is not intended to condemn the industry altogether. Rather, it strives to foster a dialogue among stakeholders—governments, mining companies, communities, and environmental activists—to enact change and explore innovative approaches that strike a balance between economic interests and environmental protection.

The future of the African Copperbelt region lies in our hands, and it is incumbent upon us to take responsibility for the harms caused by mining and seek harmonious solutions. We owe it to the people of Zambia and the DR Congo, whose lives and lands have been deeply impacted by the pursuit of valuable minerals. As you embark on this journey through the pages that follow, I implore you to approach the subject matter with an open mind and a compassionate heart. Let us not turn a blind eye to the contamination and pollution that persist in the African Copperbelt region, but rather, let us collectively work towards a more sustainable and equitable future for all those affected.

Chancellor, University of Lubumbashi (UNILU) Gilbert Kishiba Fitula
Democratic Republic of Congo

Acknowledgments

We sincerely thank Jhonnah Mundike, R. Nasilele Mutafela, P. Chishimba, C. Manase and Womba Kaumba at CBU in Zambia, and Ilunga Mwanabute and F. Ilunga Kabeya at the "Institut Superieur de Pedagogie" in the DR Congo for field and laboratory assistance in the collection and preliminary sample processing. The involvement of several township and village inhabitants for interview and provision of samples for the study is greatly acknowledged. We are also grateful to CBU in Zambia and the University of Lubumbashi in the DR Congo for hosting and executing the project, and to Professor Erik Smolders of the Division of Soil and Water Management, Department of Earth and Environmental Sciences, Faculty of Bioscience Engineering, KU Leuven, Leuven, Belgium, and Benoit Nemery of the Centre for Environment and Health, Department of Public Health and Primary Care, Faculty of Medicine, KU Leuven, Leuven, Belgium, for their invaluable support and expertise in the final sample processing and analysis as well as analyzing the data for this research. Their insightful contributions and guidance were instrumental in shaping the findings presented in this book. Most of the book photographs were taken by Ambayeba Muimba-Kankolongo while others were provided by Professors Celestin Banza Lubaba Nkulu, Jacob Mwitwa, and Florence Kampemba. The views expressed in this book are those of the authors and not necessarily those of the University of Lubumbashi and the Copperbelt University, and the various literatures which were consulted and thus cited.

Competing Interests The authors have no competing interests to declare that are relevant to the content of this manuscript.

Contents

Chapter 1
General Introduction

Keywords Mining · Copperbelt · Environment · DelPHE project · Partnership · Metallurgy · DR Congo · Zambia

1.1 Overview

The conception to prepare and write this book arose due to a growing recognition of the pervasive and often overlooked effects of mining and metallurgic industry pollution on community residents in the central African Copperbelt region. As environmental degradation and pollution levels continue to rise, more and more people are experiencing first-hand serious consequences on their health, well-being, food and nutrition security, and economic stability. Inhabitants have loudly cried out to political leaders, but not one has heard them. There is, therefore, a critical need to bring out attention to the disproportionate impact of pollution on marginalized communities, who often bear the brunt of environmental harm without adequate resources or support to mitigate the effects.

The Central African Copperbelt, also described as the Lufilian Arc, is a large area of land located in the southern part of the former Katanga Province in the Democratic Republic of Congo (DR Congo) and the Copperbelt and Northwestern Provinces in Zambia. This region is one of Africa's largest mining areas, extending over more than 500 km from Kolwezi in the DR Congo to Luanshya in Zambia (Butler 2006), and it holds substantial copper and cobalt mineral reserves (Lydall and Auchterlonie 2011). These minerals are crucial to the economies of both countries (Wolfire et al. 1998; Ashton et al. 2001; MFNP, GRZ, 2002; Ministry of Mining and Hydrocarbons 2003; Global Witness 2006; Fraser and Lungu 2007; Mwitwa and Kabemba 2007; ICMM, 2014). Mining exploitation contributes about 28% to the DR Congo's GNP and 70% of its export value (Ministry of Mining and Hydrocarbons 2003), making it the main source of foreign currencies (Global Witness 2006). The mineral deposits are found in sedimentary rocks known as the Mines Group (Lower Roan Group), which are between 1050 million and 650 million years old.

A. Muimba-Kankolongo et al., *Mining Impacts on the Environment in the Central African Copperbelt of Zambia and the Democratic Republic of Congo*,
https://doi.org/10.1007/978-3-031-96786-3_1

For the last several years, the DR Congo mainly in the Haut Katanga Province, there has been a mining surge, bringing job opportunities and economic growth to the population. The country now produces around half of the world's cobalt and is Africa's top copper producer. The copper and cobalt mined in the DR Congo are essential for various consumer electronics and industrial applications worldwide. Each year, the mining and metallurgical sectors greatly contribute to the global availability of these minerals. The increase in mining activity has created employment opportunities and stimulated economic progress for the local communities.

The DR Congo holds over half of the world's cobalt reserves and is the largest producer of cobalt, with nearly 70% of the world's cobalt mined there in 2022 (Statista 2025) and is also the third-largest producer of copper. Cobalt is a by-product of copper mining in the DR Congo, and the extraction of both minerals is crucial for the economy. In 2019, copper and cobalt accounted for 86.4% of the DR Congo's total exports, and the mining sector provided nearly a quarter of the country's employment. Despite its mineral wealth, the country remains among the poorest in the world, with 62% of its population, or about 60 million people, living below the international poverty line of $2.15 a day in 2022 (Mufungizi 2025).

The DR Congo is also home to half of Africa's forests and water resources, giving it significant environmental potential amid global climate change (Butler 2006). However, the country faces challenges like deforestation, species depletion, heavy metal pollution, and land degradation from mining activities. These environmental issues have serious implications for the Congolese people. For example, around 51 million Congolese lack access to clean drinking water, and Congolese children are among the most vulnerable globally to climate and environmental shocks, ranking ninth worldwide.

The town of Kolwezi, for instance, which is the capital city of Lualaba province and a mining town in southern DR Congo, is at the center of the country's immense cobalt industrial production. About 88% of the DR Congo's cobalt production comes from industrial mines in Kolwezi and surrounding areas, with the rest from artisanal mining. Both industrial and artisanal cobalt mining are closely linked with copper production, and these minerals are vital for the transition to cleaner energy sources. This report mainly focuses on cobalt, given its essential role in the global push for cleaner energy.

Several of the industrial mines in the DR Congo are run through partnerships between the state-owned company Gécamines and private foreign mining firms that began in the late 1990s. Over the past 15 years, Chinese companies, encouraged by their government to invest in mineral resources and infrastructure, have joined North American and European firms in the DR Congo. The 2008 China-DR Congo "minerals for infrastructure" deal saw China's state-owned banks invest $3 billion in Congolese infrastructure projects in exchange for mineral rights to copper and cobalt. Today, Chinese investment supports about 70% of the DR Congo's mining sector. Meanwhile, North American and European countries are also vying for critical minerals, developing strategies to increase their interests in Congolese copper and cobalt. The US Inflation Reduction Act of 2022 includes measures to strengthen America's clean energy supply chain, offering incentives for domestic manufacturing of EV rechargeable batteries and aiming to reduce reliance on Chinese refining and manufacturing. To support this strategy, the USA is enhancing its relationship with the DR Congo and Africa to secure access to critical minerals. For example, on December 13, 2022, the USA signed a Memorandum of Understanding with the DR Congo and Zambia to develop a value chain for EV battery production.

1.2 Context of the Book

The reasoning behind the study, which forms the basis of this book, emphasized the environmental and health challenges resulting from mining and metallurgic activities. These operations have had a profound impact on the farmland, ambient air and water, biodiversity, and the well-being of the residents in mining towns located in both Zambia and the DR Congo. Within these areas, the two countries have entered a partnership with several mainly Chinese companies such as CREC and SINOHYDRO, backed by Exim Bank, focusing on two key projects: the development of infrastructure and a joint mining exploitation venture.

Nevertheless, during mineral exploitation, acid waste has reportedly been discharged into the various Rivers near the homesteads, leading to significant harm, according to local accounts. Residents in the affected surroundings have consistently reported daily struggles with environmental degradation and health concerns. In response to community concerns in the DR Congo, numerous NGOs like AFREWATCH, ASADHO, Premi, Global Witness, etc. have conducted field investigations to document the adverse environmental effects linked to mining's and metallurgical industries' activities.

In several areas, the industrial activities of mineral exploitation significantly contribute to pollutants entering aquatic environments, impacting coastal and terrestrial ecosystems. The pollution of rivers, lakes, and reservoirs by heavy metals and hydrophobic organic compounds poses a significant environmental concern, threatening freshwater resources in the DR Congo. Mining effluent waters are major sources of toxic metal contamination, affecting both water and sediment. Mining activities and artisanal mineral exploitation cause severe contamination of surface waters, sediments, groundwater, soils, and biota. These activities also negatively impact local environments and water resources through freshwater withdrawal and pollution from mine water discharge.

Equally, the mining sector is all important to Zambia's economy, being responsible of about 70% of foreign earnings (Zambia EITI Report 2020). From the colonial period to the present time, copper (Cu) and various other important minerals, notably copper (Cu), cobalt (Co), zinc (Zn), lead (Pb), gold (Au), diamonds (C), manganese (Mn), tin (Sn), tantalum (Ta), lithium (Li), nickel (Ni), and uranium (U), have been extracted from large-scale mines in the region and processed in local mine smelters. Since the late twentieth century, artisanal mining has also significantly contributed to cobalt production in the DR Congo (Yager 2014; KPMG

Global Mining Institute 2014; Banza Lubaba Nkulu et al. 2018) and in Zambia (Kambani 2003; Ministry of Lands and Natural Resources, Government of the Republic of Zambia 2018; Hilson 2020; Nachinanga Gubula Siaciti 2022).

The World Bank's report by Broadman (2007) highlighted on the recent increase in global demand for copper raw materials, driven by rapid economic growth in Asia, which has sparked the higher commodity prices and a climb for natural resource excavation across the entire continent of Africa. This has prompted African leaders to call for increased foreign investment in their respective countries, leading to the multiplication of mining exploration and exploitation, which now significantly shape the overall continent's economy. Consequently, the DR Congo and Zambia, both located in the Central African Copperbelt, are among the world's top ten leading copper producers, with rapidly increasing production volumes and substantial global investment in the sector (Global Business Reports 2014).

Years of mining extractions and metallurgical industry operations in the Central African Copperbelt of Zambia and the DR Congo have significantly undermined environmental sustainability and social equity. The negative impacts observed at each development stage of their operations comprise habitat and vegetation losses, social displacement, land alienation, increased soil erosion and siltation of nearby water bodies, ground surface disturbance, and the discharge of hazardous and contaminating elements into the natural bodies of water. These issues pose health risks, particularly for children, and release contaminants into the air, including heavy metals, organic compounds, and sulfur dioxide, all of which have adverse health effects. Industrial activities and mineral exploitation contribute significantly to the introduction of pollutants into aquatic environments, affecting coastal ecosystems where these pollutants are deposited. The pollution of rivers, lakes, and reservoirs by various contaminants—including heavy metals, hydrophobic organic compounds, and pathogenic organisms—poses a significant environmental concern and threatens freshwater resources worldwide (Vörösmarty et al. 2010; Haller et al. 2011; Thevenon et al. 2011, 2012; Thevenon and Poté 2012).

Mining effluent waters are major sources of toxic metal contamination in aquatic environments, affecting both water and sediment compartments. Mining activities and artisanal mineral exploitation can lead to dramatic contamination of surface waters, sediments, groundwater, soils, and biota. Additionally, these activities can negatively impact local environments and water resources through freshwater withdrawal and pollution from mine water discharge (Concas et al. 2006; Liu et al. 2011).

Despite these known facts, there is limited understanding of the effects of mining contamination on human populations living near and relying on contaminated rivers for drinking water, food, and livelihood (Miller et al. 2004).

The Copperbelt region undergoes several ecological problems associated with water pollution which have negative effects on human health and biodiversity, particularly in the neighborhood of mining and metallurgical industries. Environmental degradation, such as contamination of ground and surface water from industrial discharges and waste dumps, poses significant risks to human health, food security, and energy for the population (Feeney 2001; Zambia Consolidated Copper Mines-IH (ZCCM-IH) 2005; Kříbek et al. 2011; Ncube et al. 2012). Water pollutants adversely affect the growth and well-being of the population, especially children (Yabe et al. 2015; Banza Lubaba Nkulu et al. 2018; Musa Obadia et al. 2018; Nemery and Banza Lubaba Nkulu 2018; Human Rights Watch 2019; Umpula and Van Woudenberg 2024).

Mining-induced contamination of ambient air, both surface and underground water, and soil (dry and wetlands) has been a major concern since the colonial period. During this time, African housing was deliberately built close to mining facilities, while housing for whites was located further away (Larmer et al. 2021). Industrial effluents and waste have been released across the landscape, causing water flooding and erosion due to mine waste spills from broken pipes. Toxic smoke plumes from mining and metallurgical industries further contribute to the pollution of surface and underground water (Chipundu and Kunda 1994; Zambia Consolidated Copper Mines-IH (ZCCM-IH) 2005; Ncube et al. 2012; Mwaanga et al. 2019).

Of equally significant to the two countries is the agriculture sector which provides the necessary food and nutrition to the population. However, the population adjacent to mines in the Copperbelt of Zambia and the DR Congo are food unsecured and prone to hunger, malnutrition, and poverty. The nutritional quality of food consumed is compromised by heavy metal contamination from mining. Heavy metals are toxic and can negatively impact human health by causing diseases like cancer, lung and kidney dysfunction, and heart damage. They pollute water, soil, and environment and affect the growth pattern and the well-being more especially of children. They also lead to monetary losses by farmers where the contaminated produce is rejected at marketplaces due to stringent standards of permissible limits. Hence, this proposal intends to ascertain levels of contamination of food crops to improve food security and safety and enhance the health status and nutrition of people living around mines in DR Congo and Zambia. The project is structured in a manner that will promote standards for environmental quality by mines to enhance commodity production and health status of smallholder farmers by implementing mechanisms to meet food quality standards. The main focus is on strategic approaches which would facilitate the integration of smallholder growers into the national economy and pave ways that will promote agricultural production and commercialization in the DR Congo and Zambia.

1.3 Book Structure

The book examines the significant environmental impacts of mining activities in Zambia and the DR Congo, focusing on the multifaceted challenges posed to water quality, food security, and air quality. Through a comprehensive analysis, the work highlights how mining operations have led to pervasive pollution, adversely affecting local ecosystems and the health of surrounding communities. The degradation of water resources has resulted in contaminated water supplies essential for agriculture and human consumption, exacerbating food insecurity. Additionally, the loss of forest cover poses threats to biodiversity, undermining both the ecological balance and the livelihoods reliant on these natural resources. We delve into the health implications faced by residents linked to air and water pollution. The book aims to raise awareness of these critical issues and foster discussions on sustainable practices and policies that could mitigate the negative impacts of mining, ensuring the well-being of communities while preserving the environmental integrity of the DR Congo and Zambia. Through a combination of case studies, expert testimonies, and a review of current regulations, this work seeks to contribute to the broader conversation on responsible mining practices and environmental justice in the region.

Chapter 1: Introduction—The population adjacent to mines in the Copperbelt of Zambia and the DR Congo are food insecure and prone to hunger, malnutrition, and poverty. The nutritional quality of food consumed is compromised by heavy metal contamination from mining and metallurgical processes. Heavy metals are toxic and negatively impact human health by causing diseases like cancer, lung and kidney dysfunction, and heart damage. They pollute water, soil, and the environment, thereby affecting the human well-being more especially of children. They also lead to monetary losses by farmers where the contaminated produce is rejected at marketplaces due to stringent standards of permissible limits. Hence, the book intends to ascertain levels of contamination of food crops to improve food security and safety and enhance the health status and nutrition of people living around mines in the DR Congo and Zambia. The book is structured in a manner that will catalyze the promotion of acceptable mining standards that do not compromise environmental quality, commodity production, and the health status of smallholder farmers. The main focus is on strategic approaches that facilitate the integration of smallholder growers into the national economy and pave ways that promote agricultural production and commercialization in the DR Congo and Zambia.

Chapter 2: In this chapter, emphasis is on the highlights of the collaborative efforts among various stakeholders including scientists (agriculturists, public health professionals, foresters, policymakers, and mining specialists) from the two countries underscoring the importance of unity in addressing mining pollution and inspire readers to consider the power of partnerships in solving environmental challenges. The integration of environmental science, public health, governance, and socio-economic aspects has further made the analysis multi-dimensional. These partnerships bring together diverse expertise, resources, and perspectives, enabling a comprehensive approach to tackling the issue and allowing for the integration of knowledge from fields such as environmental science, public health, engineering, and social sciences. This interdisciplinary approach ensures that solutions are well-rounded and effective. Moreover, collaborative efforts enable stakeholders to pool resources, making it possible to implement large-scale initiatives that would be unattainable for individual entities.

Chapter 3: The study was carried out in Zambia and the DR Congo from September 2010 to September 2012 in two provinces with a history of copper-cobalt extraction to enable a cross-country comparison of contamination. The study sites were Katanga Province in the south-eastern DR Congo where Gécamines, the state mining company with its subsidiary industries, have huge concessions of approximately 30,000 km^2 around the towns of Kolwezi (West Group), Likasi (Central

Group) and Lubumbashi (East Group), and the Copperbelt Province in Zambia where various companies such as Konkola Copper Mines (KCM), Mopani Copper Mines, Chambishi Metals, First Quantum and Lumwana Mines extract minerals. In addition, locations of artisanal mining were also selected. Then, three administrative districts were specifically selected within each country based of recent proliferations of mines and previous reports on environmental contamination. They were Lubumbashi City, Likasi City and Kipushi Territory in Haut Katanga Province for the DR Congo, and Chingola, Kitwe and Luanshya Districts for Zambia. Within each district, several sites were identified for sampling either in the proximity of intense mining and metallurgic activities or far away from the mines as controls. Samples of food crops, water, and urines were collected and analyzed for metal contents.

Chapter 4: The central African Copperbelt is a geologic region, also known as the Lufilian Arc, situated in the southern part of the former Katanga Province in the DR Congo and the Copperbelt and Northwestern Provinces in Zambia. This arc contains large reserves of copper and cobalt minerals which represent the mainstay of the two countries' economies. With about 28% contribution to the GNP and 70% export value, mining exploitation remains the main source of DR Congo and Zambia's foreign currency earnings. The deposits occur in a sequence of sedimentary rocks that are known as the Mines Group (Lower Roan Group) and are between 650 million and 1050 million years old. Similarly, the mining sector is Zambia's most sensitive economic sector owing to its linkage to the export market, and copper contributes about 70% of foreign earnings. Ores containing copper and various other metals—most notably cobalt, but also zinc, cadmium, and uranium—have been extracted, in both countries, from large-scale mines and then further processed and refined in local smelters since the colonial times. From the end of the twentieth century onward, artisanal mining has also contributed to a substantial share of cobalt production in the DR Congo and for several minerals in Zambia.

Chapter 5: Water is vital to the existence of all living environments, but this valued resource is increasingly being threatened as human populations grow and demand for more water of high quality for domestic and economic purpose increases. Water is the essence of life and plays a vital role in nearly every function of the body, protecting the immune system—the body's natural defenses—and helping remove waste matter. Lack of safe water, sanitation, and hygiene remains one of the world's most urgent health issues. Almost one-tenth of the global disease burden could be prevented by improving water supply, sanitation, hygiene, and management of water resources. Ensuring poor people's access to safe drinking water and adequate sanitation and encouraging personal, domestic, and community hygiene will improve the quality of life of millions of individuals. On 28 July 2010 the United Nations General Assembly through Resolution A/RES/64/292 declared safe and clean drinking water and sanitation a human right essential to the full enjoyment of life and all other human rights. Moreover, the resolution highlighted its importance of ensuring access to water and sanitation as human rights constitute an important step toward making it a reality for everyone.

Chapter 6: About 70% of the population in both countries depend on agriculture for their livelihoods. Agriculture is also a significant source of exports and intra-regional trade. In Zambia, the agricultural sector is the backbone of its rural economy and arguably the country's most important sector employing about 56% of the population. It also serves a critical function of buffering employment volatility in other sectors, remaining key to ensuring food and nutrition security for vulnerable communities. However, agriculture's contribution to GDP declined from 17.3% in 2004 to 5.3% in 2015, evidence that most Zambians remain locked into low-production subsistence agriculture which is characterized by a lack of access to key productive assets and market opportunities. Zambia is among countries with the highest rates of hunger and malnutrition. Although the incidence of stunting in children under five has declined from 45% in 2007 to 40% in 2015, it still remains high. Similarly in the DR Congo, agriculture is the primary economic sector accounting for about 40% of the national GDP and employing some 70–75% of the population. Between 2006 and 2010, the Congolese agriculture sector grew by an average of about 4.4%, and projections are that it will grow at an average rate of 4.6% between 2011 and 2015.

Chapter 7: Air pollution is one of the most common environmental problems in the world from which constant exposure poses high health risks. Several reports indicate that the major contributors to air pollution stem from industrial emissions releasing pollutants such as sulfur dioxide, nitrogen oxides, and particulate matter into the atmosphere. Additionally, the agricultural sector generates most of the methane and nitrous oxide emissions and releases carbon into the atmosphere through the burning of solid waste. Moreover, high levels of air pollution have been linked to respiratory diseases, cardiovascular problems, and premature deaths. The World Health Organization shows that an estimated 4.2 to 7 million people die from air pollution worldwide every year and in Africa. According to UNICEF, 258,000 people died as a result of outdoor air pollution in 2017, up from 164,000 in 1990. The major sources of air pollution in Zambia are manufacturing industries, motor vehicles, mining and quarrying activities, construction, and rampant burning of the savannah woodland. Mining and metallurgic industries' operations release particulates, sulfur dioxide, and other harmful substances into the air making it of poor quality. The release of pollutants contributes to numerous respiratory diseases like pneumonia, bronchitis, and asthma, especially among children and the elderly in the DR Congo.

Chapter 8: Several of the subterranean and aquatic species have reduced considerably or are threatened by mining and metallurgical industries' pollution in various terrestrial and aquatic ecosystems in the DR Congo and Zambia. The biodiversity—that is made up of ecosystems, species, and genetic diversity in the region—is steadily being eroded due to human activities. All kinds of pollution, air, water, noise, plastic, and even light pollution, affect biodiversity. Pollutants—from mining and metallurgical industries—alter the environmental conditions directly or indirectly and negatively affect the flow of nutrients and constitution of the ecosystems. This leads to changes in the abundance of species, and hence, the biodiversity declining lowers the ecosystems' productivity—the amount of food energy that is converted into biomass—and the quality of the ecosystems' services (soil quality maintenance, purifying water, supplying food and shade, etc.). In addition, biodiversity loss greatly threatens the structure and proper functioning of ecosystems and affects economic systems and human livelihoods. Humans need biodiversity for various services including provision of food, clean water and air, fertile soil, medicines, raw materials, protection from storms and floods, a stable climate, and recreation.

Chapter 9: The results of toxicological analysis of plant tissues from crops utilized as food show that concentrations of pollutants were significantly different ($p < 0.001$) between the DR Congo and Zambia and between the six locations surveyed in the two countries. The concentrations of metals were higher (per dry weight) in leafy vegetables (amaranth, pumpkin, sweet potato, and cassava leaves), tomato, and beans than in maize and tuberous roots. There were significant differences ($p \leq 0.001$) in the concentrations of all metals in water samples from the DR Congo and Zambia with higher contamination recorded in the DR Congo than in Zambia, for all elements except Zn. Urine samples revealed abnormally high contents of trace elements in the urine from children than those in the urine from adults. Almost 100% of the respondents across the surveyed villages have recognized that mining operations have negatively impacted their livelihoods. Water and air quality was compromised and was therefore unsafe for human uses. Children displayed acute symptoms of malnutrition including stunting and underweight. Most of the people in proximity to mines and metallurgical plants suffer from numerous sicknesses such as cancer and respiratory difficulties. Noise pollution from mining was reported to cause health effects on their auditory organs, blood pressure (cardiovascular system), and physiological well-being of individuals.

Chapter 10: Decades of mineral extraction in the region have caused severe damage to the environment. It is therefore urgently important for mining and metallurgic industries to implement better environmental and social responsibility activities to prevent environmental pollution and assist communities in the host areas to benefit from their sector activities. Besides the economy, the potential benefits of the Environmental Impact Assessment (EIA) include effective compliance with acceptable environmental standards, sustainable use of natural resources, better governance and management processes by involving stakeholders, improvement in the design of infrastructures, and potential reduction of operating costs. To be effective, however, an EIA mechanism needs to be grounded on well-defined legislation and guidelines that govern procedural rules where the rights and obligations of all stakeholders are clearly defined, and their enforcement is ensured through appropriate implementation and compliance monitoring procedures and other instruments. State national governments are ultimately responsible for the regulatory and institutional settings within which mining and metallurgic industries operate and their actions are critical to achieving sustainable benefits from the extractive industries for national economic development. Improvements in the two countries' governance mechanisms and policies are necessary in order to optimize the benefits from the sector for the social, economic, and environmental development.

Chapter 11: At the end, the book wraps up with the conclusion and recommendations in which the perspectives of mining sustainability to protect the environment and ensure community health and food and nutrition security in the region are presented. They serve as a critical culmination of the analysis presented in the book, synthesizing key insights on the environmental impacts of mining in Zambia and the DR Congo. The results have provided compelling evidence of mining and metallurgic operations in the Copperbelt region of Africa discharging large amounts of toxic metals in the ecosystem causing severe pollution of soils and water, food crops, and ambient air. These impacts from mining and metallurgical operations negatively affect the well-being of the communities. We observed that children were more vulnerable than adults to pollutants. Besides exploiting and processing minerals solely for financial gain and the prioritization of tax revenue, governments and mining companies have tremendously failed to pay more attention to acceptable environmental and socio-economic standards which are supposed to be safeguarded by effective monitoring and enforcement of national mining and environmental laws and regulations. There are therefore urgent needs for stringent enforcement of existing legislation and international best practices to ensure sustainable mineral resource exploitation to alleviate negative impacts on the environmental protection, food and nutritional safety, and population health. Several solutions that can help mitigate and reduce the environmental impacts of mining and mineral processing are available and only need a framework for joint implementation by the private and public sectors, international organizations, and both bilateral and multi-lateral communities. The participation of independent organizations that are neither government nor part of the mining and mineral processing entities is likely to reinforce the implementation of guidelines regulating mining and metallurgic operations to protecting the environment, ensuring food and nutritional safety, and preserving the population health in the central African Copperbelt region.

Chapter 2
Research Partnership

Keywords Collaboration · Coalition · Expertise · Inter-institutional relations · Joint working · Knowledge · Relationship · Research output · Resources sharing

2.1 Collaborative Between Commonwealth, ACU, CBU, and UNILU

The key element of this book on Mining Impacts on the Environment in the Central African Copperbelt of Zambia and the DR Congo is the development of a partnership relation between Copperbelt University (CBU) and the University of Lubumbashi (UNILU), respectively, to address common environmental challenges. Collaborative partnerships between Zambia and the DR Congo are vital in addressing the transboundary impacts of mining on the environment. Both nations are rich in mineral resources, particularly in the Copperbelt region, which spans across their borders. However, mining activities in this region have contributed to share the environmental challenges, such as soil degradation, biodiversity loss, deforestation, air and both surface and underground water contamination, and risks to community residents' well-being, which transcend political boundaries. By working together, Zambia and the DR Congo pooled their human resources, expertise, infrastructures, and political will to develop joint strategies for mitigating mining environmental pollution. This collaboration has not only enhanced the effectiveness of environmental research but also fostered regional integration and solidarity, benefiting both nations' ecosystems and communities.

The involvement of expert scientists from diverse fields, such as environmental, agriculture, public health, mining, forestry, governance, and socio-economic sciences, is equally critical in understanding and addressing the multifaceted impacts of mining on the environment. Environmental scientists bring expertise in assessing pollution levels and ecosystem health, while agricultural experts examine how soil and water quality affect crop production and food security. Public health specialists analyze the effects of mining pollutants on community health, particularly the risks

A. Muimba-Kankolongo et al., *Mining Impacts on the Environment in the Central African Copperbelt of Zambia and the Democratic Republic of Congo*,
https://doi.org/10.1007/978-3-031-96786-3_2

of heavy metals in water and food supplies. Mining experts contribute insights into sustainable extraction techniques and pollution mitigation technologies, while forestry scientists explore strategies for reforestation and biodiversity conservation. Social scientists play a crucial role in understanding the human dimension, such as the socio-economic impacts of mining pollution on communities and the effectiveness of policy interventions. This interdisciplinary approach ensured that solutions are well-rounded and effective. Moreover, collaborative efforts enabled stakeholders to be a Think-Thank pool of considerable expertise, making it possible to implement large-scale initiatives that would be unattainable for individual entities.

By fostering interdisciplinary collaboration between scientists and bilateral cooperation between Zambia and the DR Congo, the research has led to comprehensive, science-based solutions to mining-related environmental challenges. Such partnerships paved the way for innovative approaches to pollution control, sustainable land use, and improved public health outcomes. Furthermore, they empowered local communities with knowledge and tools to advocate for cleaner and safer environments, ensuring that mining activities contribute to development without compromising ecological integrity or human well-being.

Zambia and the southern DR Congo are part of the Copperbelt with a major geological structure called the Lufilian Arc stretching from Kolwezi in the DR Congo to Luanshya in Zambia. Although the School of Natural Resources (SNR) at CBU has a comparative advantage on strategic research to develop technologies and advocate for regulations to limit mining pollution and contamination of natural resource base including agricultural commodities, air, and both surface and underground water as well as causing soil degradation, biodiversity loss, and harmful health issues on community residents, it also recognizes that meaningful participation of both local and international interests is greatly important. Partners involved in this book have, in the past, created innovative processes in environmental sciences that needed to be captured for incorporation in the current study which led to this book. Nabuyanda has been conducting research to establish levels of pollution in rivers in the Copperbelt of Zambia. In 2007, Muimba-Kankolongo and Jacob Mwitwa carried out an extensive survey funded by the European Union through CIFOR on the relationship between mining, forest management, and livelihoods in the Copperbelt of the DR Congo and Zambia. They collaborated with various partners in this project. Interviewed farmers pointed out at the deterioration of crop produce and cases of intoxication as a result of mining. CBU and UNILU have then been great partners for multiple environmental investigations in the central African Copperbelt region.

There are three primary objectives guiding this book. They include (1) improvement of the livelihoods of the communities living on the fringe of the mines and metallurgical industries by enhancing food security and quality nutrition and protection of the public health and safety; (2) exploration of standard policy and governance instrument arrangements at diverse levels (community, district, national, international) to minimize environmental damage; and (3) raising awareness of the consequences of mining pollution to the public and fostering accountability concerning environmental degradation. The high level of mining activities and processing of metal ores in the Copperbelt region has impacted negatively on the environment over the years. Water is polluted, the land for agricultural production is degraded, natural vegetation and crops are affected, and human health is at risk because of pollution. The book intends to ascertain that the contamination levels resulting from trace metals in food crops in Copperbelt region are unquestionable to create awareness for the issue to improve the livelihoods of the communities adjacent to mines including risks resulting from dietary intake of contaminated foods. A comprehensive study on the relationship between mining, forest management, and community livelihood was undertaken by CBU in Lubumbashi, Likasi, and Kipushi in the DR Congo and in Kitwe, Luanshya, and Chingola in Zambia, and the results highlighted considerable food spoilage as a result of mining operations (Mwitwa et al. 2012a). Another survey by UNILU (Banza Lubaba Nkulu et al. 2009a, b) on high human exposure to metals in Katanga, a mining area of the Democratic Republic of Congo, deals with the same issue of environmental pollution. They found considerably high levels of heavy metals in human blood. This book capitalizes on the above studies from the two partner institutions collaborating in research and development including others as shown under references to enhance its content outputs.

2.2 The Commonwealth

The Commonwealth is a voluntary association of 56 independent and equal countries (https://thecommonwealth.org/about-us). The Commonwealth consists of 2.5 billion people, featuring both highly developed and developing nations. Among its 33 member states, many are small island nations. The member governments collectively aim for goals such as development, democracy, and peace, as outlined in the Commonwealth Charter. The organization originated from the British Empire, but nowadays, any nation can join. Gabon and Togo were the most recent additions, becoming members in 2022.

2.2.1 Organizations

The Commonwealth is often considered as an entity comprising several nations out of which are the following three main intergovernmental organizations:

- The Commonwealth Secretariat supports member countries to achieve the Commonwealth's aims. This website is run by the Commonwealth Secretariat.
- The Commonwealth Foundation supports people's participation in democracy and development.
- The Commonwealth of Learning promotes open learning and distance education.

The Commonwealth Secretariat and Commonwealth Foundation have their international headquarters at Marlborough House on Pall Mall in London. The diverse member countries, ranging from the world's largest to the smallest, richest, and poorest nations, are supported by a network of over 80 intergovernmental, civil, cultural, and professional organizations. Together, they strive to achieve shared goals and values.

The Secretariat constitutes the intergovernmental organization delivering needed supports to affiliated member countries in order for them to fulfill the Commonwealth's objectives of development, democracy, and peace. The member countries:

- are a voice for small and vulnerable states,
- champion young people,
- promote justice and human rights,
- address threats like climate change, debt, and inequality,
- help grow economies and boost trade,
- support decision-makers to make good laws and deliver policies,
- help strengthen governance and build inclusive public institutions,
- provide technical help and training,
- send experts and observers to countries to give impartial advice and help solve national problems,
- provide systems, software, and research for managing resources.

2.3 The Association of Commonwealth Universities (ACU) in the UK

ACU provides for a universal structure of high-learning institutions united in the quest to improve the world through international partnerships in higher education (https://www.acu.ac.uk/). As the pioneering and oldest global university network, we have been uniting universities from across the Commonwealth for over a century.

The ACU, the world's first and oldest international university network, was founded in 1913 to facilitate the exchange of information, knowledge, and ideas among universities. In 1912, the University of London initiated a meeting where representatives from 53 universities gathered in London for the Congress of Universities of the Empire. Consequently, the Universities Bureau of the British Empire was established in London in 1913. The Bureau was renamed the Association of Universities of the British Commonwealth in 1948, and in 1963, it received a Royal Charter, adopting its current name: The Association of Commonwealth Universities. In 1986, Her Majesty as Head of the Commonwealth, the Queen became the ACU's Patron. After 33 years, she passed this role to The Duchess of Sussex in 2019. In 2021, it was announced that the Royal patronage would return to Her Majesty. The Association of Commonwealth Universities (ACU) is a UK-registered charity established by the Royal Charter. The ACU Council, which includes up to 23 members who are vice-chancellors of ACU member universities from across the Commonwealth, serves as the governing body, overseeing the organization's activities and setting its strategic direction. Council members also act as Trustees and are elected by their peers within the ACU membership network.

2.3.1 ACU Current Council Members

- Chair: Professor Cheryl de la Rey, University of Canterbury, New Zealand.
- Vice-Chair: Professor Wendy Thomson CBE, University of London, UK.
- Honorary Treasurer: Professor Feridun Hamdullahpur, formerly the University of Waterloo, Canada.
- Professor Abiodun H. Adebayo, Covenant University, Nigeria.
- Professor Kabiru Aderemi Adeyemo, Lead City University, Nigeria.
- Professor Pal Ahluwalia, University of the South Pacific, Fiji.
- Professor Sir Hilary Beckles, The University of the West Indies, Jamaica.
- Professor Amanda Broderick, University of East London, UK.
- Professor Barney Glover AO, Western Sydney University, Australia.
- Professor Ms. Hina Tayyaba Khalil, Pakistan Institute of Fashion and Design, Pakistan.
- Professor Rhonda L. Lenton, York University, Canada.
- Professor Sibongile Muthwa, Nelson Mandela University, South Africa.
- Professor Barnabas Nawangwe, Makerere University, Uganda.

2.3.2 Executive Committee

- Chair: Professor Cheryl de la Rey, University of Canterbury, New Zealand.
- Professor Wendy Thomson CBE, University of London, UK.
- Professor Feridun Hamdullahpur, formerly the University of Waterloo, Canada.
- Professor Barnabas Nawangwe, Makerere University, Uganda.
- Professor Pal Ahluwalia, University of the South Pacific, Fiji.
- Dr. Rhonda L. Lenton, York University, Canada.

2.3.3 The Road to 2030

Universities play a vital role in promoting sustainable development both within the Commonwealth and globally.

2.3.4 Strategic Plan

"The journey to 2030 underscores our core belief that international collaboration is essential to addressing global challenges and highlights our shared dedication to attaining the UN Sustainable Development Goals (SDGs) through higher education."

2.3.5 *Mission*

To build a better world through higher education.

2.3.6 *Vision*

A world in which higher education transcends borders, strengthens societies, fosters innovation, and lifts the lives of people throughout the Commonwealth and beyond.

2.3.7 *Priorities*

- To champion the power of higher education to improve lives.
- To support the long-term vitality of universities.
- To engage and connect universities across borders and promote collaboration among them.
- To deliver educational opportunities that make a positive and lasting difference.
- To uphold the ACU's reputation for excellence and demonstrate its impact.

2.3.8 *Values*

Quality

ACU strives to excel in all we do and continuously reflect on our work to find new ways forward.

Collaboration

ACU considers that teamwork and sharing thoughts and notions make it stronger.

Innovation

ACU promotes innovative thinking, fosters curiosity, and encourages the exploration of new ideas.

Respect

ACU treats others in the way it would like to be treated—with fairness, integrity, and generosity.

Inclusion

ACU appreciates diversity and actively promotes and celebrates the unique perspectives of its colleagues and partners.

2.3.9 Partners

The ACU values collaboration in all its endeavors. By partnering with organizations that share our values and goals, we can exchange knowledge and ideas, expand our influence, and work toward common objectives.

2.4 The Copperbelt University (CBU)

CBU, Zambia's second largest public university, was established through Act No. 19 of Parliament in 1987, and its ongoing operation is ensured by the Higher Education Act No. 4 of 2013 (https://www.cbu.ac.zm/). The University provides a wide range of programs in Business, Built Environment, Engineering, Natural Sciences, Medicine, Mining, Natural Resource Management, and Peace/Conflict Studies at both undergraduate and postgraduate levels. Its mission is to enhance the well-being of the Zambian people through innovative, flexible, entrepreneurial, and inclusive approaches to teaching, learning, research, and service.

The University employs a multidisciplinary approach to both teaching and research, emphasizing practical applications aligned with its core values of creativity, innovation, and entrepreneurship. Through its various schools, the University tackles local, national, and global issues by applying knowledge in four key areas: training and consultancy services, capacity development, research contracts, and entrepreneurial services.

CBU led the project application and has been a key player in technology development and transfer, collaborating with numerous public and private organizations. Each partner's role and benefits are clearly outlined in the proposal. Established by Act No. 19 of 1987, CBU began operating as a separate institution that same year, after the GRZ abandoned the federal university system. Copperbelt University has now 12 faculties, also called Schools. They are:

1. School of Business
2. Built Environment
3. Engineering
4. Graduate Studies
5. Mathematics and Natural Sciences
6. Medicine
7. Mines and Mineral Sciences
8. Natural Resources
9. Directorate of Distance Education and Open Learning
10. Humanities and Social Sciences

11. Information and Communication Technology
12. Dag Hammarskjöld Institute for Peace and Conflict Studies

The SNR was founded in 1995 in response to the need for skilled professionals and technicians to manage and utilize Zambia's natural resources and conduct relevant research in fields, such as forestry, wildlife, and fisheries.

SNR is Zambia's sole higher education institution awarding degrees in Forestry. It holds a national mandate to provide quality education programs tailored to the country's needs. SNR's mission is to address the shortage of skilled professionals in Forestry, Wood Science and Technology, Wildlife and Fisheries Management, and Agroforestry while conducting research for sustainable management of these resources in Zambia and the surrounding region. With over 20 academic staff specializing in various fields, they engage in multidisciplinary research aimed at contributing to national and global economic growth. The school addresses forest protection and sustainable use of forest resources, including non-wood forest products (NWFPs), generating knowledge and technologies to mitigate potential threats to food security and the livelihoods of those reliant on forests.

CBU management stresses the importance of the University, and its various Schools, in addressing the country's current development challenges. Given the severe and the overwhelming issue of forest depletion across the country and the urgent need for a sustainable solution, this project focuses on environmentally friendly and people-centered activities. It aims to focalize on empowering stakeholders, particularly women, reducing poverty through income-generating activities for rural households, and preserving the environment by improving technologies that promote rapid forest growth and development.

2.4.1 DelPHE Project and School of Natural Resources (SNR)

The role of SNR in this project was to provide technical support to participating groups and associations in collaboration with NGO and CBO partners. SNR will provide training in sustainable management of forest natural resources and in production, post-harvest handling and processing, and marketing of mushrooms focusing on several of those associations who have a track record in technology adoption to accelerate the impact of the project. The research team will work alongside with the collaborating NGOs, CBOs, consultants, and international partners to identify viable groups to participate in the program implementation and conduct the monitoring of the project. SNR will further ensure proper financial management for efficient delivery of services and timely reporting of the implemented activities.

2.4.2 Consultancy Services at School

Consulting services are a keyway for staff at Copperbelt University to share their academic knowledge and expertise with external organizations such as government bodies, public sector entities, community groups, and businesses. The university offers consultancy services to industry, private and public institutions, and both profit and non-profit organizations in Zambia and beyond.

2.4.3 Available Experts

CBU leverages the expertise and resources of the university for commercial purposes. It enlists consultants from the university's extensive pool of staff and associate consultants from the private sector, many of whom are alumni. CBU offers one of the most comprehensive consultancy services in the country, with access to over 600 specialists across 8 academic schools and 4 directorates and institutes.

2.4.4 Consultancy Thematic Area

With its numerous experts, the university provides services across a wide range of disciplines, organized into 25 broad consultancy thematic areas.

- Actuarial sciences and epidemiological services.
- Agriculture development and value chains.
- Analytical services—mining and chemical.
- Biotechnology and bio-prospecting.
- Building, real estate, and construction.
- Business development.
- Climate change and sustainable development.
- Clinical and dental services.
- Contract administration and project management.
- Educational, professional, and executive development.
- Engineering services.
- Environmental impact assessments and management audits.
- Feasibility study assessments.
- Financial management.
- Geotechnical investigations.
- GIS and remote sensing and land surveying services.
- Health systems.
- Information and communication technology.
- Leadership, governance, and management systems.
- Monitoring and evaluation.
- Peace and conflict management and resolution.
- Social impact assessment.
- Software and information management system development.
- Solid waste management and classification.
- Statistical and mathematical modeling and analysis.

2.4.5 Training and Professional Development

CBU provides training and capacity development across various fields. These tailored in-service courses, which range from a few days to several weeks, aim to enhance knowledge, skills, and competencies for improved performance and greater productivity. Accredited and certified by relevant organizations, these courses are delivered by highly qualified professionals from CBU's schools and units. They are offered as seminars, workshops, and symposia, either on site or at the client institution or organization.

2.4.6 Research Contracts and Applied Research

The University supports both industry and the public sector through contract-based research services, leveraging its extensive multidisciplinary research capabilities to address organizational challenges.

Examples of the University's research activities include the following:

(a) Domestication and Commercialization of Wild Tubers and Bulbs: This project, funded by the Finnish and Danish governments, focuses on the propagation, commercialization, and distribution of wild tubers and bulbs (chikanda, busala, and mumbu) in Zambia's central province.
(b) Traffic Lights Research (TLR): This initiative involves research, design, product development, and commercialization aimed at enhancing traffic light technology. As a result, many new traffic lights in Zambia now utilize technology developed at Copperbelt University.
(c) Assessing Institutional and Governance Partnerships for Climate-Compatible Development in Sub-Saharan Africa: A joint research project with the University of Leeds, University of Lubumbashi, and Eduardo Mondlane University, funded by the Climate Development Knowledge Network (CDKN) and DFID, UK.
(d) Green Economy Development in Zambia—Forest Restoration in Mining Wastelands: Funded by the Swedish Science Council, this project focuses on forest restoration in mining wastelands.
(e) Situation Analysis of the State of Preparations and Preparedness for the January 20, 2015, Presidential By-Election: Funded by the International Conference on the Great Lakes Region.
(f) Preliminary Study on the Drivers of Deforestation and Potential for REDD+ in Zambia: Funded by FAO.

Additionally, the University hosts an Entrepreneurship Center dedicated to promoting entrepreneurship research and practice within the community. The center provides various services, including skills training, business incubation, and the development and commercialization of ideas and products.

2.4.7 At SNR, the Research Project Was Accommodated Within the Department of Plant and Environmental Sciences (PES)

Established in 1995 as the Department of Forest Resource Management, the Department of Plant and Environmental Sciences enrolled its first students for a Bachelor of Science Degree in Forestry in 1996. The degree programs offered by the department aim to build capacity for the sustainable use, management, and conservation of forests, water, environmental, and agricultural resources in the context of climate change. The department is well-known for its emphasis on forestry, sustainable agriculture, and the management of natural resources in a sustainable manner, as well as addressing climate change.

Its mission includes:

To build and enhance our reputation as a top-notch teaching and research unit of the institution which is recognized for its innovation, excellence, and discovery and attracts the best students and staff.

The Department offers the following programs:

- BSc Forestry.
- BSc Plant and Environmental Resources.
- BSc Sustainable Natural Resources Management and Climate Change.
- MSc/MPhil/PhD Natural Resources Management.
- MSc/MPhil/PhD Sustainable Agriculture.
- MSc/MPhil/PhD Climate Sustainable Development.

Currently, all undergraduate programs are offered on a full-time learning basis. Postgraduate programs are offered on full-time and distance learning mode. For admission to a postgraduate program, a candidate must possess a first degree in related field from a reputable and recognized institution.

Research and Teaching Interests

- Natural resources economics.
- Natural resource assessment.
- Environmental sustainability.
- Natural resources governance and policy.
- Environmental valuation.
- Soil evaluation and land use.
- GIS and remote sensing.
- Plant physiology and ecology.
- Agroforestry and sustainable agriculture.
- Plant pathology and entomology.
- Ecosystem services and climate change.
- Tissue culture.

- Plant breeding and agronomy.
- Silviculture.

Related Posts

- Department of Biomaterials Science and Technology (BST).
- Department of Zoology and Aquatic Sciences (ZAS).

SNR designated the following staff as key personnel for the research project assessing the impacts of mining and metallurgic industry pollution on the environment:

Dr. Ambayeba Muimba-Kankolongo, who initiated the concept note and prepared the proposal, for the DeLPHE project, served as Principal Project Coordinator. He has experience in agricultural production, research, and project management having worked for the last 25 years in USAID-funded projects in the DR Congo and in SADC region. He will liaise with the Dean of SNR to ensure smooth implementation of the project activities, financial accountability, and timely reporting of the project progress to various partners, the donor organization, and the CBU management. As a plant pathologist, he will also bring his experience to the project in handling the pathological aspects in various field ecosystems.

Further, Ambayeba Muimba-Kankolongo, a former Contract Instructor in the Department of Biology and Biochemistry at Carleton University in Ottawa, Canada, and Senior Lecturer in the Department of Environmental and Plant Science, School of Natural Resources at the Copperbelt University in Kitwe, Zambia, has worked for more than 30 years assisting National Agricultural Research Services (NARS) and small-scale farmers in Central and Southern Africa. His career began as a researcher in plant pathology for a USAID-funded project National Cassava Program (PRONAM) of the Ministry of Agriculture in DR Congo and then continued with the International Institute for of Tropical Agriculture (IITA) in Southern Africa. He is an agricultural engineer with MS and PhD degrees in Agriculture with specialization in Plant Pathology and minors in Plant Breeding and International Agriculture at Cornell University.

His research interest focused on disease etiology and epidemiology, mainly toward insect-pathogen pest complex as it relates to plant predisposition to pathogenic infection and disease spread within and between fields as well as identification and determination of the biology of the causal organisms. It also involved establishing mechanisms of host resistance through identification of specific traits that are incorporated in the breeding program in order to develop high yielding and disease-resistant cultivars for cultivation by farmers. Mycoflora responsible for food spoilage and related mycotoxin production and contamination have also been at the center of my effort to ensure food safety for the population. Finally, my research has also concentrated on developing sustainable crop cultivation practices to enhance food productivity and security, on agricultural crop and forest tree protection and improvement, and in the resilience of rural communities to climate change to sustain their livelihood.

Upon completion of his BS degree at IFA Yangambi, Dr. Muimba joined the DRC National Cassava Program—PRONAM—of the Ministry of Agriculture at INERA M'vuazi station in Kongo Central where he worked as a Pathologist from 1977 to 1982 assisting in the development of resistant genotypes and then successively held the positions of Head of Pathology Section from 1982 to 1992 and PRONAM Director from 1992 to 1994 overseeing overall planning and management of the Program and coordination of the project procurement and field operations.

In 1994, Dr. Muimba Joined the International Institute for Tropical Agriculture (IITA) of the Consultative Groups on International Agricultural Research (CGIAR) as a Regional Agronomist responsible for Ecologically Sustainable Plant Protection (ESPP) activities in its project Southern Africa Root Crops Research Network (SARRNET) in Southern African Development Community (SADC) region with country responsibility for research and development for Angola, Namibia, and Zambia. At SARRNET, Dr. Muimba assisted in transferring developed innovative agricultural technologies including resistant cassava varieties to growers; in planning and coordinating the training of collaborating National Agricultural Research Services (NARS), the project personnel, and farmers; developing proposals to seek additional funding to allow NARS attain their objectives; exchanging technical information to national programs to enhance their capacity and representing IITA/SARRNET to the local government and other stakeholders (donors, NGOs, UN agencies, private sector, etc.).

At the end of the IITA contract, Dr. Muimba worked as Consultant with the Cooperative League of USA (CLUSA) in Lusaka, Zambia, to organize and provide a training course on safer cassava processing procedures for food utilization to the farming communities the Organization is supporting in Southern Province of Zambia before being offered a job at Copperbelt University (CBU) as a Senior Lecturer. As such, he was in charge of crop cultivation practices including Agroforestry and Conservation Agriculture and issues related to Plant and Forest Protection through Integrated Pest Management (IPM) concepts and practices to advance the learning and knowledge of plant health. At CBU, Dr. Muimba contributed significantly to the preparation of and writing on behalf of the School grant research proposals for submission for funding to international development agencies, bilateral organizations, and foreign missions in Zambia as well as to other foundations and ensuring that the School networks with various local partners in the field of Forestry, Agroforestry, Wildlife and Fisheries, and other related fields as well as with other national, sub-regional, and international high learning institutions (Universities and Colleges) including private sectors, NGOs, and CGIAR centers to maintain a productive and effective working relationship.

At the same time, Dr. Muimba was the Director/Farm Manager of the Mukabentch Investment Ltd. farm in Kitwe, Zambia, training individual farmers and farmers' association groups and other interested farming organizations in sustainable agricultural practices.

In Canada, Dr. Muimba first joined the Corn Breeding and Pathology Program of the Agriculture and Agri-Food Canada (AAFC) at its Eastern Cereal and Oilseed Research Centre, Ottawa, in Ontario. He worked collaboratively with the Program Manager and her Assistant to learn and conceptualize the major diseases of corn in Canada and how the crop is selected and breed for resistance to these diseases at the AAFC corn breeding program in Ottawa and assisted with the evaluation of disease resistance during the harvest in fields and in the laboratory using some pathological culture techniques. Then, he moved to the Department of Biology and Institute of Biochemistry at Carleton University in Ottawa, Canada, to teach Mycology course focusing on issues related to involvement of fungi in causing plant, animal, and human diseases, importance of fungi in medicines including antibiotics, and in food industries (bread and bear fermentation, etc.).

Dr. Muimba is a member of the International Society for Tropical Root Crops (ISTRC), the American Phytopathological Society (APS), and the Society for Congolese Agricultural Professionals (SOCOAP).

Dr. Muimba has received many awards and honors such as the Copperbelt University Award for contribution to the University research, 2010; Fellow of the Office of International Program (OIP) of the American Phytopathological Society (APS); and the Foundation International Travel Award, the American Phytopathological Society in 2005 in Austin, Texas, USA. He has published several peer-reviewed articles in various scholarly journals and attended and presented papers at some international and local conferences and workshops. He is an author of three books and co-author of two other books.

2.4.8 *Professor Jacob Mwitwa*

Jacob Mwitwa was trained at professional and technical levels in Zambia, Tanzania, Switzerland, the UK, and South Africa. He has over 20 years extensive background in natural resources management, education and research, rural community work, interdisciplinary work, project and program management, curriculum development and evaluation, and human resources management. Additionally, he has the ability to meet deadlines and work productively and adapted at handling steadily increasing levels of responsibility and is persistent, patient, and willing to learn. His qualification for positions leading to major responsibilities in academic, natural resources management, as well as project and program administration and management is a tailoring value Jacob will bring to the project. He has a high desire to use various scientific and management tools in resolving academic, natural resource management, and project challenges.

Professor Mwitwa has also a solid experience of building relational capital, natural resources policy review and development, and more than 10 years as a tertiary-level educationist. The foregoing is broken down as detailed in subsequent sections.

He has been the following:

(a) The Dean of School of Natural Resources (Current): responsible for coordination and strategic planning of academic programs, faculty administration, fundraising and creating linkages with external organizations, human resources management and development including recruitment, budgeting, and procurement.
(b) Head, Department of Forest Resources Management: responsible for coordination and planning of academic programs, administration of department, human resources management including recruitment, budgeting, and procurement.
(c) Program Management (WWF, World Wild Fund for Nature), providing vision and leadership for the WWF Zambia country program and effective representation of WWF with key audiences, such as the Government of Zambia and other key stakeholders, in order to build strength and support for the Country Program, and technical supervision to implement WWF's approved Program in Zambia, within the WWF policies, procedures, and standards.
At WWF, he played a leading role in identifying opportunities for program/project development and fundraising for the Zambia Coordination Office of WWF, in conjunction with the staff of the country office and partner national organizations. Building relational capital in working with various land users and policymakers pertaining to community-based natural resources management in protected areas of Zambia.
(d) Acting Chief Forestry Research Officer for the Division of Forest Research (Ministry of Tourism, Environment, and Natural Resources) with responsibilities including planning, directing, and evaluation of research and other institutional activities for programs approved by the head of the Forestry Department.

Moreover, Professor Mwitwa has additional project management skills, having managed the IDRC (International Research Centre of Canada)-funded project for the introduction of fast-growing Australian dry zone multipurpose Acacia tree species in Zambia; assessing land tenure and use for selecting land to be used for tree crops, provision of high-quality seed to farmers and rural communities for the establishment of village woodlots; promoting and maintaining networking among countries which carried out similar projects; and dissemination of research findings to interested parties.

He managed for 5 years the CIDA/SADC Tree Seed Centre Network project with responsibilities including planning, implementation, and evaluation of seed handling and marketing programs; identification and recommendation of multipurpose tree species; financial management US$ project account; identification of skills deficiency and training institutions; planning of human resources development programs for technical and auxiliary staff; identification and procurement of equipment and machinery from local and external sources; initiating and maintaining networking within and outside the SADC region; and provision of backstopping services to CIDA field office and the Zambian government.

The management of the Mufunta project intended to facilitate the acquisition of information for the purpose of developing the Mufunta GMA project on the western side of the Kafue National Park. The assignment included negotiating with traditional, political, and governance structures, assessment of vegetation and land use of Mufunta, ecological assessments and mapping of fauna and flora, livelihood profiling through the application of the resource use assessment (RUA) tool, and development of a development and conservation project for Mufunta game management area.

He has also assumed some professional consultancies such as the Preparation of CBNRM curriculum for the diploma and certificate in forestry course modules at the Zambia Forestry College for the diploma and certificate in forestry programs at the Zambia Forestry College (WWF), strategic review of the CIFOR Strategic Plan (2008–2018), CIFOR Medium Term Plan 2009–2011, and ICRAF Medium Term Plan 2008–2010 relative to research program for dry forests of Sub-Saharan Africa; social surveys of tree species preference formulation and implementation in Central, Lusaka, Southern and Western provinces of Zambia and project Review and Formulation (SADC FSTCU Consultancy included) of research program strategies including preparation of project profiles; prioritization of research programs based on stakeholder analysis; preparation of concept notes, project profiles, and full project proposals for natural research and capacity building (inclusive of training and education and formation of SADC regional information networks); and timber supply chain survey being part of a consortium of consultants that executed a survey and analysis of the timber supply chain to Anglo-American-owned Konkola Copper Mines which included the impact of timber harvesting and transportation on the natural regeneration of Miombo woodlands and social economy of labor employed in the supply chain. The study included the social economic effects of mining technological changes from the use of timber to steel. The findings constituted part of the environmental report for the IFC/World Bank counterpart funding for the Konkola Deep Mining Project (KDMP) and the assessment of fruit and fruit processing activities and potentials in Tanzania and Zambia.

Other consultancies include the social and corporate survey funded by the Ford and Kellog Foundations. The project is executed by CP WILD (Commercial Products from the Wild) and managed by ICRAF, the Forest Stewardship Council Pre-Assessment and Main Assessment Qualified auditor for forestry investments in indigenous forestry enterprises regarding SGS designed FSC Pre-assessment and Main Assessments. The pre-assessment various components include indigenous peoples' knowledge; community rights; OHS, ILO, and WHO conventions related to forestry operations; and Zambian policies related to the environment, labor, health, and sawmilling; the FAO/MTENR National Forestry Program facility, the contribution of the forestry sector to poverty alleviation and Zambia's national economy, is a partnership research project between the FAO-NFP and the Forestry Department of MTENR. The objective of the research was to determine the potential contribution of the forestry sector to poverty reduction and the national economy and generate valid information facilitating the integration of the forestry sector into the national planning process. The study involved the collection of statistics from published national and international sources, estimating missing data; conducting household, industry, and institutional baseline surveys and data processing and analysis; and the review of the relevance to conservation targets of environmental education in Southern Africa (WWF), a consultancy implemented to review the relevance of environmental education to WWF contribution to regional conservation targets. The consultancy covered the southern African regional coverage of WWF.

2.4.9 *Dr. Misery Mulele Nabuyanda*

Misery Mulele Nabuyanda has a PhD from IHE-Delft/Wageningen University in wetlands and trace metal pollution in wetland sediments and macrophytes (2011–2024). Mulele has work experience spanning more than 28 years of managing natural resources, lecturing, examining, researching, supervising, and carrying out Environmental Impact Assessments (EIAs) and Environmental Project Briefs (EPBs). From 1991 to 1992, he was a Wildlife Biologist under the USAID Project where he was involved in curbing the poaching of elephants in the Luangwa Valley of Zambia (around Nothe and South Luangwa National Parks). Thereafter, he worked as head of the Fish Processing Unit under the Fisheries Department in Zambia (1993–1995), and from 1995 to 2000, he was transferred from the Department of Fisheries to the Natural Resources Management College (NRDC) in Lusaka Zambia as a training officer. From 2000 to 2021, when he retired, he worked at the Copperbelt University as a Lecturer.

He was instrumental in the establishment of the Environmental Engineering BSc Course, the Environmental Engineering Monitoring and Management MSc Course, as well as the School of Mathematics and Natural Science at the CBU. He also headed the Environmental Engineering Laboratory which services not only students but the water companies on the Copperbelt Province (2002–2011). He was Assistant Dean for the School of Mines and Mineral Sciences (2010–2011). He has also published several articles in various international scientific journals.

2.5 The University of Lubumbashi (UNILU)

UNILU is among the most important high learning institutions in the Democratic Republic of Congo (DR Congo) (https://www.unilu.ac.cd/). Situated in the copper-rich province of Haut-Katanga in southern Congo, the University of Lubumbashi benefits from a resource-rich environment, making it an ideal site for scientific research. The diverse fauna, flora, and geological wealth make this area exceptional.

The university, known for its philosophical orientation, vibrant academic authority, and dedicated student community, attracts many foreign students from Africa for comprehensive training and from Europe as interns and researchers. The campus hosts over 450 national tribes, creating a cosmopolitan atmosphere of researchers known as the “KASAPARDS.”

To register at the university, specific documents are required based on your personal situation, country of origin, and intended studies. It's advised to consult the registration and admission program pages for detailed information. For more detailed information, get access to the reformed course program pages as well as our Faculties and Departments' page and the university's orientations which will provide you with more information about the courses.

2.5.1 UNILU Missions

Teaching

Ensure the education of design frontrunner leaders in various sectors of both domestic and national life. The delivered curriculum is aimed at promoting scientific innovations and the development of professional skills.

Research

Conduct fundamental and applied scientific research aimed at addressing specific issues in Katanga province in particular and the whole country in general, while considering international advancements in science, technology, and techniques.

Service to the Community

Engage with the environment that nurtures the University to make a meaningful impact. This involvement supports the development of its immediate community and contributes to nation-building.

This mission is symbolized in the University's emblem. On the right, against a gold background, the initial "L" represents the city of Lubumbashi, a cross for Katanga Province, and the wavy lines of the Congo River, which originates in Katanga and flows throughout the country. On the left, against a hopeful green background, a spearhead crosses the open book of the Student, pointing to the three stars symbolizing study, research, and social teaching.

2.5.2 At UNILU, the Project Was Hosted Within the Department of Public Health

The Department is currently led by Ordinary Professor Luboya Numbi Oscar, a pioneer in public health at the University of Lubumbashi and in the DR Congo. After earning Master's degree in Public Health in Antwerp, Belgium, in the 1980s and interning with the renowned KASONGO Project, Professor Luboya spear-headed an outstanding research project on primary health care at Kapolowe. This center has trained numerous public health doctors nationwide. The department's research efforts have resulted in the creation of national curricula for undergraduate and graduate public health training and the inclusion of tuberculosis and leprosy in primary health care. He founded the UNILU School of Public Health and, with his team of medical researchers, promotes research and publication at both national and international levels, with over 150 international publications to his name. The department is supported by two secretaries (Research and Teaching) and focuses on post-graduate training for doctors specializing in public health.

Training

The training aims to build and enhance professional skills for various public health roles, including patient care, care organization within a network or local care system, family doctor, head of a multidisciplinary care team, and district or zone chief physician.

Health Promotion

Managing healthcare institutions, such as hospitals and other medical facilities, includes roles like hospital department manager, quality expert, and healthcare executives.

In the areas of risk management, research and development, decision support, and teaching, roles include epidemiologists, professionals involved in epidemic control, and researchers in fields like epidemiology, clinical practice, health systems, health economics, and bioethics. Additionally, it covers specialists in financial management.

Department Organization

The department is structured into various research units, including Biostatistics, Epidemiology and Disease Control, Occupational Medicine, Health Services Organization, Hygiene and Health Promotion, Maternal and Child Health, and General Epidemiology.

The Unit of Toxicology and Environment

The Unit of Toxicology and Environment at the School of Public Health within UNILU's Faculty of Medicine has access to global scientific databases on environmental management. This enhances biodiversity conservation and pollution mitigation technologies. Leveraging its status as the sole higher education institution in the province for public health and environmental sciences, the Unit plays a pivotal role in achieving project objectives, thereby improving agricultural and ecosystem biodiversity.

With over a decade of experience in public health, especially concerning environmental pollution and health management, the Unit's academic staff has conducted fundamental and developmental research supported by various grants. The staff involved in the project proposal were selected for their expertise, aiming to integrate research and development into a multidisciplinary core for smooth implementation of activities and outcomes. They are experienced in training and possess high-level skills in their respective fields, showing readiness to collaborate with donors, government authorities, and other partners.

The Faculty of Medicine's Unit of Toxicology and Environment nominated the following staff as vital human resources for the project.

2.5.3 Professor Célestin Banza Lubaba Nkulu

Professor Banza Lubaba was trained as a scientific research intern at the Toxicology Laboratory of the Katholieke Universiteit Leuven, from August 11 to October 18, 2009, on the assessment of exposure to heavy metals and metalloids of the populations living along the Lufira in Katanga (DR Congo) and has a certificate of participation in the International Conference on Environmental Epidemiology and Exposure in Paris, France, September. He has attended and presented papers at various workshops and symposia including the International Symposium on Mineralogy, Environment, and Health in France; the Global Youth Tobacco Survey Training Workshop in Brazzaville, Republic of Congo, organized by WHO in collaboration with CDC; the Training Workshop on policies and strategies of tobacco control in Ouagadougou, Burkina Faso, organized by WHO in collaboration with CDC; and the University Pedagogy Seminar on "the definition educational objectives."

He has been the Community Coordinator of the Protestant Convention Schools (ECP/49th CEBA) for the Katanga Province, the Prefect of Studies in Official Secondary Schools of Katanga, Assistant at the ISP of Lubumbashi, Assistant at the School of Public Health of the University of Lubumbashi, Associate Professor at the Faculty of Medicine and at the School of Public Health for the University of Lubumbashi, Head of the Toxicology Unit and Environment at ESP/UNILU, Provincial Coordinator of the National Control Program against Drug Addiction and Toxic Substances (PNLCT in acronym) for the Province of Katanga, and Research coordinator at the National Program of the fight against drug addiction and toxic substances for the DR Congo.

2.5.4 *Professor Florence Kampemba Mujinga*

Professor Florence Kampemba Mujinga is in the Faculty of Agricultural Sciences, Department of Animal Sciences, UNILU, where she currently conducts research in environmental science, animal science, and nutritional biochemistry. Florence holds a PhD in Agronomic Sciences, with a focus on animal production, since 2014, funded by an ARES-AI project between the University of Lubumbashi and the University of Liège. He is Professor at the Faculty of Agronomic Sciences of the University of Lubumbashi since February 2021, Vice-Dean in charge of tutoring at the Faculty of Architecture of the University of Lubumbashi since November 2018, Head of the Department of Zootechnics at the Faculty of Agronomic Sciences of the University of Lubumbashi from 2014 to 2018, and Advisor for networking and scientific production to the Secretary General in charge of research at the University of Lubumbashi since September 2022.

He actively participated in the NSIMA project on the development of maize varieties adapted to the edapho-climatic conditions of Katanga in collaboration with CIAT (2008–2011); in the environmental impact and public health study in the Copperbelt region funded by DelPHE (Development Partnerships in Higher Education), Round 5 (2010–2013); and in the Taking into account societal challenges in African universities' food safety and food security strategies (DAfrAli) project funded by ERASMUS +: Capacity building (2016–2018), and he is responsible for project 5 "Climate smart agriculture for sustainable food systems" of the institutional program of the Universal Cooperation (CUI), VLIR-UNILU 2022–2032.

Chapter 3
Research Methodology and Operational Overview

Keywords Research · Locations · Sites · Sample collection · Data analysis · Mining · Case studies · Residents

3.1 Background

Mining activities often negatively affect vegetation types, including forests, which are vital for rural community livelihoods. Clearing surface vegetation and soils to access minerals can have long-lasting impacts. Mines cause surface scarring, erosion, and siltation, worsened by spoil heaps, tailings dams, and disrupted water tables. Local chemical changes, such as acid mine drainage and heavy metal release, further pollute soils and waterways. Mining operations also consume and frequently pollute vast amounts of water and timber. Extensive mineral resources can lead to significant damage through large-scale open-cast mining or numerous small-scale mines.

The recent increase in FDI in extractive industries has brought significant costs. For example, mining effluents in rivers and streams are common issues in Peru and Bolivia. In Santiago, air pollution from mining incurs annual healthcare costs of US$435 million (IDRC 1996). In Zambia, increased FDI in copper mining has led to major air and water pollution problems (Boocock 2002). Air pollution from waste dumps, tailings, and smelter emissions has increased bronchial diseases. Water pollution has resulted in rivers containing copper levels 80 times higher than acceptable in the early 1990s. Social impacts of FDIs in the extractive industry are often underreported, focusing mainly on their contribution to foreign earnings and GDP.

A. Muimba-Kankolongo et al., *Mining Impacts on the Environment in the Central African Copperbelt of Zambia and the Democratic Republic of Congo*,
https://doi.org/10.1007/978-3-031-96786-3_3

3.2 Research Locations

The study determined specific locations based on maps and previous studies on mining pollution's impact on communities' livelihoods (Mwitwa et al. 2012a) and vulnerability assessments (ASADHO-Katanga 2007; Nyasulu et al. 2011). Additional reports on food security and mining pollution concerns (FAO/WFP 2005; Ashton et al. 2001; Global Witness 2006) helped prioritize vulnerable and underprivileged areas.

3.2.1 Study Areas

From September 2010 to September 2012, scientists from CBU and UNILU embarked on an ambitious journey through the central African Copperbelt, spanning the DR Congo and Zambia to probe into the depths of copper-cobalt extraction's legacy across these two provinces. This study provided a comprehensive cross-country comparison of ambient air quality, the safety of food crops, and contamination levels in both surface and underground water. By examining these areas, rich in copper-cobalt history, they aimed to uncover the true environmental impact of mining activities and operations of metallurgic industries.

- Zambia: In Zambia's Copperbelt Province (Lat. 13°03′25.23″S, Long. 27°32′58.50″E, Alt. 1247 m, Fig. 3.1), a hub of mineral wealth, giants like Konkola Copper Mines (KCM), Mopani Copper Mines, Chambishi Metals, First Quantum, and Lumwana Mines dominate the landscape, tirelessly extracting precious minerals. Amidst these industrial titans, numerous artisanal mining sites have also been chosen, showcasing the vibrant, albeit challenging, tapestry of small-scale mining operations woven throughout the region. Here, Luanshya, Chingola, and Kitwe Districts were specifically targeted.
- The DR Congo: In the southeastern DR Congo, Gécamines, the state mining giant, along with its subsidiary industries, holds vast mining concessions that span an impressive 30,000 km^2. These concessions stretch across strategic locations around the bustling towns of Kolwezi (West Group), Likasi (Central Group), and Lubumbashi (East Group), painting a picture of extensive industrial footprint and mining prowess in the region. Selected locations comprised Kipushi territory, and Likasi and Lubumbashi cities in Haut-Katanga Province (Lat. 11°39′52.84″S, Long. 27°28′46.38″E, Alt. 1239 m, Fig. 3.1).

Within each location, we pinpointed several strategic sites for sampling. Some were nestled near the heart of intense mining and metallurgical activities, while others stood as control points, far from the mining chaos. The selection was methodical, considering past farmland degradation, varying levels of exposure, easy access, and crucially steering clear of health hazards due to potential irradiation from

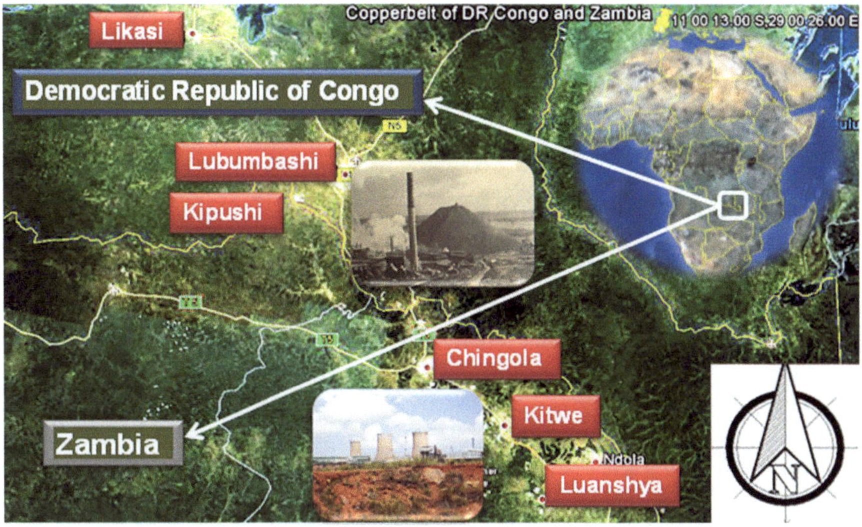

Fig. 3.1 A geographic map of different locations with mineral mining cities and districts surveyed in the Copperbelt region with Kipushi Territory, Likasi District, and Lubumbashi City in the DR Congo and Chingola, Kitwe, and Luanshya Districts in Zambia for sample collection and interview with community members

hazardous mining zones. This approach ensured a comprehensive and safe study of the environmental impact.

3.2.2 *Methodology*

Households near mining areas were interviewed about the impacts on livelihoods and agricultural production. Samples of crops and produce from fields and marketplaces were collected for laboratory analysis to detect nine heavy metals including Mn, Co, Ni, Cu, Zn, As, Cd, Pb, and U. GPS readings determined geographic positions and pollution sources.

3.2.3 *Mining Companies Involved*

In addition to formal mining listed above, informal artisanal mining is widespread and often non-compliant with environmental regulations (Kambani 2003; IPEC 2008; Tsurukawa et al. 2011).

To ensure a broad distribution of sampling sites across the six locations in the DR Congo and Zambia, the study selected ten sites per country. In each town, one site was located within 100 m of a mine, followed by sites more than 1 km but less than

20 km away. The farthest site was chosen in an area assumed not to be affected by mining but within at least 20 km from a mine.

Sampling Sites

The DR Congo

- Lubumbashi: Kabecha, Tshansansa, Kasongo.
- Likasi: SNCC-Shituru, Gécamines-Shituru, Panda cities, Kansalabwe village.
- Kipushi: Mungotshi, Douane areas, Betty city, Kampemba farm, Nkumanwa village.

Zambia

- Chingola: Shimulala, Heleni-Kasenji, Kalipenta, Pule, Chama villages.
- Luanshya: Roan area, Mpatamatu, Mwansa villages.
- Kitwe: Chankalamu village, Wusakile, Nkana-West compounds.

Study Purpose

This study aimed at assessing levels of trace metal contamination of the environment leading to poor quality ambient air and water, food spoilage, soil degradation, biodiversity loss, and residents' illnesses in the fringe of copper and cobalt mining and metallurgical facilities. The goal was to raise public awareness and influence policy dialogue for long-term solutions on sustainable mining management in Zambia and the DR Congo.

Community Engagement

The study included discussions with locals and workshops with policymakers, mining company representatives, environmental agencies, civil society, academia, and community members in Lubumbashi (DR Congo) and Ndola (Zambia). These sessions gathered perceptions on the environmental and social impacts of industrial activities on communities and sought to improve local livelihoods.

Focus Areas

- Characterizing heavy metals in food: Investigating the presence of heavy metals in food crops and their impact on population health and agricultural production.

- Identifying favorable factors: Understanding factors that influence heavy metal occurrence and plant uptake to prescribe mitigation measures.
- Improving nutrition and health: Promoting good quality food to enhance household income, nutrition, and health, particularly for vulnerable groups like HIV/AIDS patients.
- Supporting agricultural practices: Encouraging alternative agricultural practices to reduce heavy metal loads in food, leading to healthier farmers and children, reduced medical expenditures, and improved livelihoods.

Expected Outcomes

The study anticipated that healthier farmers and children would lead to less spending on health care, more productive lives, and increased self-sustainability. Improved knowledge of health risks from food contamination could boost family incomes through sales of quality products, fostering productivity and enhancing household food security.

By addressing these critical issues, the study aimed to create a positive impact on communities affected by mining activities and contribute to sustainable development in the region.

3.3 Descriptions of Locations

3.3.1 Zambia

The Copperbelt Province is a bustling urban sprawl, encompassing the interconnected mining hubs of Chingola, Luanshya, Kitwe, and Ndola. This region hosts the largest and continually growing population in the country. For years, the mining sector has been the powerhouse, contributing to roughly 80% of the nation's foreign earnings and a significant chunk of its GDP. Since the liberalization of the economy in the early 1990s, the mining of copper and cobalt has surged as the dominant industrial activity.

Miombo woodlands blanket much of the province, but unfortunately, they've been heavily exploited, primarily for timber to support mining activities. Early retrenchments and increased pollution from the mines have also played a part, resulting in substantial depletion and degradation of these precious natural forest resources.

Looking ahead, project activities will roll out in Bwanamukubwa, Maposa, Kalulushi, Masaiti, Lufwanyama, and the rural areas of Ndola, aiming to address these critical issues and foster a more sustainable future.

3.4 Chingola District

Chingola, a vibrant city in Zambia's Copperbelt Province, boasted a population of 147,448 as per the 2000 census. Located about 400 km north of Lusaka, the nation's capital, Chingola is renowned as the home of the Nchanga Open Pit Mine, the second-largest open-cast mine in the world. Founded in 1943 to serve the burgeoning copper-mining center at Nchanga, the city has grown around this industrial giant. The Nchanga Open Pit stretches in an 11 km arc around the west and north of the town, covering nearly 30 km^2, with its deepest point plunging 400 m below the surrounding plateau. Additionally, Chingola hosts an underground mine with a tailings dam that spans 24 km^2, located about 7 km south of the city.

Strategically, Chingola's main road to Lubumbashi in the DR Congo via Chililabombwe is the sole major link between Zambia and the DR Congo, connecting the two nations to the broader Southern African Development Community (SADC) region. Another key road branches off from Konkola, diverging from the main Copperbelt highway, which runs southeast from Kitwe and northwest to Solwezi.

Beyond the urban center, the surrounding rural areas are sparsely populated and face significant environmental degradation due to deforestation, charcoal burning, and uncontrolled subsistence farming. The privatization and subsequent change in ownership of the mines have brought about various land use changes. While large tracts of land remain under KCM's ownership, areas previously owned by Zambia Consolidated Copper Mines (ZCCM) have been handed over to the government. These lands are now extensively used for agricultural production and other purposes, such as charcoal burning, by local residents and former mine workers who have settled in the Chingola District.

The boundaries of the study area in Chingola start at the confluence of the Mwitimpi Stream with the Kafue River, following downstream to the Mbafi River confluence and then upstream to the eastern boundary of National Forest No. 12: Luano. The boundary then extends southwest for about 6 km to the Musakashi River and continues upstream to its source and then northwest to Beacon CG500 of Lot No. 1311/M. It traces along the northern and western boundaries through beacons CG506, CG505, and CG504 to the most western corner and southwest across the Zambia Railways Strip Reserve for about 1 km to the source of an unnamed tributary of the Musenga River. The boundary follows this tributary to the northwestern boundary of the Pitanda Special Grant No. 524 m and along the boundaries of the Pitanda and Mwambeshi North Special Grants to Beacon M219. It continues south along the western boundary of the Mwambeshi South Special Grant to the Mwambeshi River, upstream to its source, then south to the source of the Chisanga River, downstream to the West Katembula River confluence, and upstream to its source. Finally, the boundary moves northwest for approximately 16 km to a point about 2 km north of the Chipopo Stream source, northwest for about 13 km to the Solwezi-Chingola Road, and further northwest passing between Muchishi East Stream to the north and Milopa and Miwale Streams to the south, to a point on the northern side of Chingola Road, approximately 20 km south of the Mwitimpi Stream source, and back north to the Mwitimpi Stream confluence with the Kafue River, where it began.

3.5 Kitwe District

Kitwe, one of Zambia's fastest-growing cities, owes its rapid development to the opening of some of the country's largest and most productive mines. The city has become a bustling hub of construction, trade, and services, all clustered around the mining sector. Businesses in Kitwe have thrived, ensuring that the value created by mine supply chains remains within the region. Kitwe's diverse economic landscape, driven by mining and bolstered by manufacturing, finance, and commerce, underscores its significance as a rapidly growing urban center in Zambia.

Starting at the point where the western edge of the Kitwe-Mufulira Road Reserve crosses the Kafue River, the boundary follows the Kafue River downstream to its confluence with the Mutupa Stream. It then moves up the Mutupa Stream to the southern edge of the Mufulira-Ndola Road Reserve. From there, it heads southward for approximately 5 km along the road reserve's edge to the Nakolwe Stream, then down this stream to its confluence with the Mwekera Stream. It continues down the Mwenkwenta Stream to an unnamed south bank tributary, approximately 1.5 km north of Beacon 01356 on the Zambia Railways Strip Reserve (Nkana Branch Line). The boundary then moves up this tributary to its source, southward to Beacon 01356, across the Railway Reserve to Beacon 01357 near Mile Peg 23, and then southwest along the southern edge of the Railway Reserve, skirting the eastern, southern, and western boundaries of Farm No. 1386 (Mabote Siding Reserve South). It continues to a point where the eastern boundary of Lot No. 573/M intersects the southern edge of the Railway Reserve, then southward to Beacon BA56, the south-eastern corner Beacon BA57, and further south to the north-eastern corner of Farm No. 2265 near Beacon YR1466.

The boundary proceeds southward along the farm's eastern boundary to Beacon KL180, its southernmost corner beacon, and then southward across the Kitwe-Ndola Road Reserve through Beacons JB216, L541, JB224, and KL169 to the point where the western boundary of Local Forest No. 4 (Maposa) crosses a northern tributary of the Baluba River. It follows this tributary to its confluence with the Baluba River and down the Baluba River to its confluence with the Kafue River. The boundary continues down the Kafue River to its confluence with the Kalulushi River and then up the Shibolo Stream to its source. From there, it moves northeast to Beacon M209, the southernmost corner beacon of the Kalulushi Special Grant No. 437m, and then along the eastern boundary of this grant to Beacon M208. It then heads northwest to the source of the Kasombo Stream, down the Kasombo Stream to its confluence with the Chibuluma Stream, up the Chibuluma Stream to Beacon U420 near the northern point of Chibuluma, northward to Beacon U419, and westward to Beacon U418. The boundary continues along the northwestern boundary of

Nkoma Special Grant Area C2 No. 772m to Beacon RU183, the northwestern corner beacon of Farm No. 2493 (now part of National Forest No. 8: Ichimpe), and northeast along the southern boundary of this national forest for approximately 7 km to where it intersects the southern edge of the Chingola-Kitwe Road Reserve. From there, it moves northwest along the road reserve edge to its junction with the Kitwe-Mufulira Road Reserve, then along the western edge of the Kitwe-Mufulira Road Reserve to the point where it crosses the Kafue River, the starting point.

3.5.1 Economic Sectors

Mining: Since the late 1800s, mining has been Kitwe's dominant economic activity, particularly copper and cobalt extraction (Infrastructure and Cities for Economic Development (ICED) 2018). Despite a decline in mining employment from a peak of 65,000 to 30,000 due to privatization, the sector remains vital. However, the reduction in public services previously provided by the mines has impacted the community.

Manufacturing: Manufacturing is crucial in Kitwe, accounting for 26% of formal employment, while services cover 74% (Census of Business Establishments 2010). The city specializes in manufacturing metals and machinery, with a location quotient of 1.79. Although the capacity utilization of manufacturing firms is 62.3%, above the national average of 56.2%, it has declined by 6.6 percentage points since 2007.

Finance and Commerce: Kitwe's financial sector is robust, with almost all Zambian banks having branches in the city. Major supermarkets, along with Chisokone and Obote markets, serve as retail and wholesale hubs. Despite this, 14 of the 21 markets operate below 50% capacity, with 5 deserted and only 2 functioning at 80% capacity.

3.6 Luanshya District

Luanshya, a town nestled in Zambia's Copperbelt Province near Ndola, boasts a population of 117,579 as per the 2008 census (CSO 2008). Historically under Chief Mushili of the Lamba people, Luanshya was founded in the early twentieth century. Its inception is quite the tale—two prospectors, including William Collier, shot a Roan Antelope on the banks of the Luanshya River. Remarkably, the fallen antelope's head rested on a rock with a visible seam of copper ore. This serendipitous discovery led to the formation of the "Roan Antelope Copper Mines Ltd," marking the beginning of significant copper mining in the area.

Throughout the twentieth century, Luanshya thrived on copper mining, but by the century's end, the mining had become less economically viable, plunging the town into a severe recession. Despite this downturn, there's still plenty of copper underground, and the town's future hinges on the efficient extraction and sale of this resource. Interestingly, the Roan Antelope Rugby Club once held a Guinness World Record for the highest goal posts at 110 ft, 6 inches (Butler 2006).

A local legend speaks of a river snake blamed for early development issues at the Roan Antelope Mine. Workers faced high illness and mortality rates from diseases like malaria, blackwater fever, typhus, dysentery, and pneumonia, and the Lamba people believed a giant snake caused these troubles by wriggling along the river and mine shafts, leading to floods, cave-ins, and poisonous gas releases (Lydall and Auchterlonie 2011). By 1928, construction of the plant and township began following the discovery of approximately 20 million tonnes of copper reserves. The Luanshya River was eventually channeled, and its swamps drained during a 1929–1932 anti-malaria campaign, resolving the flooding issue.

Despite the social challenges brought by the mines, Luanshya has experienced a revival in employment rates. The population is steadily increasing, and new economic activities are emerging, including the installation of a new soya solvent extraction plant and the expansion of a well-known milling company. The agricultural sector is also flourishing, with small-scale farmers boosting their crop outputs to meet growing demand. Key crops include maize, wheat, and soybeans.

3.6.1 The DR Congo

In the DR Congo, the study focus was on the Haut-Katanga Province. Over the past decade, the province has undergone a dramatic transformation, fueled by a surge in mining activity. This boom has sparked hope among local communities, offering the promise of employment and economic upliftment in a region long marked by hardship. The DR Congo has emerged as a global powerhouse in mineral production, supplying nearly half of the world's cobalt and standing as Africa's leading producer of copper. These minerals are not just commodities— they are the backbone of modern technology. From smartphones and electric vehicles to industrial machinery, copper and cobalt extracted from Congolese soil play a vital role in powering the global economy. Each year, the mining sector contributes significantly to the international supply chain, reinforcing the DRC's strategic importance. Yet, this progress comes with a heavy cost. The environmental and social consequences of mining are profound and increasingly difficult to ignore. As the world confronts the challenges of climate change, biodiversity loss, and waste management, the sustainability of mining practices has come under intense scrutiny. The need to balance economic development with environmental stewardship has never been more urgent.

3.7 Lubumbashi City

Lubumbashi, once known as Élisabethville, stands as the second-largest city in the DR Congo, just after the capital, Kinshasa. Positioned near the Zambian border, Lubumbashi serves as the vibrant heart of the southeastern region and the capital of the prosperous Katanga Province, formerly Shaba. Covering 747 km^2, the city's population is estimated to be around 1.2 million.

Founded by the Belgians in 1910, the city flourished with the rise of the copper mining industry. Lubumbashi has grown into a significant commercial and industrial center, hosting a variety of manufacturing industries including textiles, food products, beverages, printing, bricks, and copper smelting. Despite underinvestment making the railway lines to Ilebo, Kindu, Sakania, and Kolwezi unreliable, the city remains a key transportation hub. It also boasts the modern Luano International Airport and acts as a distribution center for minerals like copper, cobalt, zinc, tin, and coal.

From a past viewpoint (http://en.wikipedia.org/wiki/History_of_Katanga), the earliest known permanent settlements in Katanga belong to the ancestors of the Luba people, who established themselves in the Upemba Depression near Lake Upemba. By the sixth century, the Luba were engaged in ironworking and traded salt, palm oil, and dried fish. These goods were exchanged for copper, charcoal, glass beads, iron, and cowrie shells from the Indian Ocean. The region's wealth in ores facilitated the development of iron and copper technologies, as well as a primitive but extensive commercial network stretching over 1500 km to the Indian Ocean.

In the late sixteenth century, the Luba Empire and Lunda Kingdom controlled the province, triggering migrations of warriors and tribes into neighboring regions. Many of Zambia's Bemba, Kazembe-Lunda, Kanongesha-Lunda, and Lozi people trace their origins back to Katanga. On October 1, 1933, the Belgian Government renamed the province Elisabethville, after its capital (now Lubumbashi). Moreover, Gordon (2021) explores how various tribes contributed to the copper trade in Copperbelt during the early nineteenth century. Traders from different directions—west, east, and north of Katanga—played key roles in expanding copper production. The Nyamwezi, originating from present-day central Tanzania in the east, were instrumental in the copper trade, utilizing copper for symbolic purposes and trading it for ivory bound for the east coast. Western traders, including the Chokwe, Ovimbundu, and potentially the Lunda, pursued commercial opportunities in the region. From the north, the Luba exchanged salt and tools for Katanga copper. Within Katanga, the Sanga community notably adapted to growing trade demands by producing sizable copper ingots.

The Eastern Lunda Kazembe kingdom, situated around the Luapula River and Lake Mweru, emerged as a hub for these nineteenth-century trade routes. Portuguese accounts from the late eighteenth to early nineteenth centuries portrayed this polity as dominated by a wealthy and occasionally authoritarian Mwata Kazembe leader. Yet, these descriptions often neglect the complex reciprocal systems underpinning the kingdom's political and economic framework. The wealth of the Kazembe kingdom derived from its varied resources, such as salt, fish, copper, and the fertile Luapula Valley. While Portuguese records emphasize powerful rulers and ivory commerce, oral traditions illuminate networks of mutual exchanges, including connections with copper-producing elites.

Oral histories reference early chiefs involved in copper production in Katanga, such as the Kipimpi, Kibuye, and Kasongo titleholders. These accounts tie their origins to migration narratives of the Eastern Lunda tradition under Mwata Kazembe. These leaders reportedly formed alliances, often through marriage, with local landowners and skilled copper artisans. Whether these chiefs were descendants of Lunda migrants in the eighteenth century remains uncertain. Nevertheless, the narratives highlight the trade and clientage relationships of the early nineteenth century. Copper served as a unifying force, linking Katanga's autonomous communities to the Eastern Lunda Kazembe political hierarchy through lasting kinship and mutual ties.

3.8 Kipushi Territory

Kipushi, in the capital of Haut-Katanga Province in the DR Congo, lies close to the Zambian border and is predominantly a mining town. It is one of Gécamines' major production centers, alongside Lubumbashi, Likasi, Kambove, and Kolwezi. Kipushi is strategically served by national road 37, situated 30 km southwest of Lubumbashi.

3.8.1 Historical Significance

Kipushi's mining heritage dates back to 1924 when Union Minière began operations, exploiting an 18% copper deposit from a surface open pit. At that time, it was the world's richest copper mine. The mine later transitioned to Africa's richest underground mine for copper, zinc, and germanium. In 1967, the state-owned Gécamines took control of Kipushi, operating it until 1993 when it was placed on care and maintenance due to economic and political challenges. Over 69 years, Kipushi produced 6.6 million tonnes of zinc and 4.0 million tonnes of copper from 60 million tonnes of ore, along with 278 tonnes of germanium and 12,673 tonnes of lead (Butler 2006).

3.8.2 Modern Developments

Since Ivanhoe Mines acquired its interest in 2011, drilling campaigns have significantly upgraded and expanded Kipushi's mineral resources. The current estimates are as follows:

- Zinc: 11.78 million tonnes grading 35.34%.
- Copper: 0.80%.
- Silver: 23 g/tonne.
- Germanium: 64 g/tonne.

These resources contain 9.2 billion pounds of zinc, 8.7 million ounces of silver, and 24.4 million ounces of germanium.

3.8.3 *Strategic Metals*

Germanium is crucial for electronic devices, flat-panel display screens, LEDs, night vision devices, optical fibers, and solar arrays. Gallium, another strategic metal found in Kipushi, is essential for manufacturing semiconductor wafers used in integrated circuits and optoelectronic devices like laser diodes, LEDs, photodetectors, and solar cells.

3.8.4 *Ivanhoe Mines*

Ivanhoe Mines, a Canadian company, focuses on advancing three principal projects in Southern Africa:

1. Kamoa-Kakula Copper Complex: Expansion in the DR Congo.
2. Platreef Project: Palladium-nickel-platinum-rhodium-copper-gold project in South Africa.
3. Kipushi Mine: Restart of the historic ultra-high-grade zinc-copper-germanium-silver mine in the DR Congo.

The mine also explores new copper discoveries across its 2400 km^2 of exploration licenses and newly acquired joint venture licenses in the Western Foreland, adjacent to the Kamoa-Kakula Copper Complex.

3.9 Likasi City

Likasi, a lively city in the Haut-Katanga Province of the DR Congo, has a fascinating past and serves as a major industrial, particularly mining, hub. Once called Jadotville (in French) and Jadotstad (in Dutch), the city offers much to explore. Boasting rich history and stunning natural beauty, it stands as a unique and captivating tourist destination. Its location between Lubumbashi and Kolwezi makes it an ideal stop for travelers exploring the Katanga mining region.

3.9.1 *Historical Significance*

Likasi's history stretches back to the Belgian colonial era when it emerged as a vital mining center. Situated 20 km west of Likasi is the Shinkolobwe mine, once described as the world's most significant uranium deposit in a 1943 report related to the Manhattan Project. Uranium from this mine was critical in creating the atomic bombs used in Hiroshima and Nagasaki in 1945. In 1961, during the Katanga conflict, Irish

UN troops stationed in Jadotville were besieged and later surrendered to forces loyal to Katangese Prime Minister Moïse Tshombe. Visitors can delve into this storied past at the Likasi Geology and Mining Museum, which showcases an impressive collection of minerals, rocks, and exhibits detailing the region's mining history.

3.9.2 *Natural Beauty*

Surrounded by lush green hills, Likasi is renowned for its breathtaking natural beauty. Among its many attractions is the Lofoi Waterfall, Africa's highest, with a dramatic 340-m drop. For adventure enthusiasts, Likasi offers hiking and cycling in the hills, canoeing on the Lufira River, and visits to nearby game reserves teeming with wildlife like elephants, lions, and antelopes.

3.9.3 *Mining Heritage*

Located midway between Lubumbashi (114 km) and Kolwezi (190 km), Likasi is at the heart of the Katanga mining region. As far back as the sixteenth century, the "copper eaters" handcrafted copper crosses, which served as currency. These artifacts, along with other historical objects, can be admired at the city's mineralogical museum in the Gécamines center. Likasi remains a vital industrial center, particularly for mining, and acts as a transport hub for the region. Nearby are deposits of copper and cobalt, with refineries and mines supporting the city's economy. In early 2019, a $17 million crushing plant in Kamatanda began operations. Ore from this plant is processed at the Heap Leach Unit in Panda, and the copper solution is transported to the Shituru Factories in Likasi, where high-grade copper electrodes are made. While an old gold mine in Likasi is commercially depleted, it still attracts artisanal miners.

3.9.4 *Development and Modern Day*

The city's discovery dates to the Belgian expedition led by Jules Cornet in 1893. Cornet's geological map, which remains relevant today, coined the term "Katanga is a geological scandal" due to the region's abundant mineral wealth. Likasi was officially founded in 1917 following the opening of its namesake mine and the development of an industrial complex, including the Panda Foundry and Shituru factories. In 1931, the city was named Jadotville, honoring Belgian engineer Jean Jadot, and experienced significant urban and landscape developments, earning it the nickname Coquetteville.

3.9.5 Present Day

Today, Likasi retains some of its historical charm but has lost much of its past splendor. Though it remains an important copper processing center, the city often appears to be running in slow motion, with many installations reflecting a bygone era. Its large avenues, colonnade buildings, and orange facades, now covered in dust, give it a ghost town feel. This atmosphere, reminiscent of Havana's old chic register, adds to Likasi's unique character, where time seems to stand still.

3.9.6 Adventure Activities

Likasi is not just a city with a rich history—it is also renowned for its breathtaking natural scenery. Surrounded by lush green hills, Likasi boasts several impressive waterfalls, including the magnificent Lofoi Waterfall. With a staggering drop of 340 m, it's the tallest waterfall: For those who love the great outdoors, Likasi offers plenty of exciting activities. You can hike or cycle through the picturesque hills or take a thrilling canoe ride along the Lufira River. Nearby game reserves provide opportunities to spot incredible wildlife such as elephants, lions, and antelopes. The city offers a variety of outdoor activities for adventure seekers. Hiking and cycling in the surrounding hills, as well as canoeing on the Lufira River, provide exciting ways to explore the natural landscape. Nearby game reserves allow visitors to see wildlife such as elephants, lions, and antelopes.

3.9.7 Cultural Significance

Likasi holds cultural importance with its unique artifacts, including the handcrafted copper crosses from the sixteenth century, which served as currency. These pieces can be admired at the city's mineralogical museum. In addition, Likasi's culinary delights combine African and European flavors. You'll find local specialties like fufu, a mash made from cassava or banana, and moambe, a palm nut-based chicken stew. For international tastes, the city hosts restaurants serving Italian, French, and Belgian cuisine, offering something for everyone.

3.9.8 Urban Charm

The city's urban planning and architecture from the Belgian era, with wide avenues lined with trees and gardens, contribute to its charm. The remnants of its past glory give Likasi a distinctive atmosphere that feels both historic and timeless.

3.10 Study Sites

Following the identification of several sampling sites, both near-intense mining and metallurgical activities and further away to serve as control sites, various criteria were considered. These included past farmland degradation, degrees of exposure, accessibility, and avoiding health risks from potential irradiation. The selected sites encompassed:

- Lubumbashi: Kabecha, Tshansansa, Kasongo.
- Likasi: SNCC-Shituru, Gécamines-Shituru, Gécamines Panda, Kansalabwe village.
- Kipushi: Mungotshi, Douane, Betty city, Kampemba farm, Nkumanwa village.
- Chingola: Shimulala, Heleni-Kasenji, Kalipenta, Pule, Chama villages.
- Luanshya: Roan, Mpatamatu, Mwansa villages.
- Kitwe: Wusakile, Nkana-West compounds, Chankalamu village.

3.10.1 Sample Villages and Households

Aligned with previous studies (CSO 1990, 2000), the survey targeted 10,000–12,000 households as a representative sample. Adjustments were made based on the 2000 population and household census (CSO 2003; INS 1984), considering budget, distances between districts and villages, population size, language diversity, study duration, and village population.

The survey assumed that collected data from the sample population could be generalized to the target population in various locations. This meant that the data from those interviewed would roughly reflect the broader population's experiences.

3.10.2 Diverse Actors and Interviewees

Interviewees were selected across village, district, provincial, and national levels (Muimba-Kankolongo et al. 2022). They included the following:

- Provincial/National: Local government officers (health, environment, forestry), mining representatives.
- District: Political party leaders, mine workers' affiliations, residents, environmental institutions, NGOs, CBOs.
- Local Communities: Sample households were randomly selected at regular intervals (10–50 km) from town fringes and industrial mining areas for group discussions. This diverse input aimed to capture community opinions on mining pollution impacts.

3.10.3 Participatory Rural Appraisal

A participatory rural appraisal approach was employed, focusing on both men and women, traders, and artisanal miners, acknowledging their differing perceptions of mining benefits and impacts. Adults, children, and those with disabilities in targeted urban, suburban, and rural communities participated in semi-structured interviews, integrating open-ended questionnaires and face-to-face discussions (Fig. 3.2). This structured approach ensured a comprehensive understanding of mining impacts on communities, highlighting diverse perspectives and aiming for better-informed policy decisions.

Fig. 3.2 Community members were selected near and far away from mining and metallurgic industry operations for interview

3.10.4 Questionnaire

The questionnaire was carefully structured into thematic components tailored to relevant actors, aiming to gather the most reliable information about the mining impacts on the surrounding population and natural resources. Key areas of focus included the following:

Thematic components and key areas

1. Negative impacts on farmland, food, and air and water quality: Questions were designed to gauge the community's awareness of how mining activities affect farmland, food, and water quality and the potential illnesses that residents face due to these impacts.
2. Suggestions for mining management and governance: Respondents were asked for their thoughts on how to manage and govern mining activities to better protect the environment.
3. Face-to-face interviews: Conducted with stakeholders at households, workplaces, or in the fields, these interviews avoided the participation of mining and industrial companies to ensure unbiased information. Topics included the availability of alternative water sources, the importance of water for livelihoods, consumption of home-grown food, food and livelihood security, time spent in the area, and the corporate social responsibilities of mining companies.

Specific Questionnaire Topics

- **Water sources and usage**: Questions about alternative water sources, the impact on normal life if water sources were contaminated or cut off.
- **Food security and livelihood**: Inquiries into the consumption of food grown around homesteads or villages, factors contributing to food security, and the impacts of mining on livelihoods.
- **Environmental concerns**: Opportunities for respondents to voluntarily express concerns about environmental pollution were provided.

The response rate to the interview was an impressive 100%, reflecting high interest from inhabitants and ensuring that the data collected were representative of the entire community population.

Published Literature Review

A thorough review of grey and published literature was conducted to understand global and regional trends in mining impacts, particularly in the Copperbelt region of Zambia and the DR Congo. This review included official documents, publications, reports, and various articles to identify the key drivers of environmental degradation in copper mining and mineral processing.

Findings from Existing Literature

The review highlighted both the ecological, social, and economic benefits and the negative impacts of mining and metallurgical industries. Key negative impacts included the following:

- Loss of vegetation cover: Widespread deforestation due to mining activities.
- Destruction of water bodies: Significant damage to water bodies from mining pollutants.
- Loss of biodiversity: Negative effects on local biodiversity.
- Land use changes and food insecurity: Alterations in land use and resulting food insecurity.
- Increased social vices and conflicts: Mining-related social issues and conflicts.
- High cost of living and air pollution: Economic strain and environmental pollution impacting communities.

3.11 Sample Collection and Preparation

3.11.1 Sampling Food Crops

Before collecting samples, local residents, village headmen, and community members were approached, usually at their homes, to brief them about the study. This included both women and men, focusing on the potential health effects of long-term exposure to hazardous chemicals. The briefings aimed to gauge their knowledge about the negative impacts of mining on agricultural lands, food and water quality, and the potential illnesses they might face due to these impacts.

Sample Collection

After obtaining consent, various edible leaves, tubers, bulbs, fruits, grains, and seeds were sampled from crops in farms and gardens (Table 3.1) including the following:

- *Amaranthus hybridus* (amaranth).
- *Allium cepa* (onion).
- *Capsicum frutescens* (pepper).
- *Colocasia esculenta* (spinach).
- *Cucurbita maxima* (squash).
- *Hibiscus trionum* (hibiscus).
- *Ipomoea batatas* (sweet potato).
- *Manihot esculenta* (cassava).
- *Ocimum basilicum* (basil).

Table 3.1 Types of food crops cultivated in farms or çi = n home gardens for domestic consumption in the Central African Copperbelt of Zambia and the DR Congo

			DR Congo (Swahili, French)	Zambia (Bemba)		
Amaranth	*Amaranthus hybridus* L.	Amaranthaceae	Mchicha or Lengalenga; Epinard du Sudan	Bondwe	Leaves	Boiled with condiments
Basil	*Ocimum basilicum* L.	Lamiaceae	?;?	?	Leaves	Fresh or dried in soups
Cabbage	*Brassica oleracea* (L. var. *capitata*)	Brassicaceae (mustard)	Sukuma wiki; Choux	?	Leaves	Boiled, fried, sautée or stuffed, in salads
Cassava	*Manihot esculenta* Crantz	Euphorbiaceae	Mhogo, Kakonte, Sombe: Manioc	Tute, Katapa	Tubers, leaves	Tubers: boiled, fried, grilled, raw, flour, bread, beverage Leaves: boiled to cook with condiments
Common beans	*Phaseolus vulgaris* L.	Fabaceae	Maharagwe or Mahalaki; Haricot	Chilemba, Chimpapila	Leaves, grains	Leaves: cooked with condiments Grains: boiled to cook with condiments; whole immature pod boiled
Hibiscus?	*Hibiscus trionium* L. var. *trionium*	Malvaceae	Ngayingayi; Oseille	Lumanda	Leaves	Fried, mushed, and boiled
Maize	*Zea mays* L.	Poaceae, Gramineae	Mahindi; Maïs	Mataba	Grains	Fresh boiled, grilled, or dried milled in flour to cook a stiff pasty mush known as fufu or ugali; Nsima
Onion	*Allium cepa* L.	Lilliaceae	Kitungulu; Oignon	?	Bulbs, leaves	In soups as condiment, in salads

(continued)

Table 3.1 (continued)

			DR Congo (Swahili, French)	Zambia (Bemba)		
Pepper	*Capsicum frutescens* L.	Solanaceae	Pilipili; Piment	?	Fruits	Condiment for soups
Pumpkins, squash	*Cucurbita maxima* (Dutch)	Cucurbitaceae	Malenge; Courge	Chibwabwa, Chipushi	Leaves, fruits, seeds	Leaves: chopped and boiled with condiments Fruits: boiled Seeds: inner part steamed, pounded, and cooked with condiments
Sorrel	*Rumex acetosa* L.	Polygonaceae	?; Oseille also	?	Leaves	In salads or puree in soups or sauce
Sweet potato	*Ipomoea batatas* L.	Convolvulaceae	Viazi vya tamu; Patate douce	Fyumbu, Kalembula	Tubers, leaves	Leaves: boiled to cook with condiments Tubers: boiled, fried, raw
Taro, cocoyams	*Colocasia esculenta* L.	Araceae	?; Ignames	?	Leaves, tubers	Leaves: chopped and boiled with condiments Tubers: boiled
Tomato	*Lycopersicon esculentum* Mill.	Solanaceae	Nyanya; tomate	?	Fruits	As a condiment in soups, ripe raw in salads

- *Phaseolus vulgaris* (beans).
- *Rumex acetosa* (sorrel).
- *Solanum lycopersicum* (tomato).
- *Zea mays* (maize).

Harvesting was done with community assistance, following local practices. Samples were immediately placed in labeled paper bags and stored in a refrigerator. A total of 48 root and tuber crops, 143 leafy vegetables, 51 fruits, 5 bulbs, and 33 other plants were collected. Of these, 205 samples were from areas with intense mining activities, while 75 were from remote areas.

Pre-treatment Process

The samples were pre-treated in laboratories at Copperbelt University in Zambia and the Institut Supérieur Pédagogique in Lubumbashi, the DR Congo. They were rinsed in tap water, followed by three to four rinses with distilled water, sliced with a sterile knife, oven-dried at 70 °C to a constant weight for 48 h, and then homogenized using sterile porcelain mortars and pestles. The weight of each powdered sample was recorded before packing in Minigrip sealed bags for further analysis. Unless otherwise specified, samples were shipped to Belgium for laboratory analysis. The thorough and systematic approach adopted in the study ensured the collection of comprehensive data for a better understanding of the impacts of mining and metallurgical industries on agricultural produce and community well-being.

3.11.2 Water Samples

Every effort was made to maintain a high degree of cleanliness for all equipment used in water sample collection, including bottles, containers, ropes, and filtering equipment. Water samples were collected from various sources:

- Residential faucets/taps: 23 samples.
- Domestic boreholes: 38 samples.
- Rivers and natural springs: 18 samples.

Collection Process

- **Residential faucets/taps**: Water was left to run for a few minutes to clear the line. Samples were slowly filled into 40 mL sterile polystyrene bottles (Plastiques-Gosselin, Hazebrouck, France) and securely capped.
- **Boreholes/wells**: Water was drawn using cut tins or plastic containers attached to long canes with wire or rope and then transferred to polystyrene bottles.
- **Rivers**: Samples were collected from the surface by pointing open bottles in the direction of the river flow, avoiding floating debris.

Out of 79 water samples, 42 were from Zambia and 37 from the DR Congo. Samples were pipetted into 1.5 mL Eppendorfs, immediately stored at −20 °C, and shipped for analysis to Belgium in sealed cooler boxes with ice packs.

3.11.3 Urine Sample Collection

Urine samples were only collected in the DR Congo due to administrative issues with Zambia's Ministry of Health. Donors included the following:

- Infants and children (2+ years): 9 samples.
- Adolescents: 4 samples.
- Adults: 13 samples.
- Elderly: 3 samples.

Collection Process

Following procedures by Banza Lubaba et al. (2009), urine samples were collected throughout the day with local health personnel's assistance. Participants were informed about the study's need for urine samples and provided instructions to avoid external contamination. Their participation was purely based on their willingness. An aliquot of 5N hydrochloric acid was added to prevent decomposition and loss of metals. Samples were then stored in a cooler box and later in a refrigerator, before being shipped to Belgium for analysis.

Additional Information

For each participant, the following data were recorded to track possible levels of intoxication:

- **Anthropogenic indices**: Birth date, age, sex, weight.
- **Demographic indicators**: Time spent at site, past and present occupation.
- **Medical history**: Symptoms, general health status.
- **Vital signs**: Heart rate, pulse.
- **Lifestyle factors**: Smoking and alcohol intake patterns.

3.12 Sample Analysis

3.12.1 Food Crops

Sample Preparation and Analysis

According to Muimba-Kankolongo et al. (2021), ground oven-dried powder samples of various food crops were first crushed and then sequentially digested in nitric acid (HNO_3). The process involved three successive runs of dissolution in HNO_3 (1 mL suprapur 70%), followed by evaporation to dryness for 4 h at 110 °C. This was complemented by microwave heating at different power levels (250 W for 5 min; 400 W for 5 min; 650 W for 10 min; 250 W for 5 min). After cooling, the contents were filtered through Whatman filter paper No. 42 and kept in a final solution of 2% hydrochloric acid (HCl). The samples were then analyzed for total metal content using inductively coupled plasma/optical emission spectroscopy (ICP-OES,

Perkin-Elmer Optima 3300 DV, Norwalk, CT, USA) at the Louvain Centre for Toxicology and Applied Pharmacology (Université catholique de Louvain, Belgium).

Measurement of Trace Elements

For concentrations below 5 µg Co/L in the digest, measurements were made by ICP-MS (Thermo X-series I) in He mode, using Ga as an internal standard. Certified reference material NIST SRM-1573a (tomato leaves, National Institute of Standards and Technology, Gaithersburg, Germany), with a certified value of 0.57 mg Co/kg and a measured value of 0.54 ± 0.01 mg Co/kg, was included in duplicate with each analysis.

Determination of Trace Metals and Metalloids

The analysis determined the contents of 20 different trace elements and metalloids, with a focus on those of major concern in foods, such as manganese (Mn), nickel (Ni), lead (Pb), zinc (Zn), cobalt (Co), arsenic (As), uranium (U), cadmium (Cd), and copper (Cu) (FAO/WHO 1993; Musa Obadia et al. 2018; Nemery and Banza Lubaba Nkulu 2018). Metals and metalloids released by digestion were added to an ammonium iodide (NH_4I) solution, extracted by a MIBK-TBP mixture (80:20), and then stripped with H_2O_2 (5%), HNO_3 (0.5%), and ammonium dihydrogen phosphate ($NH_4H_2PO_4$) (0.1%).

Quantification

Quantification was achieved by injecting aliquots of sample solutions (10 µL) into a carbon rod atomizer of an atomic absorption spectrophotometer, set at a variable detector with 0.5 absorbance unit full scale. Results were expressed in mg/kg of dry matter for each crop species.

3.12.2 Surface, Underground, and Tap Water Samples

Preparation and Analysis of Water Samples

Before analysis, water samples were filtered using a 0.45 µm micropore membrane filter to prevent clogging the atomizer burner. The trace metal contents were then quantified as described by Muimba-Kankolongo et al. (2021). Specifically, the analysis was performed using inductively coupled argon plasma mass spectrometry (ICP-MS) with a detection limit as low as 0.5 parts per billion (ppb) (Perkin-Elmer Optima 3300 DV, Norwalk, CT, USA).

Quality Control and Detection Limits

For quality control, a certified reference water sample from the National Institute of Standards and Technology (NIST) was included (certified value = 27 μg Co/L; measured value = 26 ± 1 μg Co/L). The detection limits for all samples were calculated as twice the standard deviation of six preparation blank sample measurements, as outlined by Cheyns et al. (2014). For values falling below the detection limits, half of the detection limit for each trace element was considered, and the results were expressed in micrograms per liter (μg/L) of water.

3.12.3 Urine Samples

Sample Preparation

Urine specimens (500 μL) were diluted quantitatively (1 + 9) with a solution containing 1% nitric acid (HNO_3), 0.5% hydrochloric acid (HCl), and internal standards (Sc, Ge, Rh, and Ir). Additionally, another solution containing 2% w/v 1-butanol, 0.05% w/v EDTA, 0.05% w/v Triton X-100, and 1% w/v ammonium hydroxide (NH_4OH) with the same internal standards was also used. Mercury (Hg) and palladium (Pd) were analyzed using no-gas mode.

Analysis of Solubilized Urine Samples

Solubilized urine samples were analyzed at the Louvain Centre for Toxicology and Applied Pharmacology (Université Catholique de Louvain, Belgium), using blind analysis as previously published by Banza Lubaba Nkulu et al. (2009a, b) and Hoet et al. (2013). The analysis quantified 24 trace metals using inductively coupled argon plasma mass spectrometry (ICP-MS) on an Agilent 7500ce instrument.

Detection Limits and Quality Control

The limits of detection (LoD) and quantification (LoQ) were calculated as three and nine times, respectively, the standard deviation of the blank signal. The laboratory has achieved successful results in external quality assessment schemes organized by the Institute for Occupational, Environmental, and Social Medicine of the University of Erlangen, Germany (G-EQUAS program), and by the Institut National de Santé Publique, Québec (PCI and QMEQAS programs). For the QMEQAS 2012-01 round, all results, except for Li, In, and Pd (not included in the QMEQAS program), were associated with a z'-score between −1.00 and +1.00.

Laboratory Certification

The laboratory holds ISO15189 certification for the measurement of 20 metals in urine, including Be, Al, V, Cr, Mn, Co, Ni, Cu, Zn, As, Se, Mo, Cd, Sn, Sb, Te, Ba, Tl, Pb, and Bi.

Creatinine Determination

Creatinine is a metabolic waste product generated by the body, which serves as an indicator of kidney function. The kidneys filter creatinine and other waste products out of the blood, expelling them through urination. Therefore, the concentration of creatinine in urine can be used to measure the levels of various chemicals present in the urine (Santopinto et al. 2003; Galea et al. 2015). The urinary creatinine was determined using a Beckman Synchron LX 20 analyzer (Beckman Coulter GmbH, Krefeld, Germany) to correct for urine dilution and assess kidney function. Creatinine is a waste product of metabolism filtered out of the blood by the kidneys. The final metal concentrations were corrected for creatinine by dividing the concentration of urinary trace elements of each heavy metal and metalloid by the creatinine concentration, and the results were expressed in µg/g creatinine.

3.13 Data Analysis

The collected data were compiled and input in different excel spreadsheets, double-check the accuracy of the data entered to ensure consistency and completeness, and remove unnecessary data points and highlighted outliers. They were then analyzed using a Statistical Package for Social Sciences (SPSS version 19, Chicago, IL, USA), and Fisher's test ($i \leq 0.05$) for a one-way analysis of variance (ANOVA) was used to compare variances between trace metal concentrations across countries, districts, and sites in districts within each country and their interactions. Some data were logarithmically transformed for statistical analysis after testing the original data for normality. Because the data did not follow a normal distribution, results are shown as medians along with the 5th, 25th, 75th, and 95th percentiles. Non-parametric methods, including the Mann-Whitney and Kruskall-Wallis tests as well as Spearman correlation analyses, were used for statistical evaluation. The Mann-Whitney test was used to compare the differences between two independent groups by assessing whether one group tends to have higher or lower values than the other. That of Kruskall-Wallis evaluated whether there were significant differences in the median ranks among the groups. As a non-parametric method, the Spearman correlation analysis was used to assess the strength and direction of the relationship between two variables. It is particularly useful when the data are ordinal (ranked) or non-normally distributed. The Spearman correlation coefficient (ρ) ranges from −1 to 1:

A value close to 1 indicates a strong positive relationship.
A value close to −1 signals a strong negative relationship.
A value near 0 suggests no correlation.

This method is widely applied in scenarios where assumptions about normality or linearity are not met, making it a versatile tool for statistical analysis. These analyses were performed using GraphPad Prism version 8.0.0 for Windows (developed by GraphPad Software, San Diego, CA, USA). Statistical significance was defined as a p value of less than 0.05, using a two-sided approach.

Chapter 4
Mining and Metallurgic Industry Sectors in the Region

Keywords Mining industry · Copperbelt · Economy · Artisanal mining · Historical context · Regulations

4.1 Background

Mining activities in Zambia and the R Congo have profound environmental impacts, shaped by the unique challenges and opportunities in the central African Copperbelt region. Although Zambia's mining sector, particularly in the Copperbelt province, has been a cornerstone of its economy, it has also come with significant environmental costs. Historical and ongoing mining operations have led to pollution of air, soil, and water, as well as land degradation. For instance, Kabwe, once home to industrial lead and zinc mining, is now one of the most lead-polluted places globally. Decades of mining have left behind 6.4 million tons of lead-bearing waste, exposing up to 200,000 people to toxic lead dust (Human Rights Watch 2019; Blacksmith Institute report 2007). Efforts like the Copperbelt Environment Project (World Bank 2015b) have aimed to address these issues, yet challenges persist due to inadequate implementation of environmental regulations and lack of technical capacity. In the DR Congo, the extraction of critical minerals like cobalt and coltan has fueled global technological advancements but at a steep environmental and human cost. Industrial cobalt mining in the southern provinces like that of Haut Katanga has led to severe air, soil, food, and water pollution, harming local ecosystems and communities and agriculture (Banza Lubaba Nkulu et al. 2009b; Mwitwa et al. 2012a; Muimba-Kankolongo et al. 2021, 2022). Communities living near these mines face health risks, including reproductive health issues linked to toxic contamination. Additionally, artisanal mining practices often result in deforestation and soil erosion, further exacerbating environmental degradation. While international demand for these minerals continues to rise, both Zambia and the DR Congo grapple with balancing economic benefits against the environmental and social toll.

A. Muimba-Kankolongo et al., *Mining Impacts on the Environment in the Central African Copperbelt of Zambia and the Democratic Republic of Congo*,
https://doi.org/10.1007/978-3-031-96786-3_4

4.2 Mine and Metallurgic Industries in the Region

The Central African Copperbelt, also described as the Lufilian Arc, is a large area of land located in the southern part of the former Katanga Province in the Democratic Republic of Congo (DR Congo) and the Copperbelt and Northwestern Provinces in Zambia. This region is one of Africa's largest mining areas, extending over more than 500 km from Kolwezi in the DR Congo to Luanshya in Zambia (Butler 2006), and it holds substantial copper and cobalt mineral reserves [Lydall and Auchterlonie 2011]. These minerals are crucial to the economies of both countries [Wolfire et al. 1998; Ashton et al. 2001; MFNP, GRZ, 2002; Ministry of Mining and Hydrocarbons 2003; Global Witness 2006; Fraser and Lungu 2007; Mwitwa and Kabemba 2007; ICMM 2014]. Mining exploitation contributes about 28% to the DR Congo's GNP and 70% of its export value [Ministry of Mining and Hydrocarbons 2003], making it the main source of foreign currencies [Global Witness 2006]. The mineral deposits are found in sedimentary rocks known as the Mines Group (Lower Roan Group), which are between 1050 million and 650 million years old.

Equally, mining sector is all-important to Zambia's economy, being responsible of about 70% of foreign earnings [Zambia EITI Report 2020]. From the colonial period to the present time, copper (Cu) and various other important minerals, notably cobalt (Co), zinc (Zn), lead (Pb), gold (Au), diamonds (C), manganese (Mn), tin (Sn), tantalum (Ta), lithium (Li), nickel (Ni), and uranium (U), have been extracted from large-scale mines in the region and processed in local mine smelters. Since the late twentieth century, artisanal mining has also significantly contributed to cobalt production in the DR Congo (Yager 2012; KPMG Global Mining Institute 2014; Banza Lubaba Nkulu et al. 2018) and in Zambia (Kambani 2003; Ministry of Lands and Natural Resources, GRZ 2018; Hilson 2020; Nachinanga Gubula Siaciti 2022).

The World Bank's report by Broadman (2007) highlighted on the recent increase in global demand for copper raw materials, driven by rapid economic growth in Asia, which has sparked the higher commodity prices and a climb for natural resource excavation across the entire continent of Africa. This has prompted African leaders to call for increased foreign investment in their respective countries, leading to the multiplication of mining exploration and exploitation, which now significantly shape the overall continent's economy. Consequently, the DR Congo and Zambia, both located in the Central African Copperbelt, are among the world's top ten leading copper producers, with rapidly increasing production volumes and substantial global investment in the sector (Global Business Reports 2014).

Years of mining extractions and metallurgical industry operations in the Central African Copperbelt have significantly undermined environmental sustainability and social equity. The negative impacts observed at each development stage of their operations comprise habitat and vegetation losses, social displacement, land alienation, increased soil erosion and siltation of nearby water bodies, ground surface disturbance, and the discharge of hazardous and contaminating elements into the natural bodies of water. These issues pose health risks, particularly for children, and release contaminants into the air, including heavy metals, organic compounds, and sulfur dioxide, all of which have adverse health effects.

The region undergoes several ecological problems associated with water pollution which have negative effects on human health and biodiversity, particularly in the neighborhood of mining and metallurgical industries. Environmental degradation, such as contamination of ground and surface water from industrial discharges and waste dumps, poses significant risks to human health, food security, and energy for the population (Feeney 2001; Zambia Consolidated Copper Mines-IH (ZCCM-IH) 2005; Kříbek et al. 2011; Ncube et al. 2012). Water pollutants adversely affect the growth and well-being of the population, especially children (Yabe et al. 2015; Banza Lubaba Nkulu et al. 2018; Musa Obadia et al. 2018; Nemery and Banza Lubaba Nkulu 2018; Human Rights Watch 2019; Umpula and Van Woudenberg 2024).

The recent alarming report from Zambia by DownToEarth (2025, available at https://www.downtoearth.org.in/africa/environmental-catastrophe-in-zambia-after-acid-leak-at-chinese-operated-copper-mine) indicates that Zambian authorities and environmental advocates are grappling with the severe consequences of an acid spill at a Chinese-owned mining operation, which has contaminated a key river and poses a potential risk to millions of people. Traces of pollution have been found over 100 kilometers downstream, raising fears of a major ecological disaster. The spill occurred on February 18 when a dam holding acidic waste from a copper mine in northern Zambia collapsed. Investigators from the Engineering Institution of Zambia estimate that approximately 50 million liters of waste, including strong acids, heavy metals, and dissolved solids, entered a tributary connected to the Kafue River, the country's most critical water source.

Environmental activist Chilekwa Mumba, based in Zambia's Copperbelt, described the event as a catastrophic environmental crisis with far-reaching consequences. This incident highlights China's prominent role in Zambia's copper mining sector, as the nation is one of the top global producers of copper—a resource crucial for modern technology. To mitigate these environmental impacts, the country's President has called for expert support to address the emergency, describing it as a threat to both human and wildlife populations along the Kafue River, which spans more than 1500 kilometers across Zambia. Efforts are underway to assess the full scale of the damage.

An Associated Press journalist revealed dead fish along the riverbanks nearly 100 kilometers away from the mine, operated by Sino-Metals Leach Zambia, a company primarily owned by the China Nonferrous Metals Industry Group. Zambia's Ministry of Water Development and Sanitation confirmed extensive damage, including destroyed farmlands along the riverbanks. There is a growing concern that toxic waste is infiltrating groundwater, potentially contaminating surrounding areas and further disrupting the river's ecosystem. Local residents have observed drastic changes, with one individual, Sean Cornelius, lamenting that the river, once vibrant, has become lifeless almost overnight.

The Kafue River supports 60% of Zambia's population, providing water for domestic use, agriculture, and industry. It is a vital drinking water source for five million people, including residents of Lusaka, the capital city. As a result of the spill, the water supply in Kitwe city, home to roughly 700,000 people, has been suspended. In response, the Zambian government has mobilized the air force to release lime into the river to neutralize the acid and mitigate further harm. Speedboats are also dispersing lime to affected areas. The Government spokesperson stressed the gravity of the situation and confirmed that Sino-Metals Leach Zambia would fund the cleanup effort.

Mining-induced contamination of ambient air, both surface and underground water, and soil (dry and wetlands) has been a major concern since the colonial period. During this time, African housing was deliberately built close to mining facilities, while housing for whites was located further away (Larmer et al. 2021). Industrial effluents and waste have been released across the landscape, causing water flooding and erosion due to mine waste spills from broken pipes. Toxic smoke plumes from mining and metallurgical industries further contribute to the pollution of surface and underground water (Chipundu and Kunda 1994; Zambia Consolidated Copper Mines-IH (ZCCM-IH) 2005; Ncube et al. 2012; Mwaanga et al. 2019).

Our survey in the Chingola district of Zambia revealed drastically poor water quality, with muddy water being common (Figs. 4.1 and 4.2). Streams draining mining areas empty into the Kafue River or its tributaries, the principal water source for the Copperbelt Province. The DDCC in Chingola recognized general cleanliness and water quality as common problems. According to KCM statistics, drinking water standards were breached for several water quality indices, although heavy metal levels fell within Zambian standards.

Likewise, in the DR Congo, numerous bodies of rivers contributing to the several sources of drinking water and the areas where the water table feeds into these rivers have profusely become polluted. In Lubumbashi, for instance, it has been noted that artisanal miners frequently pollute rivers by washing and sieving the soil they

Fig. 4.1 Quality of water from some boreholes sank by KCM in Mushishima village around the mine area, Chingola District, Copperbelt Province of Zambia

Fig. 4.2 One of the DelPHE project experts—Professor Banza Lubaba Nkulu—taking a photo of a severely polluted river in Luanshya District (above) and vegetation and soil degradation of the valley which was once utilized by the inhabitants to grow vegetables in Chingola District (below) in the Copperbelt Province of Zambia

collect to extract minerals, particularly around mining areas. This results in contaminated and dirty drinking water, undermining the right to good health and safer living conditions for communities near the mines.

Rivers affected by mining activities in the country include Kafubu, Kapolobwe, Kasapa, Kimilolo, Naviundu, and Panda. Notably, cadmium contamination is prevalent in the Kimilolo and Naviundu Rivers, while the Kafubu River has been previously contaminated with arsenic. Analysis of the samples of water from the Naviundu River in Lubumbashi showed that quality standards were breached for minerals like cadmium, manganese, mercury, and sodium (Luboya et al. 2004).

The contamination of water and its marine resources, such as fish, lobsters, turtles, crabs, shrimps, and oysters, which are important for nutritional security contributes to increased toxic buildup in people's bodies. Asthma and other respiratory problems are widespread due to mining activities, particularly affecting children and the elderly. Health hazards also affect workers in the mining industry. According to annual ILO statistics, mining and quarrying are among the most dangerous and accident-prone industries, with over 15,000 deaths per year.

Considering the extensive environmental damage including that to the ecosystems resulting from mineral extraction and operations of metallurgic industries which produce hazardous wastes containing dissolved toxic metals over the years, caused by mining activities in Zambia and the DR Congo, households in the fringe of mines and metallurgical industries have raised significant concerns about the poor quality of water available for domestic use, prompting the urgent needs for several measures to be taken to mitigate the negative impacts and promote sustainable practices.

4.3 Mining Industries in the Copperbelt of Zambia

According to the Ministry of Mines and Minerals Development (2022), Zambia has a long history of mining, dating back to 1920. Mining activities began long before colonial times, with indigenous people mining surface deposits for copper and fashioning the metal into ingots for tools, weapons, and trade, whereas major expansion occurred between 1924 and 1969, driven by foreign investment, particularly from the USA and South Africa. During the period under British rule, all mineral rights were controlled by the British South African Company (BSAC), which focused primarily on profit generation with little regard for environmental impacts. This led to the creation of numerous unattended mine waste dumps that have become significant pollution sources, particularly affecting nearby communities. After gaining independence in October 1964, Zambia enacted the Mining Ordinance (Amendment) Act of 1965, which transferred all mineral rights to the President of Zambia. This act was later amended and replaced by the Mines and Minerals Act of 1976, leading to the formation of the Zambia Consolidated Copper Mines (ZCCM), which merged the Roan Copper Mines (RCM) and the Nchanga Consolidated Copper Mines (NCCM) (Fig. 4.3).

Zambia is the largest copper producer in Africa and seventh largest globally and depends heavily on its mining industry, which is a major employer in a country with 13 million citizens (Global Business 2014). It reported further that in 2012, Zambia's copper output, estimated at 800,000 mt/y, accounted for 4.8% of global copper

Fig. 4.3 Nchanga copper mine, previously the Chingola copper mine, is an underground and open-pit cobalt and copper mine in Chingola, in the Copperbelt Province of Zambia

production. The country's copper reserves are estimated at 21 million metric tons, 6% of the world copper reserve (https://theconversation.com/climate-change-action). While copper remains the backbone of Zambia's economy, it is not the only mineral wealth waiting to be discovered in this relatively unexplored nation. Proper management and exploitation of its vast mineral resources, including gold, diamonds, coal, and various industrial minerals, could provide a strong economic safety net. Zambia also boasts an estimated 11 million metric tons of the highest-grade zinc and lead deposits globally and contains 20% of the world's emerald supply, with the world's second-largest emerald deposit and Africa's largest deposits of aquamarine and amethyst.

Zambia's large-scale copper mines are mainly located in the Copperbelt and North-Western Provinces (ICMM, 2014). The Copperbelt is the traditional mining hub, hosting large copper mines previously operated by ZCCM (Table 4.1). Together with the capital Lusaka, the Copperbelt is the most developed and urbanized province in Zambia. North-Western Province, referred to as the "New Copperbelt" due to significant mining investments over the last decade, is largely rural with limited infrastructure and is the least populated of Zambia's nine provinces. Solwezi, the provincial capital, is where current mining activities are centered. At the time of the study, two large mines were operational in the region, with several others in various stages of development.

A recent World Bank report highlighted the mining sector's importance to Zambia's economy and its likely continued significance (Reconstruction and Development/World Bank, 2015a). Zambia has considerable, vast, and multifaceted potential for further mineral discoveries due to its high prospectivity. They comprise:

Table 4.1 Some of the major mining companies in Zambia

Details of some of operational mines in Zambia as of May 20, 2013			
Mine name	Location	Ownership	Commodities mined
Baluba center copper mine	Luanshya	China non-ferrous Mining Corp ltd. (80%); ZCCM investments holdings PLC (20%)	Copper, cobalt
Chambishi copper operation	Chingola	China non-ferrous Mining Corp ltd. (85%); ZCCM investments holdings PLC (15%)	Copper, cobalt
Chibuluma copper/cobalt mine	Kitwe	Metorex Pty ltd. (85%); ZCCM investments holdings PLC (15%)	Copper, cobalt
Collum colliery	Lusaka	Government of Zambia (100%)	Coal
Kagem emerald mine	Luanshya	Gemfields PLC (75%); government of Zambia (25%)	Gemstones
Kansanshi copper/ gold mine	Solwezi	First quantum minerals ltd. (80%); ZCCM investments holdings PLC (20%)	Copper, gold
Kariba (Mapatizya) amethyst mine	Livingstone	Gemfields PLC (50%); government of Zambia (50%)	Amethyst
Kasempa copper mine	Solwezi	H and S Mining ltd. (100%)	Copper
Konkola copper/cobalt operation	Chingola	Vedanta resources PLC (79%); ZCCM investment holdings PLC (21%)	Copper, cobalt
Lilayi talc mine	Lusaka	Unspecified (100%)	Talc
Lubambe copper mine	Chililabombwe	Vale S.A. (40%); African rainbow minerals (40%); Zambia consolidated copper mines investment holdings (20%)	Copper, cobalt
Lumwana copper mine	Lubumbashi	Barrick gold corporation (100%)	Copper, cobalt, gold
Maamba colliery	Chroma	Nava Bharat Pte ltd. (65%); ZCCM investments holdings PLC (35%)	Coal
Mopani copper/cobalt operation	Kitwe	Glencore Xstrata PLC (73%); first quantum minerals ltd. (17%); ZCCM investments holdings PLC (10%)	Copper, cobalt
Muliashi copper operation	Luanshya	China non-ferrous Mining Corp ltd. (80%); ZCCM investments holdings PLC (20%)	Copper, cobalt

Source: KPMG International Cooperative, 2014

4.3.1 Key Minerals

1. **Copper**: Zambia is one of the world's largest producers of copper. The Copperbelt and North-Western Provinces are home to extensive copper mines, both operational and under development.
2. **Cobalt**: Often found alongside copper, Zambia also has significant cobalt reserves, crucial for various industrial applications.

3. **Gold**: The country has considerable gold deposits that are yet to be fully explored and exploited.
4. **Emeralds**: Zambia holds 20% of the world's emerald supply, with substantial deposits of aquamarine and amethyst as well.
5. **Zinc and lead**: Zambia has an estimated 11 million metric tons of the highest-grade zinc and lead deposits globally.
6. **Other minerals**: The country also possesses coal, diamonds, and a variety of industrial minerals.

4.3.2 *Economic Impact*

- **Foreign direct investment**: The mining sector attracts significant foreign direct investment, contributing to the country's economic growth.
- **Government revenue**: Mining tax revenues form a substantial portion of total government revenue.
- **Employment**: The mining sector is a major source of formal employment, both directly within industry and indirectly through related services and industries.

4.3.3 *Potential for Growth*

- **High prospectivity**: Zambia's geology indicates high potential for new mineral discoveries, suggesting that the mining sector can expand further.
- **Infrastructure development**: Continued investment in infrastructure, such as roads and power supply, will support the growth of the mining sector.
- **Technological advancements**: Adoption of modern and sustainable mining technologies can improve efficiency and reduce environmental impacts.

The mining sector is a major contributor to foreign direct investment and mining tax revenues, which form a significant portion of government revenue. It is also a key source of formal employment, both directly and indirectly, consistent with more mature mining economies.

4.4 Mining Industries in the Copperbelt of the DR Congo

The DR Congo boasts significant mining potential, with an estimated 1100 different mineral substances (Yearbook rapport économique 2023–24). The country holds the world's largest reserves of cobalt and substantial reserves of tantalum, tin, gold, and diamonds (Geenen and Marysse 2016). The diversity of minerals extends from copper to coltan (Global Business, 2014). The mining sector, representing over 20% of the GDP, is a major economic pillar, with all provinces claiming mineral wealth mainly, for example, copper, cobalt, zinc, manganese, uranium, cassiterite (tin ore), coal, gold, silver in Haut-Katanga; diamond in Kasaï-Oriental and Grand Kasaï; gold, cassiterite (tin ore), and coltan in Maniema and North and South Kivu; and copper, cobalt, zinc, manganese, uranium, cassiterite (tin ore), coal, gold, and silver in Lualaba.

4.4.1 Some of the Key Minerals Include the Following

1. **Cobalt**: The DRC is the world's largest producer of cobalt, making over 70% of the world's cobalt, a critical component for the production of batteries for electric vehicles and other electronic devices. This has led to increased global demand for Congolese cobalt.
2. **Copper**: The country holds some of the largest copper reserves in the world. The Katanga region is particularly known for its high-quality copper deposits.
3. **Diamonds and gold**: In addition to base metals, the DRC is also rich in precious minerals. Artisanal mining of diamonds and gold plays a significant role in local economies, although it is often informal and unregulated.
4. **Coltan**: The DRC is a significant source of coltan, essential for the manufacture of capacitors in electronics.
5. **Lithium**: Which is used in rechargeable lithium-ion batteries for electronics and electric vehicles and also very important for manufacturing of glass and ceramics.
6. **Iron ore**: Primarily used in the production of steel, which is essential for construction and manufacturing industries.
7. **Manganese**: Used in steel production to improve hardness and durability and also in batteries and as a chemical oxidizer.
8. **Zinc**: Used for galvanizing steel to prevent corrosion and is very important in the production of alloys and batteries.
9. **Lead**: Historically used in batteries, radiation shielding, and in certain types of glass but more importantly in construction materials and ammunition.

The DR Congo's rich mineral deposits not only contribute significantly to its economy but also play vital roles in various industries worldwide.

According to the US Geological Survey (USGS) in 2010, the DR Congo accounted for 51% of the world's cobalt production, 25% of industrial diamonds, 14% of tantalum, 5% of gem-quality diamonds, and 3% each of copper and tin (KPMG International Cooperative, 2014). The DR Congo has some of the largest deposits of non-ferrous metals globally, including about 3% of the world's copper reserves, 45% of global cobalt reserves, approximately 25% of global diamond reserves, and significant reserves of gold and tantalum. The DR Congo is the largest global producer of cobalt, holds the tenth highest gold reserves worldwide, and has the largest diamond reserves in Africa. The country's vast mineral resources are expected to attract significant international interest, with the mining sector's value anticipated to reach $2.6 billion in 2017, primarily driven by copper and gold production.

Industrial mining in the country is predominantly concentrated in Haut-Katanga Province, where minerals such as copper, cobalt, zinc, manganese, uranium, cassiterite (tin ore), coal, gold, and silver are produced (Geenen and Marysse 2016). Additionally, mining activities are expanding in North and South Kivu, Maniema, and East Province (gold). In Katanga, the national company Gécamines (Fig. 4.4), which once contributed 85% of the country's export earnings and 42% of public revenues, saw production decline from 500,000 tonnes of copper in the 1970s to about 30,000 in 1990 due to a lack of maintenance and investment.

The mining sector attracts substantial foreign investment (Table 4.2) and contributes significantly to government revenue and employment. This influx of foreign investments in the country significantly boosted copper and cobalt production from 2006, reaching historic levels of over one million tonnes in 2014 (Banque Centrale du Congo, 2013 as cited by Geenen and Marysse 2016). This development positioned the DRC as Africa's leading copper exporter and the world's top cobalt exporter. Key companies involved in this sector include Sicomines (20% Gécamines, 80% Chinese companies), Kamoto Copper Company (KCC, 20% Gécamines, 80% other, including Glencore/XStrata), Tenke Fungurume Mining (TFM, 80% Freeport-McMoRan and Lundin Mining, 20% Gécamines), Frontier (95% ENRC Congo, 5% Congolese state), and Mutanda Mining (MUMI, 60% Glencore, 40% other) (EITI, 2014: 12–14 and 147–8).

Fig. 4.4 A general view of the Gécamines industrial plant in Lubumbashi city, Haut-Katanga Province, the Democratic Republic of Congo

Table 4.2 The directory of some mining businesses and quarries in the DR Congo

Name	Origin	Exploited mineral	Address
DE NOVO CONGO			KINSHASA—Gombe
AEL Mining services			Lubumbashi
Boss mining sprl			Lubumbashi
Congo cobalt corporation (CCC)			Lubumbashi
Entreprise Générale Malta.			
Forrest			Lubumbashi
GECAMINES			Lubumbashi
Golden African Resources SPRL			Lubumbashi, Likasi
Kamoto Copper Company (KCC)			Lubumbashi, Kolwezi
Katanga Metals SPRL			Lubumbashi, Kolwezi
Kinsenda Copper Company (KICC)			Lubumbashi
Mining Company Katanga Group			Lubumbashi
Phelps Dodge Congo SPRL			Lubumbashi
Rulvis Congo			Lubumbashi
Sase Mining SPRL			Lubumbashi
Swanmines & carrieres – Amenagement Minier & Transport SPRL			Lubumbashi
Tenke Fungurume Mining (TFM) Trade Services SPRL			Lubumbashi
Société Anhui Congo d'Investissement Minier (SACIM)	China	Industrial diamonds	Kinshasa,
Proximity Mining Expertise (Proximiex SPRL)			Lubumbashi
Robinson Int'l Afrique			Lubumbashi
SAFRICAS			Lubumbashi, Kinshasa –
AFRITEC			Kinsahsa
CARRIGRES			Kinshasa
Congo Roche SPRL			Lubumbshi
CORETEK Drilling Solutions SPRL			Lubumbashi
Entreprise Générale MALTA Forrest			Lubumbashi
La Nouvelle Commerciale—CGD			Lubumbashi
Trade Services SPRL		Analyses laboratories	Lubumbashi
Congo Diam SPRL			Kinshasa
De Beers SPRL			Kinshasa
MARGAUX			Kinshasa
SICOMINES	China	Mines cobalt and copper	Kolwezi, Kinshasa

(continued)

Table 4.2 (continued)

Name	Origin	Exploited mineral	Address
MIBA mine		Industrial diamonds	Mbujimayi, Kinshasa
M ACEC Congo			Kinshasa
Africa Commodities SPRL			Lubumbashi, Likasi

Source: Ministère de l'économie nationale, 2022

The dramatic increase in copper and cobalt production has shifted export composition. In 2004, nearly 57% of exports were diamonds, but by 2013, copper and cobalt accounted for 85% of all exports. Additionally, exports from the eastern provinces have gained importance. In 2011 and 2013, Banro Corporation (South Kivu) and Kibali Gold (East Province) began industrial gold production, increasing gold exports to 6125 kg, compared to 2546 kg in 2012 and less than 500 kg from 2006 to 2011 (Ministry of Mines, 2013). Other international companies, primarily Chinese and South African, are exploring and initiating production in several other eastern provinces.

4.5 Artisans and Small-Scale Mining in Zambia and the DR Congo

Artisanal and small-scale mining (ASM) denotes mining activities by non-registered individuals or organizations which are formal, informal, or illegal depending on the regulations in the area (International Labor Organization, 2020). Artisanal mining is characterized by small-scale mining activities that typically use rudimentary methods and tools. It is distinguished from small-scale mining, which can be semi-mechanized and more capital-intensive. Although artisanal mining typically involves rudimentary technologies like pickaxes, shovels, and pans, small-scale mining can be mechanized, semi-mechanized, and have a higher degree of capitalization. Together, these activities are categorized as ASM. Globally, about 40 million people work in ASM across 80 countries, with 150 million people relying on ASM practices, including miners' families and the broader community. Artisanal miners contribute to 20% of the global supply of gold and diamonds, 80% of sapphires, 26% of tantalum, and 25% of tin. Artisanal mining plays a significant role in the global mineral supply chain:

- **Global supply contribution**: Artisanal miners contribute to about 20% of the world's gold and diamond supply, 80% of the world's sapphire supply, 26% of the global tantalum supply, and 25% of tin production.
- **Economic impact**: With approximately 40 million people engaged in ASM across 80 countries and an additional 150 million people dependent on it, arti-

sanal mining supports a substantial portion of the population, especially in developing nations.

Artisanal mining involves a diverse group of participants, often driven by economic necessity:

1. **Local miners**: The majority of artisanal miners are local individuals or families who engage in mining as a primary or supplementary source of income.
2. **Small enterprises**: Some small-scale enterprises participate in artisanal mining, seeking to capitalize on local mineral resources.
3. **Women and children**: In many regions, women and children also partake in artisanal mining, either directly or by supporting activities such as transporting materials or processing minerals.
4. **Migrants**: Migrant workers from other areas or countries may come to mining regions in search of employment opportunities.
5. **Community members**: The broader community often depends on artisanal mining, with local businesses and services supporting the miners and their families.

The rising global demand for cobalt since 2000 has triggered a cobalt mining boom in Katanga Province, the DR Congo (Wilburn 2011). In Katanga, cobalt is also extracted in small artisanal mines (Fig. 4.5) due to its abundance in surface deposits, mainly as heterogenite, a mixed oxide and hydroxide of cobalt (CoOOH), often found in thin, friable layers (De Putter et al. 2011). This has led to widespread artisanal mining, with thousands of miners extracting heterogenite under precarious and hazardous conditions (Cuvelier 2014; Elenge and De Brouwer 2011; Tsurukawa et al. 2011; Amnesty International 2018).

Artisanal mining, while vital for many communities and integral to the livelihoods of millions of people, poses numerous environmental challenges due to the informal and often unregulated nature of these activities that need to be addressed through better regulations and support systems.

Some key environmental issues associated with artisanal mining include the following:

4.5.1 Deforestation and Habitat Loss

- **Tree clearing**: Artisanal miners frequently clear large areas of forest to access mineral deposits, leading to significant deforestation.
- **Biodiversity impact**: The loss of forest cover disrupts local ecosystems and habitats, threatening plant and animal species.

Fig. 4.5 Artisan miners digging out minerals with primitive tools such as picks, shovels, buckets, large pans, and sieves in Lubumbashi city, Haut-Katanga Province, the DR Congo (Source: Courtesy of Professor Banza Lubaba Nkulu, Lubumbashi University, DR Congo)

4.5.2 Soil Erosion and Degradation

- **Topsoil removal**: The mining process often involves stripping away the topsoil, which can lead to erosion and degradation of fertile land.
- **Loss of soil quality**: The soil structure is disrupted, reducing its ability to support plant life and agriculture.

4.5.3 Water Pollution

- **Mercury and cyanide use**: Artisanal miners often use mercury and cyanide to extract gold, which can contaminate rivers and groundwater.
- **Sedimentation**: Mining activities increase sediment in water bodies, affecting aquatic life and water quality.

4.5.4 Air Pollution

- **Dust emissions**: The excavation and processing of minerals generate dust, which can affect air quality and respiratory health.
- **Toxic fumes**: Burning of minerals or improper disposal of waste can release toxic fumes into the atmosphere.

4.5.5 Land Degradation

- **Unrehabilitated mining sites**: After mining activities cease, the land is often left barren and scarred, with little to no rehabilitation efforts.
- **Landscape alteration**: The physical alteration of landscapes can lead to unstable landforms and increased vulnerability to natural disasters.

4.5.6 Water Resource Depletion

- **Overuse of water**: Artisanal mining operations often require substantial amounts of water, which can deplete local water resources.
- **Lowered water tables**: Intensive water use can lower the water table, impacting both ecosystems and human water supplies.

4.5.7 Chemical Contamination

- **Toxic runoff**: Chemicals used in mining can seep into the soil and water, causing long-term contamination.
- **Bioaccumulation**: These chemicals can accumulate in plants and animals, entering the food chain and affecting human health.

4.5.8 Waste Management Issues

- **Improper waste disposal**: Waste from mining activities, including tailings and scrap, is often disposed of improperly, leading to soil and water pollution.
- **Heavy metal contamination**: Waste products can contain heavy metals, which are harmful to the environment and human health.

4.5.9 Flooding and Water Diversion

- **Altered water courses**: Mining activities can divert natural water courses, leading to flooding and changes in local hydrology.
- **Increased flood risk**: The removal of vegetation and soil destabilization can increase the risk of flooding during heavy rain.

Throughout the twentieth century, cobalt extracted as a by-product of other metals was sufficient to meet market demands. However, the surge in cobalt demand over the past decade has led many ASM to be involved in the extraction of the heterogenite, a common cobalt ore, in southeastern of the DR Congo. This could have presented an important occasion for considerable revenue generation for the country and for local development. Regrettably, due to the multitude of challenging coexistence of industrial and artisanal mining, coupled with inadequate governance, severe negative social impacts have emerged in Katanga, a region that accounted for half of the global cobalt production in 2010 (Tsurukawa et al. 2011).

In Zambia, the ASM sector is more formalized. Individuals, organization groups, or companies are required to operate under a legal mining title, such as an Artisan's Mining Right, a Small-Scale Exploration License, or a Small-Scale Mining License. Activities outside these regulations are considered illegal mining under Zambian law (Ministry of Lands and Natural Resources 2018). Formalizing ASM helps to ensure that these contributions are recognized and taxed appropriately, and by legalizing it, ASM allows the government to regulate and monitor mining activities, ensuring safer working conditions and environmental protection.

Chapter 5
Water Resources in the Region

Keywords Water quality · Hydrology · Surface and ground water · Safety · Sanitation · Pollution · Management

5.1 Contextual Framework

The environmental impacts of mining activities in Zambia and the DR Congo play a pivotal role in shaping the quality of the available water resources. In this region, mining is a cornerstone of economic development, with substantial operations extracting copper, cobalt, and other minerals. However, these activities exert considerable pressure on local water systems, making water management a critical aspect of environmental sustainability.

Mining processes often consume large volumes of water, whether for mineral processing, dust suppression, or cooling machinery. In Zambia and the DR Congo, this demand often leads to competition with other vital water uses such as agriculture and drinking water supply, particularly in communities near mining sites. Moreover, poor management of water resources results in over-extraction, leaving local ecosystems and residents vulnerable to water scarcity. This underscores the need for stringent regulations and sustainable practices to ensure that water remains available for both mining operations and community needs.

Another major concern is water contamination caused by mining. The release of tailings, which often contain toxic substances like heavy metals, seep into rivers, groundwater, rivers, and lakes. In the region such as in Zambia's Copperbelt Province or the DR Congo's Katanga Province, this contamination has led to biodiversity loss and health risks for populations dependent on these water resources. Effective water resource management, including proper treatment of wastewater and monitoring of water quality, is crucial to mitigate these environmental and social impacts.

Furthermore, water resources are integral to the resilience of local ecosystems. Wetlands, rivers, and aquifers in Zambia and the DR Congo support diverse flora

A. Muimba-Kankolongo et al., *Mining Impacts on the Environment in the Central African Copperbelt of Zambia and the Democratic Republic of Congo*,
https://doi.org/10.1007/978-3-031-96786-3_5

and fauna, many of which are sensitive to disruptions caused by mining. Preserving these ecosystems is vital not only for biodiversity but also for the livelihoods of communities that rely on natural resources for fishing, farming, and tourism. Integrating conservation efforts into mining operations can help strike a balance between economic development and environmental stewardship.

5.2 Background

Water is fundamental to life and a crucial component for sustaining livelihoods. It is essential for all living environments, but this critical resource is increasingly under threat as human populations grow and demand more high-quality water for domestic and economic activities (United Nations Environment Program Global Environment Monitoring System (GEMS)/Water Program 2008). Water is fundamental to life because without it, humans as well as livestock and several other living organisms can only survive for a few days. It plays a crucial role in nearly every bodily function, including digestion, nutrient absorption, and temperature regulation and protecting the immune system and aiding in waste removal (World Health Organization 2003). Lack of safe water, sanitation, and hygiene remains a pressing global health issue (UN-Water Decade Program 2015).

For sanitation, water is necessary for personal hygiene practices such as bathing, handwashing, and cleanness of numerous facilities, which help prevent the spread of infectious microorganisms.

Nearly one-tenth of the global disease burden could be prevented by improving water supply, sanitation, hygiene, and water resource management. Ensuring access to safe drinking water and adequate sanitation for the poor and promoting hygiene at personal, domestic, and community levels will enhance the quality of life for millions. Reports from the United Nations Environment Program Global Environment Monitoring System (GEMS)/Water Program (2008), WHO (2003), UN-Water Decade Program (2015), and UN-Water (2011) indicated that ensuring sustainable access to water resources is vital for several reasons, all of which contribute to the protection and enhancement of the quality of life for both current and future generations. They are the following:

5.2.1 Health and Well-Being

- **Clean drinking water**: Access to safe and clean drinking water is fundamental for maintaining good health. Contaminated water can lead to waterborne diseases such as cholera, dysentery, and typhoid, which can be fatal, particularly among vulnerable populations such as children and the elderly.
- **Sanitation and hygiene**: Adequate water supply is crucial for sanitation and hygiene practices. Proper sanitation facilities and hygiene practices help prevent the spread of infectious diseases, significantly improving public health.

5.2.2 Economic Development

- **Agriculture**: Water is essential for irrigation and livestock farming, which are critical components of food production. Ensuring a sustainable water supply supports agricultural productivity, food security, and economic livelihoods for farming communities.
- **Industry**: Many industries rely on water for manufacturing processes, cooling, and cleaning. Sustainable water management supports industrial growth, job creation, and economic stability.

5.2.3 Environmental Sustainability

- **Ecosystem services**: Freshwater ecosystems provide vital services, such as water filtration, flood control, and habitat for biodiversity. Protecting water resources helps maintain these ecosystems, ensuring they continue to support human and environmental health.
- **Climate resilience**: Sustainable water management practices can enhance resilience to climate change by improving water storage, reducing vulnerability to droughts and floods, and supporting adaptive agricultural practices.

5.2.4 Social Equity

- **Access for all**: Ensuring equitable access to water resources helps reduce social inequalities. Communities, especially marginalized ones, benefit from reliable water access, which supports their development and inclusion.
- **Empowerment**: Access to water empowers communities by allowing them to participate in decision-making processes regarding water management, leading to better resource allocation and governance.

5.2.5 Intergenerational Benefits

- **Sustainable resource use**: By managing water resources sustainably, we ensure that future generations have access to the water they need for their own health, economic activities, and environmental sustainability.
- **Education and innovation**: Promoting water conservation and sustainable practices encourages innovation and education, equipping future generations with the knowledge and tools to manage water resources effectively.

5.2.6 Global Perspective

- **Achieving SDGs**: Sustainable access to water is aligned with several United Nations Sustainable Development Goals (SDGs), particularly Goal 6: Clean Water and Sanitation. Achieving this goal supports broader efforts to eradicate poverty, promote good health and well-being, and foster economic development.

On July 28, 2010, the United Nations General Assembly, through Resolution A/RES/64/292, declared access to safe and clean drinking water and sanitation a human right indispensable for the gratification of living and all other human rights (UN-Water Decade Program 2015). This declaration emphasizes that:

- Access to safe water and basic sanitation is statutory, not a generous service.
- Strong energies should be dedicated to achieving rapidly basic and improved levels of access.
- Those lacking them should be prioritized, thus lessening discrimination disparity.
- Communities and vulnerable groups should be trained in issues of safe water and sanitation to fully contribute in decision-making processes.
- United Nations human rights mechanisms should be used to monitor nations' progress in realizing the right to water and sanitation and to hold governments accountable.

Clean, safe, and adequate fresh water is crucial for the survival of living organisms and the proper functioning of ecosystems, communities, and economies (UN-Water 2011). Declining water quality is a global concern as human populations increase, industrial and agricultural activities expand, and climate change threatens to alter the hydrological cycle.

5.3 Water Sector in Zambia

In their report on Zambia's environment, Chipundu and Kunda (1994) detailed the country's water resources, noting their adequacy to meet both short- and long-term needs for the growing population. Zambia boasts a well-distributed system of perennial rivers, streams, lakes, and swamps, all being essential for Zambia's social, economic, and environmental well-being.

The importance of water resources to the country are several including the following:

5.3.1 Health and Sanitation

- **Clean drinking water**: Access to safe and clean drinking water is critical for public health. It helps prevent waterborne diseases such as cholera, dysentery, and typhoid.

- **Sanitation**: Adequate water supply is necessary for sanitation practices like handwashing, bathing, and the operation of sanitation facilities, which are vital for disease prevention.

5.3.2 *Agriculture*

- **Irrigation**: Water is crucial for irrigation, supporting crop production and food security. Efficient irrigation systems help maximize agricultural productivity, especially in arid regions.
- **Livestock**: Water is essential for maintaining the health of livestock, which are a significant source of livelihood for many rural communities.

5.3.3 *Hydroelectric Power*

- **Energy production**: Zambia relies heavily on hydroelectric power, which is generated from its major rivers and man-made lakes, such as the Zambezi River and Lake Kariba. This renewable energy source is crucial for the country's electricity supply and economic development.

5.3.4 *Industry*

- **Industrial processes**: Many industries depend on water for manufacturing, cooling, and cleaning processes. Reliable water supply is essential for the smooth operation of these industries, which contributes to economic growth and job creation.

5.3.5 *Environmental Sustainability*

- **Ecosystem services**: Water bodies like rivers, lakes, and wetlands support diverse ecosystems. They provide habitat for wildlife, contribute to biodiversity, and help maintain ecological balance.
- **Flood control and water filtration**: Wetlands and natural water systems help regulate flooding and filter pollutants, ensuring cleaner water and reducing disaster risk.

5.3.6 *Economic Activities*

- **Fisheries**: Lakes and rivers in Zambia are vital for fisheries, which provide food and livelihood for many communities.

- **Tourism**: Water-based tourism activities, such as boating, fishing, and wildlife viewing in national parks, contribute to the economy and promote conservation.

5.3.7 Social and Cultural Significance

- **Cultural practices**: Water bodies hold cultural and spiritual significance for many communities. They are integral to traditional practices and ceremonies, fostering social cohesion and identity.

As a landlocked country, all of Zambia's water resources are freshwater sources, which are largely classified as follows:

5.3.8 Surface Water

Surface water is the predominant water source in Zambia, featuring six major basins, four natural lakes, two man-made lakes, and several swamp areas.

River Basins

1. **Zambezi River Basin**: Covering about 261,000 km^2, the Zambezi River Basin is Zambia's largest water resource. Spanning eight countries (Angola, Botswana, Malawi, Mozambique, Namibia, Tanzania, Zambia, and Zimbabwe), the Zambezi River Basin is the fourth largest in Africa, covering 1.32 million km^2 and serving nearly 40 million people. In 2004, these countries established the Zambezi Watercourse Commission (ZAMCOM) to foster cooperation over shared water resources, aiming to boost agricultural yields, hydropower production, and economic opportunities.
2. **Kafue River Basin**: Encompassing 152,000 km^2 and entirely within Zambia, the Kafue River Basin accounts for 17% of the Zambezi Basin. The Kafue River originates from the DR Congo-Zambia border and flows southwest through the Copperbelt, Central, Southern, and Lusaka Provinces. It supports various economic activities such as mining, hydroelectricity, agriculture, and municipal water supply. The Lower Kafue produces 50% of Zambia's national hydroelectricity and supplies 44% of Lusaka's water, employing over 900,000 individuals (World Wide Fund for Nature (WWF) 2016).
3. **Luangwa River Basin**: Spanning 165,000 km^2, the Luangwa River Basin in northeastern Zambia borders Malawi, Mozambique, and Zimbabwe. It is a subcatchment of the Zambezi River system, draining a significant part of south-central Africa. The Luangwa River, flowing 850 km southwest from the Mafinga

Hills to the Zambezi confluence, supports diverse wildlife, local farming communities, hunting, and tourism industries.

Natural Lakes

1. **Lake Kariba**: Africa's largest man-made dam, Lake Kariba generates substantial electric power for Zambia and Zimbabwe and supports a thriving commercial fishing industry. It offers stunning views, fishing, boating, and water sports.
2. **Lake Bangweulu**: Known as "The Place Where the Water Meets the Sky," Lake Bangweulu primarily supports fishing, with an annual harvest of about 57,000 metric tons. Its fisheries are among the largest in Zambia, supporting high population densities.
3. **Lake Mweru**: Although off the beaten path, Lake Mweru supports Zambia's fishing industry and is rich in culture and trade activities.
4. **Lake Tanganyika**: Shared with several countries, Lake Tanganyika stretches 677 km and is home to over 350 fish species. It provides abundant plankton and protein for local and export markets and supports the local fishing industry.

Zambia is lagging on several Millennium Development Goal (MDG) targets, particularly those related to water and sanitation (International Bank for Reconstruction and Development/The World Bank 2015). According to the report, only 63% of Zambians had access to clean drinking water, falling short of the MDG target of 75% by 2015 set by the UNICEF/World Health Organization (WHO) Joint Monitoring Program (JMP). A notable disparity exists between urban (about 90% access in 2010 per the UN) and rural areas (46% access). In urban settings, 41% of people have water connections at home or in their yards, while 49% rely on water kiosks and standpipes. These kiosks, managed by private individuals in collaboration with water utilities and community organizations, sell bulk piped water at around one US cent per 20 L. In 2013, Lusaka had approximately 270 water kiosks, including 100 constructed under the project. The National Water and Sanitation Council (NWASCO) reported that 97% of water samples from urban areas met drinking water standards that year. However, in Copperbelt, mining pollution affects water quality, with manganese levels sometimes exceeding treatable limits, necessitating water delivery by tankers to several mining towns.

NWASCO was established under the Water Supply and Sanitation Act No. 28 of 1997 (amended by Act No. 10 of 2005) to regulate water supply and sanitation services for efficiency and sustainability (https://www.nwasco.org.zm/index.php/about-nwasco/who-we-are). Its main roles include the following:

- Licensing providers.
- Advising the government on water supply and sanitation issues.
- Establishing and enforcing sector standards and guidelines.
- Advising providers on handling consumer complaints.
- Disseminating information to consumers on water supply and sanitation issues.

NWASCO provides water supply services to urban and peri-urban areas and limited sewerage services (NWASCO 2021). In 2004, it implemented key reforms to address poor performance in service provision (NWASCO 2004). Factors contributing to poor performance included unclear roles and responsibilities, low investment, inadequate cost recovery, overstaffing with underqualified personnel, and more. Prior to the reforms, the water sector faced the following:

- Lack of comprehensive policy or strategy.
- Unclear roles leading to duplication or gaps.
- Deteriorating infrastructure from poor maintenance and investment.
- Erratic and insufficient government funding.
- Increasing pollution, particularly in mining areas.
- Outdated and incomplete legislative framework.
- Lack of stakeholder involvement and consumer ownership.

Specific challenges in Water Resources Management (WRM) included the following:

- Inadequate preparation for negotiating shared water courses and treaties.
- No legislative framework to control groundwater abstraction.
- Lack of modern river-based management principles in legal and institutional frameworks.

5.4 Water Sector in the DR Congo

The DR Congo is rich in water and aquatic ecosystems. Effective management, protection, and development of these resources face challenges posed by sustainable development, poverty alleviation, and climate change. Ensuring access to drinking water remains a significant challenge, necessitating new policies and efficient management strategies. Water should be considered not only as an economic resource but also as a social good, vital for preserving life.

The country is traversed by the Congo River, the second largest river in the world by flow (average flow, 40,000 m^3/s). The country receives abundant rainfall, averaging 1500 mm annually, which helps recharge the water tables. Surface water sources like rivers, ponds, and lakes are easier to capture but often require extensive, costly treatments to make them potable. Thus, projects often opt for groundwater to supply networks through source catchments or deep boreholes, providing clean drinking water without additional treatments. However, before commissioning, the water's physico-chemical and bacteriological quality is ensured.

A United Nations Environment Program (UNEP) report (2011) on the DR Congo's water sector highlights the country's significant water resources, which include 52% of Africa's surface water reserves (rivers, lakes, and wetlands) and the highest water richness in Africa (Fig. 5.1). The DR Congo also accounts for 23% of Africa's internal renewable water resources. With an average annual precipitation of around 6000 billion m^3, rainfall in the DR Congo is regular and abundant (average 1545 mm/year), though it varies across regions (800–1800 mm). The country enjoys

Fig. 5.1 Map of the DR Congo groundwater resources. (Source: UNEP 2011) (Reproduced with permission from UNEP)

substantial water autonomy, with 70% (900 km^3/year) of its total actual renewable water resources (estimated at 1283 km^3/year) generated internally from rainfall. The country's extensive forests, spanning over 155.5 million hectares, play a critical role in water resources, as most of the population relies on springs located in dense forests for their water supply.

The report underscores the importance of springs and groundwater for drinking water, estimating that springs provide up to 90% of the DR Congo's rural population with drinking water. No comprehensive inventory of springs exists at the national or provincial levels. These sources typically involve simple, capped springheads used widely in dispersed villages and peri-urban areas. Large-scale water production from springs also serves many cities, including Mbuji-Mayi, Lubumbashi, Kisingani, Bunia, Beni, Gemena, and Lisala. The DRC has approximately 1000 deep-drilled wells, most built between the 1960s and 1980s during the International Decade for Water Supply and Sanitation. Their yield ranges from 15 to 80 m^3/h, sometimes reaching up to 300 m^3/h. Since the 1990s, well drilling has developed slowly, though international assistance has accelerated it recently. Groundwater is mainly exploited using dug wells and hand or mechanical pumps, supplying about 10% of the drinking water.

The DR Congo possesses over 50% of Africa's surface water reserves and about one-quarter of the continent's water resources, making it the most water-abundant country in sub-Saharan Africa (USAID/Sustainable Water Partnership 2021). Most of these renewable water resources come from surface water. The country has extensive river networks stretching over 20,000 km and lakes and rivers covering approximately 86,080 km^2, which is about 3.5% of the country's land area. The Congo River Basin, covering 98% of the country and spanning 3.8 million km^2 across 8 countries, includes 62% of the basin within the DR Congo. The Congo River is Africa's largest by volume, the second largest basin in the world, and the world's deepest river. It maintains consistent flow throughout the year, replenished by 20 major tributaries in 4 main sub-catchments: Lualaba/Tanganyika in the southeast, Kwa-Kasaï in the southwest, Oubangui in the northwest, and the main Congo flowing through the center.

The drinking water sub-sector in the DR Congo receives support to ensure sustainable development in line with existing legislative texts and Decree No. 15/039 of December 14, 2015, which established the Committee National Action for Water, Hygiene, and Sanitation (CNAEHA) (ANAPI 2021). Operating under the Ministry of Planning, CNAEHA coordinates all sub-sector activities to guarantee equitable and affordable access to drinking water for the Congolese population, aiming to achieve the Sustainable Development Goals (SDG) by 2030. This initiative is crucial, given that only 50.4% of the population had access to drinking water in 2015, with significant disparities between urban and rural areas despite the country's abundant freshwater resources.

The DR Congo enjoys an annual average rainfall exceeding 1500 mm, which is ten times the precipitation of countries like Niger and 50% more than neighbors such as Tanzania and Zambia (World Bank 2015). The Congo River's daily discharge into the Atlantic Ocean could supply all Congolese with sufficient water for several years. The Congo River Basin is the largest in Africa and the second largest in the world after the Amazon, draining all surface water in the country. It has several major tributary rivers and numerous streams equipped with dams for hydroelectric energy production.

A significant challenge in the DR Congo and Zambia's water sector is the declining water quality due to contamination and pollution from decades of mining activities, including construction, exploitation, and maintenance, which have severely damaged regional ecosystems. These activities have affected ground and surface water through discharges, waste dumps, and untreated effluents, posing public health risks due to increased concentrations of trace metals and pollutants like mercury, lead, cadmium, and copper (Feeney 2001; ZCCM-IH 2005; Ncube et al. 2012).

The DR Congo is notably the most water-abundant country in sub-Saharan Africa, holding over 50% of the continent's surface water reserves and about one-quarter of Africa's total water resources (USAID/Sustainable Water Partnership 2021). Most of these renewable water resources come from surface water. The country has an extensive network of rivers stretching over 20,000 km and water bodies (lakes and rivers) covering approximately 86,080 km^2, which is about 3.5% of the DRC's land area.

The Congo River Basin, which spans 3.8 million km^2 across 8 countries, covers 98% of the DR Congo, with around 62% of the basin located within the country. The Congo River is the largest in Africa by volume and the second-largest basin globally. It is also the world's deepest river. The Congo River maintains a consistent flow throughout the year, replenished by 20 major tributaries distributed across 4 main sub-catchments: Lualaba/Tanganyika in the southeast, Kwa-Kasaï in the southwest, Oubangui in the northwest, and the main Congo flowing through the center.

The Nile Basin makes up 2% of the DR Congo's land area on its eastern border and drains into the Semliki River watershed. Despite its small size in the DR Congo, this watershed contributes approximately 4600 million cubic meters (MCM) per year, accounting for about 20% of the White Nile River's flow. The DR Congo shares several significant lakes with its neighboring countries in the Nile Basin. These shared lakes include the following:

- **Lake Albert**: Shared with Uganda.
- **Lake Edward**: Shared with Uganda.
- **Lake Kivu**: Shared with Rwanda.
- **Lake Tanganyika**: Shared with Tanzania, Burundi, and Zambia.

In addition to these shared water bodies, the DR Congo has designated 12% of its land as protected areas, which includes nearly 12 million hectares of Ramsar wetland sites. These important conservation areas include the following:

- Parc national des Mangroves.
- Parc national des Virunga.

These protected lands and shared water resources play a crucial role in the region's biodiversity, environmental health, and cooperation among neighboring countries.

The DR Congo is home to significant water resources and infrastructure, including the Bassin de la Lufira and Ngiri-Tumba-Maindombe, the largest transboundary Ramsar site in the world. The country also boasts 18 large dams and the greatest hydroelectric potential in Africa, with hydroelectricity accounting for 99.7% of the country's electricity. The two largest dams are the Inga Dams, built in 1972 and 1982 on the Congo River, located just south of Kinshasa. These dams have a generation potential of 1774 MW but currently produce only 650–750 MW due to malfunctioning turbines. The proposed Grand Inga Dam project aims to increase the number of dams to 8, raising potential capacity to 44,000 MW. The project is still in the design phase, and construction of the third Inga Dam (Inga III) was announced to start in 2021. Another significant dam is the N'Seke on the Lualaba River with a capacity of 260 MW, and construction began on the 240 MW Busanga Hydropower Dam on the same river in 2016.

5.4.1 Groundwater Resources

The DR Congo is also rich in groundwater, although data on availability, yields, and aquifer sizes is limited. There are five main hydrogeological units:

1. **Alluvial sedimentary aquifers**: Found in the Cuvette Centrale (a region of wetlands and forests in the center of the Congo Basin) and Oubangui sub-catchment in the northwest.
2. **Sandstone aquifers**: Also located in the Cuvette Centrale.
3. **Low potential Quaternary-Tertiary aquifers**: Located in the Batékés Plateau and Kasaï in the west and south.
4. **Sedimentary aquifers**: Found in Katanga and Lubumbashi Dolomites in the south.
5. **Basement and metasedimentary aquifers**: Underlie the eastern and western borders of the country.

Groundwater yields are highest in the sedimentary aquifers of the Cuvette Centrale and Oubangui, where levels can reach up to 120 m.

5.4.2 Drinking Water Supply

Most rural populations, approximately 90%, depend on springs for drinking water, although data on springs, boreholes, and wells are limited. Deep-drilled wells and hand-dug wells are common, with some regions more dependent on unprotected wells, increasing contamination risks. Urban demand for groundwater is highest in cities like Kinshasa, Lubumbashi, and Matadi, accounting for 79% of REGIDESO's customer base. Despite abundant groundwater, extreme poverty, rapid population growth, and recurrent conflicts limit water access and infrastructure development, resulting in unmet safe water demand. The rapid overgrowing population in some cities such as Beni and Butembo in the mountainous areas of the eastern DR Congo has led to over-exploitation of water leading to localized shortages. Some wells in densely populated peri-urban areas in the southern Katanga region have also run dry due to over-abstraction and soil salinization (Fig. 5.2).

Fig. 5.2 Areas adjacent to Gécamines mines in Lubumbashi which had flourishing trees, grasses, and flowers in the past but no longer bear profuse plant growth due to soil intoxication

5.5 Water Quality

Mining activities significantly impact human health through food, water, soil, and ambient air pollution. According to UNEP (2011) and USAID/Sustainable Water Partnership (2021), the main challenge for water resources in regions like southern Lubumbashi in the DR Congo is pollution from industrial operations. Groundwater in this area is typically alkaline, and mining effluent from the Katanga Copperbelt region and artisanal gold mining in the east contaminate it with several hazardous trace metals. In various areas near Lubumbashi, groundwater often exceeds WHO drinking water quality guidelines for lead, aluminum, iron, and cadmium. Additionally, Kinshasa's groundwater shows high nitrate levels, likely due to agricultural and industrial pollution.

Mining of copper and cobalt in the Copperbelt Province of Zambia has resulted in significant waste, leading to contamination of rivers and streams over the years. ZCCM's early investigations in (ZCCM-IH 2005) identified multiple pathways through which mines pollute surrounding waterways, including drainage of mine water, cooling and wash water from smelters, stormwater runoff, and erosion from waste dumps. Toxic metal-rich solutions containing elements like cadmium, copper, cobalt, lead, nickel, zinc, and uranium are often discharged into streams and rivers, contaminating drinking water sources. In Zambia, most streams from mining areas flow into the Kafue River or its tributaries, the primary source of drinking water for towns in the Copperbelt Province.

In Chingola, it was observed that the water is often muddy. The quality of water provided by various companies for domestic use in Zambia's Copperbelt is summarized in Table 5.1. According to KCM statistics, which are outdated and not independently verified, drinking water standards were breached for 4 of the 9 water quality indices and by 10 of the 13 water providers. Despite this, heavy metal levels tested fell within Zambian water quality standards. Similarly, in the DR Congo, several rivers that supply drinking water and areas where the water table feeds into these rivers have been contaminated.

The DR Congo, despite having Africa's largest freshwater resources, suffers from an acute drinking water supply crisis. Only 46% of the population had access to an improved drinking water source as of 2012, and sanitation coverage was even lower at 31% (USAID/Sustainable Water Partnership 2021). Soil and water contamination results from discharges from broken pipes, new site developments, waste dumps, and untreated effluents from copper processing. About half of the respondents in Zambia cited decreased water quality as a direct consequence of mining activities (e.g., Chingola and Mushishima streams and Kafue River). Both Mulonga Water and Sewerage Company in Zambia and REGIDESO in the DR Congo pump domestic water from the Kafubu River. This water has been reported to have poor quality and contain harmful heavy chemical elements. However, regular water analysis in Chingola by KCM indicates no harm for human use based on WHO standards (table below). Stakeholders emphasize the need for regular sampling and independent laboratory cross-checks due to rapid and extreme fluctuations in water quality from industrial pollutants.

KCM (Konkola Copper Mines) frequently contaminates freshwater ecosystems with untreated mine waste, which is high in suspended solids and lowers water pH. For example, in August 2006, a pipe burst, contaminating the Mushishima stream, a tributary of the Kafue River. While water turbidity was kept within WHO standards after liming, nitrate and magnesium concentrations remained above WHO standards (Table 5.1). High nitrate levels can lead to an increase in nitrogen and ammonia, which suppress pH levels. According to Taiz and Zeiger (1998), nitrogen and magnesium are available in the slightly acidic pH range of 5.5–6.5. This aligns with WHO's recommendation for a water pH of 6.5 to ensure the availability of nutrients for riparian vegetation in the Copperbelt region.

However, high water quality indicated by physical and chemical properties did not match observations made during a study conducted approximately 2 km from the water treatment plant (Table 5.2). The water was significantly polluted, with the source of pollution visually identified as mine waste often diverted into land near rivers. The stream was dredged to create a path for the waste to flow into the Kafue River. Characteristics of the discharged water included the folowing:

1. Direct flow of waste into the headwaters of the Kafue River.
2. Clear discoloration downstream from the mine waste outlet.
3. Local farmers are unable to farm the land around the water pathway due to soil contamination, partly due to growing artisanal mining (Fig. 5.3).

Industrial pollution in Zambia, particularly from mining, leads to heavy metal contamination and frequent breaches of water quality standards. The Copperbelt region, for instance, experiences contamination from mining effluents, leading to elevated levels of heavy metals like lead, aluminum, iron, and cadmium in groundwater (UNICEF Zambia 2024). Additionally, nitrate levels in Kinshasa's groundwater exceed WHO recommendations, likely due to agricultural and industrial pollution (UNEP 2011).

Mining activities in the DR Congo have significantly impacted several rivers, including the Kafubu, Kapolobwe, Kasapa, Kimilolo, Naviundu, and Panda. Cadmium contamination is prevalent in the Kimilolo and Naviundu Rivers, while the Kafubu River has previously been contaminated with arsenic. Water analysis of the Naviundu River in Lubumbashi revealed that the quality standards for cadmium, manganese, mercury, and sodium were breached (Table 5.3).

5.6 Health Impacts of Water Contamination

Survey results in Lubumbashi (Luboya et al. 2004) and earlier studies in Zambia (ZCCM 2005) indicated heavy metal contamination in human blood. According to the Chingola Health Board, residents near mining areas frequently suffer from eye irritation, coughing, and respiratory issues, especially children. Cadmium, a highly

Table 5.1 Quality of water distributed by various water utility companies for domestic use in the Copperbelt Province of Zambia (KCM 2001)

Waterworks	Turbidity (NTU)	NO_3 mg/L)	PO_4 (mg/L)	T/coli (/100 mL)	F/coli (100 mL)	Co (μg/L)	Cu (μg/L)	U (μg/L)	Zn (μg/L)
Zambian drinking water standard	≤10[a]	400	<0.01	0	0	≤500	≤1000	15[b]	5000
USEPA environmental standards (for aquatic life)	–	10	–			–	9	30	120
Background values for Kafue River	2.5	0.3	–	0	0	0.12	21	0.03	1
Ndola-Kafubu WS (boreholes)	8	5	0.004	0	0	–	–	–	–
Ndola-Kafubu WS (Kafubu R)	4	5	**0.026**	**50**	0	<1	5	<1	77
Luanshya AHC-MMS (Kafubu R)	3	1	0.004	0	0	4	67	6	58
Luanshya-Kafubu WS (Kafubu R)	7	8	**0.019**	0	0	<1	3	1	43
Mufulira-Mulonga WS (Kafubu R)	**13**[c]	8	0.007	**1**	0	<1	24	4	14
Mufulira-AHC-MMS WS (Kafubu R)	**59**	11	0.004	**TNTC**[c]	**30**	<1	24	4	10
Kitwe-Nkana WS (Kafue R)	2	3	0.005	0	0	–[d]	–	–	–
Kitwe-AHC-MMS (Kafue R)	**17**	3	0.008	0	0	<1	11	4	15
Chambishi-Nkana WS (mine water)[a]	**35**[a]	12	0.002	**100**[a]	**20**[a]	–	–	–	–
Chingola-Mulonga WS (Kafue R)	**11**	6	0.007	0	0	–	–	–	–

Chingola AHC-MMS (Kafue R)	**65**	15	0.003	0	0	248	680	1	82
Chililabombwe AHC-MMS (Kafue R—mine water)	**58**	10	0.005	**TNTC**[c]	**2**	1	26	2	93
Chililabombwe-Mulonga WS (boreholes)	**14**	4	0.007	0	0	–	–	–	–

[a] Note: According Nkana Water and Sewage Company's own data (August 2004–January 2005) for the Chambishi treatment plant (i.e., treated water), the monthly average turbidity values ranged from 8.1 to 21.3 NTU, and coliform bacteria were in no cases detected. Thus according to these data, the water quality should be deemed acceptable

[b] WHO standards have been used where Zambian standards are lacking

[c] Bold font = breach of quality standards

[d] Too numerous to count

Table 5.2 Physical and chemical properties of water from a KCM contaminated freshwater ecosystem at Mushishima stream in Chingola District, Copperbelt Province of Zambia

Parameter	Test results	WHO recommended standard
pH	7.11–7.7	6.5–8.5
Total dissolved solids	170–198 mg/L	1000 mg/L
Conductivity	340–396 us	3000 us
Turbidity	2.02–7.4 NTU	5.0 NTU
Nitrate	0–0.99 mg/L	50.0 mg/L
Chloride	16.7–25.05 mg/L	250 mg/L
Sodium	6.11–6.22 mg/L	200 mg/L
Potassium	4.75–4.76 mg/L	12.0 mg/L
Calcium	39.14–39.2 mg/L	200 mg/L
Magnesium	34.2–35.95 mg/L	30 mg/L
Lead	Nil	0.01 mg/L
Cadmium	Nil	0.003 mg/L

Source: Data from government's District Health Board in Chingola District of Zambia

toxic trace metal mobilized by mining activities, causes skeletal and renal tubular damage (Järup et al. 1998). High cadmium exposure can lead to diabetic glomerulopathy in animals (Bernard et al. 1991) and nephrotoxicity, with early effects seen as kidney damage (WHO 1992). Diabetic patients may be more susceptible to cadmium's toxic effects (Buchet et al. 1990). Wildlife has been also found threatened by cadmium in the western US "ore belt," causing kidney and liver dysfunction, bone decay, and reproductive issues (Larison et al. 2000).

Lead contamination poses significant health risks, with exposure levels as low as 100–200 μg/L affecting developmental and IQ tests in children. Lead exposure during pregnancy can decrease pregnancy duration, lower birth weights, and increase miscarriage risks (Wang et al. 1992). Cadmium and lead have long half-times in the body, meaning they accumulate with prolonged exposure, causing adverse health effects at even slightly elevated levels (WHO 1992, 1995). Cobalt, though essential in small amounts, is toxic in high doses, leading to respiratory issues like asthma and interstitial lung diseases (Midtgård and Binderup 1994). Manganese, another essential trace element, becomes toxic in high doses, causing "manganism," a condition with mental, emotional, and motor disturbances (ATSDR 2012). Sodium, when in contact with water, converts to sodium hydroxide, becoming poisonous.

Fig. 5.3 Artisanal diamond miners digging close to a water treatment plant near Mbuji-Mayi, Kasai Orientale. (Source: UNEP 2011) (Reproduced with permission from UNEP)

Table 5.3 Results of Naviundu River water analysis in Lubumbashi, Katanga Province, in the Democratic Republic of Congo

No	Parametres analyzed	Results	Standards (OMS and FAO)	Commentary
01	pH	8	6.5–8.5	Nothing to indicate (NTI)
02	Co [mg/L]	<0.01	1	NTI
03	Cu [mg/L]	0.02	1	NTI
04	Fe [mg/L]	0.10	0.5	NTI
05	Ni [mg/L]	<0.01	0.05	NTI
06	Pb [mg/L]	<0.01	0.05	NTI
07	Al [mg/L]	0.2	0.2	LIMITED VALUE
08	As [mg/L]	<0.01	0.05	NTI
09	**Cd [mg/L]**	**<0.01**	**0.005**	**IN EXCESS**
10	Cr [mg/L]	<0.01	0.05	NTI
11	Zn [mg/L]	<0.16	3	NTI
12	**Mn [mg/L]**	**0.80**	**0.5**	**IN EXCESS**
13	Ca [mg/L]	28	100	NTI
14	Mg [mg/L]	37	50	NTI
15	**Na [mg/L]**	**860**	**200**	**IN EXCESS**
16	K [mg/L]	1.3	10	NTI
17	**Hg [mg/L]**	**<0.01**	**0.001**	**IN EXCESS**
18	P [mg/L]	0.05	0.4	NTI
19	So	249	250	NTI
20	Se [mg/L]	<0.01	0.01	LIMITED VALUE
21	**TDS [mg/L]**	**2690**	**50**	**IN EXCESS**

Note: Standard values are maximum concentrations in mg/L of water. TDS represents solids in suspension (total dissolved solids)
Bold indicates values exceeding the standard limits

Chapter 6
Agriculture and Food Production in the Region

Keywords Agriculture · Food security · Land use · Soil quality · Farming · Crop production · Mining impacts

6.1 Agricultural Context Note

Agriculture plays a crucial role in supporting the livelihoods of populations in both Zambia and the DR Congo. These nations have vast agricultural potential due to their favorable climates, fertile soils, and abundant water resources, making farming a cornerstone of economic activity and food security. In Zambia, agriculture is vital for both rural and urban communities. It contributes significantly to employment, with the majority of rural populations depending on farming as their primary source of income. The sector supports food production, supplying staple crops like maize, cassava, and groundnuts, which are integral to the national diet. Additionally, agriculture drives export opportunities through crops such as tobacco and cotton, boosting the country's economy. In the DR Congo, agriculture is a lifeline for millions, especially in rural areas where other economic opportunities are scarce. It ensures food availability through the cultivation of essential crops like cassava, plantains, and maize, which form the staple diet. The sector also aids in poverty reduction by providing income for smallholder farmers. Moreover, given the DR Congo's rich biodiversity and natural resources, agriculture holds significant potential for sustainable development and economic diversification.

The description of agricultural practices for various food crops which are grown in the region for the population livelihoods and negatively impacted by mining pollution undoubtedly provides a holistic view of how mining pollution affects food crops through soils which are managed using several cultivation methods including ploughing, planting, fertilization, irrigation, weeding, and harvest as well as an

A. Muimba-Kankolongo et al., *Mining Impacts on the Environment in the Central African Copperbelt of Zambia and the Democratic Republic of Congo*,
https://doi.org/10.1007/978-3-031-96786-3_6

understanding of the biogeochemistry of trace elements in the soil-plant system, allowing readers to realize the complexity of plants' contamination pathways.

The scientific literature expressing the views on the roles of food cropping cultivation practices in mediating pollutants' contamination of plants are numerous (Kabata-Pendias 2011; McGrath et al. 2014; FAO and UNEP 2021). Plants are essential in the circulation of trace elements within ecosystems and their eventual incorporation into the food chain. Studies conducted across various settings have underscored this significant dual role. With their ability to adjust to diverse environmental chemical conditions, plants can take in and store trace metals in or on their tissues through the roots that are anchored in soil. Consequently, they serve as a bridge for transferring these elements from the soil, air, and water to humans and animals (Ambika Asati et al. 2016; Khan et al. 2015; Bhat et al. 2019; Buzea and Pacheco 2020; Angon et al. 2024). The absorption of heavy metals by plants depends on factors such as the type of soil, the specific plant species, variations within the same species, and the area within the plant's tissues where the metals accumulate. Leafy vegetables tend to absorb most of the heavy metals, while root vegetables show a moderate level of uptake, and legumes accumulate the least.

Therefore, detailing soil preparation, planting methods, fertilization, weeding, and harvesting practices emphasizes the interconnectedness of agricultural practices and environmental health, highlighting the necessity of sustainable farming methods. A thorough examination of farming techniques guides community farmers, researchers, policymakers, and other environmental stakeholders in developing strategies to mitigate contamination, thereby promoting safer agricultural practices. Addressing these topics raises awareness among consumers and communities about potential risks associated with contaminated food crops, empowering them to make informed decisions regarding food consumption.

They are as follows:

1. **Soil preparation and contamination**: The quality of soil plays a key role in successful crop cultivation. Elements such as soil texture, nutrient content, and pH levels have a direct effect on crop growth and productivity. To ensure ideal conditions for farming, it's important to implement soil management practices like testing and improving the soil. Mining activities alter the soil's natural composition, leading to an accumulation of hazardous metals like lead, mercury, arsenic, and cadmium. Highlight how contaminated soil acts as the primary source of metal uptake in crops through the roots. Incorporate data or observations from the samples to emphasize variations in contamination levels across different planting locations.
2. **Planting methods and impact**: Tillage is a farming technique used to loosen and aerate the soil. This involves using mechanical tools to turn and break the soil into smaller particles, improving its structure and allowing better air and water flow. These conditions support the metabolic activities of plants and microorganisms, enabling chemical and biological processes to occur more efficiently. As a result, organic matter decomposes more quickly, making essential

nutrients readily available for crops to thrive. Certain planting techniques may inadvertently expose crops to contaminated deeper soil layers.

3. **Fertilization practices**: Soil fertilization, while beneficial for crop growth, can introduce contaminants that may negatively affect food crops. Fertilizers often contain substances like heavy metals, nitrates, and phosphates. If applied excessively or improperly, these components can accumulate in the soil and eventually make their way into plants. For instance, heavy metals such as cadmium and arsenic present in some fertilizers can be absorbed by crops, contaminating the food produced. Similarly, excessive nitrates from fertilizers can leach into groundwater, posing risks to both human health and the environment. Over time, the buildup of these harmful substances can disrupt soil health and affect the safety of the harvested crops.
4. **Weeding techniques**: Weeding significantly influence heavy metal contamination in plants, both directly and indirectly. By removing weeds, farmers reduce competition for nutrients, water, and space among crops. However, if heavy metals are present in the soil, the crops might absorb higher concentrations, as they now have greater access to the contaminated soil. Weeding processes, especially those involving uprooting, disturb the soil, causing heavy metals that are bound to deeper soil layers to resurface, potentially increasing their availability for uptake by crops. In some cases, weeding improves soil aeration, which can enhance the chemical reactions in the soil, making certain heavy metals more soluble and accessible to plants. Some weeds naturally accumulate heavy metals and act as "metal sponges." Their removal might expose crops to more heavy metal uptake in the absence of these barriers.
5. **Harvesting methods and crop exposure**: Local harvesting techniques affect the levels of contamination detected in food crops. For instance, improper handling during harvest or storage may lead to secondary contamination from mining dust or residues in nearby environments.

6.2 Background

Agriculture is the lifeblood of many communities in Zambia and the DR Congo, with about 70% of the population in both countries relying on it for their livelihoods (SADC/FANR 2011).

6.2.1 Zambia's Agricultural Landscape

Apart from mining, agriculture forms the backbone of Zambia's rural economy, although it has been underperforming. The sector employs 56% of the Zambian population and plays a crucial role in cushioning employment volatility in other sectors. It is key to ensuring food and nutrition security, especially for financially vulnerable communities. Despite the high employment, agriculture's contribution to GDP fell from 17.3% in 2004 to 5.3% in 2015, indicating that many Zambians are still engaged in low-productivity subsistence farming with limited access to productive assets and market opportunities (World Bank 2018).

Food Security Challenges

Zambia faces significant food security challenges, with high rates of hunger and malnutrition. Although stunting in children under 5 decreased from 45% in 2007 to 40% in 2015, it remains a pressing issue (World Bank 2018).

6.2.2 *Agriculture in the DR Congo*

In the DR Congo, agriculture is the primary economic sector, contributing about 40% to the national GDP and employing 70–75% of the economically active population. Related agro-industries employ another 10% (CAADP 2013; Nutrition Smart Agriculture 2020). Between 2006 and 2010, the agriculture sector grew by an average of 4.4%, with projections of 4.6% growth between 2011 and 2015 (CAADP 2013).

Import and Export Dynamics

Despite its potential, the DR Congo remains a net food importer. From 2012 to 2015, it imported nearly 8 million tons of food, with cereals accounting for 51% of these imports. Fish and sugars were the next most imported products, making up 14% and 12% of food imports, respectively (CAADP 2013). During the same period, national food production increased, with 174 million tons of food produced and livestock numbering over 108 million animals. Cassava alone constitutes 79% of all food production in the country.

6.2.3 *Food Crop Production*

Smallholder farmers, primarily in rural areas, produce most of the food crops, relying on rain-fed and traditional farming methods. Key crops include root and tubers, plantains, maize, rice, groundnuts, beans, sorghum, millets, and vegetables (SADC/FANR 2011; Muimba-Kankolongo 2018). In urban and suburban areas, food crops are commonly grown in kitchen gardens (Peša 2020a, b).

6.2.4 *Environmental and Social Challenges*

Recurrent and prolonged droughts have exacerbated food insecurity in the region (Chapoto et al. 2018; Muimba-Kankolongo 2018). Additionally, environmental degradation from mineral extraction and metallurgical industries has further compromised food security (Mwitwa et al. 2012b).

Agriculture remains a cornerstone of the economies in both Zambia and the DR Congo, despite facing significant challenges from environmental degradation and food insecurity.

Since its inception in 1980, the Southern African Development Community (SADC) has made remarkable strides in fostering regional cooperation and integration. With a population of 350 million spanning 16 Member States, SADC has transitioned from a coordinating conference to a dynamic regional development community. This transformation positions SADC as a pivotal building block of African unity through its collaboration with the African Union (AU).

Progress in Regional Integration

SADC's commitment to regional integration is evident through the 33 protocols and numerous declarations, charters, and memoranda of understanding signed by Member States since 1992. These agreements cover a wide range of sectors, including:

- Trade
- Mining
- Finance and investment
- Illicit drugs
- Forestry
- Shared watercourses
- Empowerment of women and youth

Of the 33 protocols, 25 have entered into force, having been ratified by at least two-thirds of the signatory Member States. These commitments have significantly contributed to improving the quality of life for SADC citizens.

Contributions to SADC Citizens

The various protocols and agreements signed by SADC Member States have had a profound impact on the region's development. These initiatives have promoted the following:

- **Economic growth**: Through enhanced trade and investment opportunities.
- **Social development**: By addressing issues such as drug trafficking, women and youth empowerment, and environmental sustainability.

- **Infrastructure Development**: Facilitating better connectivity and resource management across the region.

6.3 Climate Change and Food and Nutrition Security

Over the past 15 years, climate change and extreme weather events have severely impacted the Southern African Development Community (SADC) region, and these effects are expected to worsen in the coming decades.

6.3.1 Temperature Changes

Figure 6.1 below illustrates the changing temperatures in cropping areas during 2015–2016, compared to a 15-year period. The data reveals significant shifts that have critical implications for agricultural productivity.

6.3.2 Severe Droughts and Rainfall Variability

During the 2018–2019 rainfall season, several SADC countries experienced severe drought, receiving the lowest levels of rain since 1981. This drought affected Angola, Botswana, Namibia, South Africa, Zambia, and Zimbabwe, prompting Angola, Botswana, and Namibia to declare drought emergencies. Additionally, unpredictable rainfall patterns disrupted crop production.

6.3.3 Floods and Cyclones

Extreme weather events also resulted in flooding across various SADC countries. Tropical storms and cyclones exacerbated the situation in Comoros, Madagascar, Malawi, Mozambique, and Zimbabwe.

6.3.4 Impact on Food Security

These extreme weather events have had significant negative impacts on food security. Data from 13 Member States indicate that 41.2 million people were food insecure in 2019–2020. This represents a 7.4% increase compared to the 2016–2017 El Niño-induced drought. The countries with the highest increases in food insecurity are as follows:

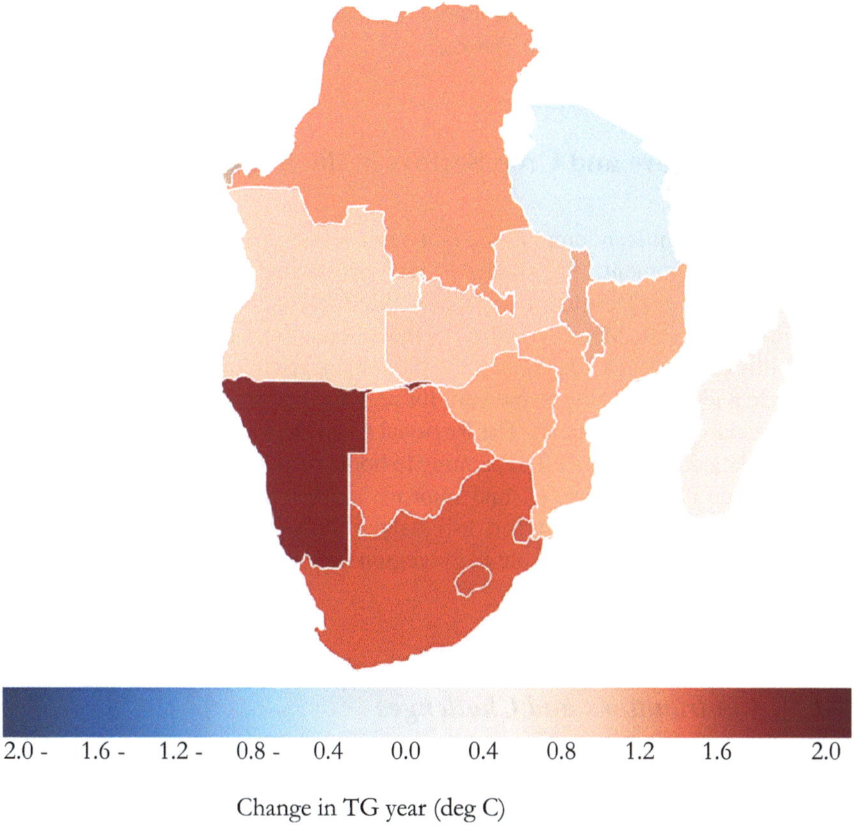

Fig. 6.1 Changing temperature in cropping areas in 2015–2016, juxtaposed with a 15-year period. (Source: Food and Agriculture Organization (FAO). (2018): State of food security and nutrition in the world)

- **DR Congo**: 80%
- **Mozambique**: 85%
- **Eswatini**: 90%
- **Zimbabwe**: 128%
- **Zambia**: 144%

6.3.5 Malnutrition and Health Issues

A high number of children in the SADC region suffer from malnutrition due to cereal-based diets lacking essential micronutrients like vitamin A, zinc, iron, folate, and iodine. In 2019, seven Member States reported that 5% of children under 5 were considered too thin for their height (wasting rate). Conversely, the region also faces a prevalence of overweight and obese individuals, particularly in Botswana, Comoros, Seychelles, and South Africa.

6.4 Agriculture and Crop Sectors in the Region

SADC places significant emphasis on the crop sector as a cornerstone of regional economic development. This focus is well articulated in the SADC Regional Indicative Strategic Development Plan (RISDP 2020–2030), which outlines the aspirations of SADC Member States to attain a common future (SADC 2020). As part of this effort, SADC developed the Regional Agricultural Policy (RAP 2014) to provide a legally binding framework for regional cooperation in agriculture and natural resources management. The Regional Crop Development Program (RCDP 2019) is a key component of RAP, aiming to boost sustainable agricultural production, enhance competitiveness, and improve market access for crops and their products. The RCDP is designed to catalyze the agricultural sector's growth, ensuring it plays a significant role in the region's economic development and food security.

6.4.1 Contributions and Challenges

Agriculture contributes between 4% and 27% to the Gross Domestic Product (GDP) of the SADC region, with crop production accounting for an estimated 61% of this contribution. Crops serve as the main source of food, employment, and income for the region. However, the sector faces numerous challenges including climate change, land degradation, rapid population growth, and urbanization, which pressure agriculture to deliver beyond current productivity levels.

6.4.2 Enhancing Agricultural Performance

To address these challenges and improve crop production, farmers need better access to inputs, finance, markets, and an enabling policy environment. The RCDP provides a framework to stimulate increased crop production, productivity, and competitiveness. It promotes sustainable crop production intensification based on Good Agricultural Practices (GAP) and agricultural production systems. The RCDP also emphasizes the importance of inputs, mechanization, irrigation, adoption of appropriate crop technologies, and reduction of post-harvest losses.

6.4.3 Regional and International Dynamics

The RCDP aligns with both regional and international dynamics in the crop sector and agriculture. It aims to implement harmonized, area-specific interventions that significantly contribute to regional socio-economic growth. Additionally, the program provides guidance on addressing plant health and phytosanitary issues as outlined in the Regional Strategy on Plant Health.

6.4.4 Investments and Support Systems

Successful implementation of the RCDP requires increased public and private investment in research and development, infrastructure, market and trade conditions, institutional and farm support systems, and access to factors of production. These investments are crucial for ensuring a competitive and vibrant crop sector that supports the economies of SADC Member States.

6.4.5 Regional Plant Health

Transboundary plant pests and diseases pose a continuous threat to regional food and nutritional security in the Southern African Development Community (SADC). These issues not only affect productivity levels and food safety but also threaten biodiversity and natural resources, which are crucial for agricultural economic development, particularly in developing countries that rely heavily on agricultural exports.

Regional Strategy on Plant Health

SADC's efforts to manage these threats are guided by the Regional Strategy on Plant Health (2019), which aims to:

- Enhance the ability of Member States to increase agricultural productivity.
- Improve market access.
- Promote a common approach to managing transboundary pests and diseases.

This strategy also ensures coordination, preparedness, and harmonized control interventions. It supports the implementation of trade obligations under the World Trade Organization-Sanitary and Phytosanitary (WTO-SPS) Agreement and the SADC SPS Annex to the Trade Protocol. Additionally, it helps safeguard against Invasive Alien Species, which pose significant threats to the region.

6.4.6 Specific Pest and Disease Management Strategies

Beyond the Regional Strategy, SADC has developed specific management strategies for high-impact pests and diseases affecting food security and trade. Five priority plant pests and diseases have been identified:

1. Fall armyworm
2. Tuta absoluta
3. Fruit flies
4. Lethal maize necrosis disease (MLND)
5. Banana fusarium TR4

6.4.7 Currently Available and Approved Strategies

Strategies for managing lethal maize necrosis disease (MLND) and banana fusarium oxysporum TR4 are in progress.

1. Regional Strategy for the Management of Fruit Flies.
2. Regional Strategy for Management of the Fall Armyworm.
3. Regional Strategy for Tuta absoluta Management.

6.4.8 Southern Pesticides Regulators Forum (SAPReF)

The Regional Strategy on Plant Health promotes ecosystem-friendly control methods through the principles of Integrated Pest Management (IPM). In recent years, pest outbreaks have necessitated the use of pesticides. The Southern African Pesticides Regulators Forum (SAPReF), a sub-committee of the SADC Plant Protection Technical Committee (SPPTC), provides technical and policy guidance on pesticide management. This includes insecticides, herbicides, fungicides, biopesticides, and other formulations useful in managing plant pests and diseases. SAPReF, comprising heads or representatives of national pesticide regulatory authorities and registrars, has facilitated the registration of new pesticides in response to pest and disease introductions, such as the fall armyworm.

6.4.9 Migrant Pests

The region continues to battle migrant pests like African migratory locusts, brown locusts, red locusts, Quelea birds, and African armyworms. These pests cause significant damage to crops. Guided by the Regional Migrant Policy, Member States continuously monitor and manage these pests. In recent years, locust prevalence has increased, leading SADC to launch regional appeals to contain their spread and promote a coordinated approach, while also building Member States' capacity to manage these pests effectively.

6.4.10 Promoting Access to Quality Inputs

Improving crop sector performance requires access to quality inputs such as land, water, seeds, fertilizers, and pesticides. However, sub-standard inputs often reach farmers. To promote a common approach, SADC developed regional guidelines for registering high-quality seeds, fertilizers, and pesticides. The seed sector is guided by the Harmonized Seed Regulatory System (HSRS), which establishes regulatory standards for seed variety release, certification, quality assurance, and quarantine measures. The HSRS aims to increase agricultural productivity and food security by standardizing seed quality across Member States.

6.4.11 Irrigation and Mechanization

SADC promotes the use of appropriate technologies to adapt to climate change through smart agricultural practices. A regional irrigation policy framework is being developed, recognizing the Water-Energy-Food (WEF) nexus to maximize water use while considering other needs. This policy aims to promote a coordinated approach among all water users for energy generation, agriculture, domestic, and industrial purposes.

6.4.12 Challenges and Opportunities for Smallholder Farmers

According to Muimba-Kankolongo (2018), most Southern African farmers rely on small land holdings for subsistence, using rudimentary methods for crop production. Typical farm holdings per household rarely exceed 1–5 ha, where they grow a variety of crops such as maize, groundnuts, cowpea, beans, sweetpotato, sorghum, millets, and cassava. Smallholders also engage in fishing, livestock keeping, and exploiting natural resources, especially during prolonged droughts.

Smallholder productivity is often constrained by limited access to technology, inputs, services, and credit, as well as an inability to bear risks. These farmers struggle to meet market demands and sanitary standards. Land preparation is usually manual, and recycled seeds are commonly used without knowing how to ensure the best quality. Minimal use of fertilizers results in low soil fertility and yields, averaging 1 ton/ha, compared to the potential 10 tons/ha achieved at research stations (Tittonell et al. 2007).

6.5 Food Crop Production in Zambia and the DR Congo

The food crop production for cereals, root and tuber, and leguminous crops and vegetables reported below are portions reproduced from Chaps. 8, 9, 10, and 11 published in the book *Food Crop Production by Smallholder Farmers in Southern Africa: Challenges and Opportunities for Improvement*, by Ambayeba Muimba-Kankolongo, 382 Pages, Copyright Elsevier Inc., 2018.

6.5.1 Cereal Production

Introduction

According to a regional paper presented at the extraordinary summit on agriculture and food security held in Tanzania in 2004 (SADC-FANR 2004), the total cereal production in the SADC region, encompassing both commercial and small-scale farmers, remained relatively stagnant from 1990 to 2003 (Table 6.1). Production estimates were reported at 22,062 metric tons in 1990 and slightly increased to

Table 6.1 SADC cereal production for each crop from 1990 to 2003

Total cereal production by crop ('000 tons)					
Year	Maize	Wheat	Rice	Sorghum/millet	Total
1990	17,348	2516	603	1595	22,062
1991	15,474	2154	491	1642	19,761
1992	6783	2365	311	1247	10,707
1993	17,137	1778	523	2079	21,517
1994	20,074	2231	562	1838	24,705
1995	11,097	2490	603	2049	16,239
1996	20,345	3170	529	2776	26,820
1997	17,054	2763	533	2032	22,382
1998	15,835	2201	922	1905	20,863
1999	17,052	2032	633	1586	21,303
2000	20,501	2729	730	2124	26,084
2001	15,880	3053	745	1909	21,587
2002	16,323	2629	595	1999	21,546
2003	18,418	1888	767	1680	22,753

Reproduced from SADC-FANR (2004). Enhancing Agriculture and Food Security for Poverty Reduction in the SADC Region: Key Issues Paper for Extraordinary Summit. Dar-es-Salaam, Tanzania. 37 p

22,753 metric tons by 2003. Meanwhile, the population of the region grew significantly, from 152 million in 1990 to 212 million in 2003. This discrepancy led to a noticeable decline in per capita food consumption over this period.

Currently however, the cereal production status in SADC countries shows some positive trends. According to the latest report by Luxon Nhamo et al. (2019), cereal production in the region has been on an upward trajectory. Despite challenges posed by water scarcity, cereal production continues to rise steadily, thanks to increased investments in irrigated agriculture and the development of improved crop varieties. This increase is a step toward improving food security and reducing poverty in the region.

The Southern African Development Community (SADC) faces significant challenges in food crop production, primarily due to droughts, floods, poor input availability, soil erosion, nutrient depletion, plant diseases, pests, inadequate irrigation, and insufficient extension services. Despite these obstacles, the region cultivates a diverse range of crops. These include other key crops such as root and tuber produce, including cassava, sweetpotato, and yams, along with bananas, which are predominantly grown in the northern parts of the region. These crops have the potential to significantly contribute to regional economic growth and poverty reduction, given that many people rely on agriculture for their livelihoods. Food crops in SADC hold tremendous potential for enhancing regional economic development and reducing poverty, particularly as the majority of the population depends on agriculture. However, achieving this potential requires addressing the current challenges in crop production.

6.5.2 Maize Cultivation

Maize (*Zea mays* L.) is the principal food crop in SADC, particularly in the southern part of the region (Fig. 6.2). As a member of the Poaceae family, maize is a staple food for millions and a crucial source of dietary carbohydrates and foreign exchange. Its cultivation is encouraged across various agro-ecological zones, even in areas less suited for its production.

Importance of Maize

Maize plays a crucial role in Southern Africa, serving multiple purposes from human consumption to industrial applications.

Human Consumption and Uses

For human consumption, maize is primarily prepared as a thick porridge or cooked fresh by boiling or roasting green maize kernels. The kernels are also versatile, being used as feed for livestock and poultry, which accounts for over two-thirds of

Fig. 6.2 A field of lush green rows of maize plants during the growing season in one of smallholder farmers in Copperbelt Province of Zambia

the total trade in feed grains. Additionally, maize kernels are raw materials for various industrial products, including starch, oil, syrup, organic liquids, alcoholic beverages, and several other nutrients (Table 6.2).

Subsidiary Uses

Mature maize plants are utilized for animal feed, and silage maize is a prominent crop in industrialized western countries, where specific cultivars and production technologies have been developed. Crop residues, such as stalks, are used for fuel or compost. The inner husks and fiber from stalks are valuable materials in the papermaking industry. Additionally, boiled or roasted kernels from unripe ears can be consumed as a vegetable.

Advantages of Maize in Southern Africa

Maize remains one of the most important cereals in Southern Africa due to its numerous beneficial features:

- **High yield per man-hour**: Maize gives the highest yield per man-hour of invested labor.
- **Protection against birds and rain**: The husks provide natural protection.

Table 6.2 Maize nutrient values per 100 g for edible portion of raw material

Nutrient	Unit[a]	Value/100 g	# of data points	Std. error	Cup 145 g
Water	g	76.05	103	0.863	110.27
Energy	kcal	86	–	–	125
Protein	g	3.27	8	0.185	4.74
Total lipid (fat)	g	1.35	8	0.118	1.96
Ash	g	0.62	47	0.042	0.90
Carbohydrate, by difference	g	18.70	–	–	27.11
Fiber, total dietary	g	2.0	2	–	2.9
Sugars, total	g	6.26	2	–	9.08
Energy	kcal	86	–	–	125
Starch	g	5.70	2	–	8.27
Calcium, Ca	mg	2	53	0.983	3
Iron, Fe	mg	0.52	93	0.022	0.75
Magnesium, Mg	mg	37	89	1.700	54
Phosphorus, P	mg	89	79	3.654	129
Potassium, K	mg	270	92	5.026	392
Sodium, Na	mg	15	92	2.935	22
Zinc, Zn	mg	0.46	58	0.040	0.67
Manganese, Mn	mg	0.163	92	0.014	0.236
Vitamin C, total ascorbic acid	mg	6.8	7	0.568	9.9
Vitamin B-6	mg	0.093	7	0.008	0.135
Vitamin A, RAE	µg	9	–	–	13
Vitamin K (phylloquinone)	µg	0.3	–	–	0.4
Fatty acids, total saturated	g	0.325	–	–	0.471

Reproduced from USDA Agricultural Research Service (ARS) (2008). Nutrient Data Laboratory. Available at: https://www.ars.usda.gov/northeast-area/beltsville-md/beltsville-human-nutrition-research-center/nutrient-data-laboratory/

[a] g, gram; IU, international unit; kcal, kilocalorie; mg, milligram; µg, microgram

- **Ease of harvest and storage**: Maize is easy to harvest and store, and it does not shatter.
- **Extended harvest period**: It can be harvested over a long period, starting with immature ears and later mature ones.
- **Temperature tolerance**: Maize can tolerate a wide range of temperatures.
- **Increasing demand**: The demand for maize as food and feed is always increasing.

Despite its significance, maize yields and profitability in Southern Africa can be substantially improved. The low yields are primarily due to:

- **Rainfed cultivation**: Maize is mostly grown as a rainfed crop, making it susceptible to droughts and unreliable rainfall.
- **Slow adoption of improved cultivars**: High costs prevent many small-scale farmers from adopting improved cultivars and advanced cropping techniques.

Factors Influencing Maize Cultivation

These yield issues are tied to several factors, including marketing, prices, transportation, drying, storage, processing, and usage of the crop. Efficient distribution of improved maize seed is often lacking, and farmers need better access to inputs like improved seed, fertilizers, and crop protection advice. Improved cultivars may not suit the mixed cropping systems typical of smallholder farms, highlighting the need for research focused on farmers' needs.

Time of Planting

Planting maize typically occurs between October and early January, depending on the start of the rainy season in specific locations. Planting during this period helps avoid yield reductions and ensures the crop reaches its critical development stages—flowering and ear filling—under suitable environmental conditions. For instance, farmers in Mtwara and Nachingwea, Tanzania, begin planting in December and January, respectively, based on rainfall patterns. Early planting offers several advantages:

1. Higher yields from better soil nutrient usage.
2. Improved plant stands.
3. Reduced soil evaporation and weed growth due to earlier soil shading.
4. Appropriate pollination period that may avoid mid-season drought.
5. More sunlight during the grain filling period.

Spacing and Plant Density

Optimal maize spacing is about 75–90 cm between rows and 20–30 cm within rows, resulting in a plant population of 44,400–55,500 plants per hectare, depending on the variety's maturity. However, small farms often exhibit high plant density due to inconsistent planting patterns.

Planting Methods

Uniform seedling emergence depends on:

- Seed quality (viability and purity).
- Seed treatment (against soil-borne pathogens).
- Proper seeding methods (depth and spacing).
- Correct seed quantity per pocket.

Common planting methods include broadcasting seeds, dibbling in furrows behind ox-drawn ploughs, and hoe-digging. Small-scale farmers often overplant, placing two to three seeds per pocket to ensure adequate plant emergence, then thin to one plant per pocket about 4 weeks after emergence.

Planting Depth and Fertilizer Requirements

Maize is typically planted 3–5 cm deep. Planting too deep results in poor emergence, while shallow planting risks bird predation and wash-away by rain. Maize requires significant nutrient inputs, particularly nitrogen (N). Nutrient uptake peaks during ear formation and tasseling. High nitrogen levels should be applied in three doses: at planting, when the crop is about 50 cm tall, and at silking. Phosphorus (P) is also essential but less readily absorbed by maize.

Weed Control

Effective weed control is crucial, as weeds significantly reduce yields. Hand weeding is common but labor-intensive. Chemical weed control is gaining popularity due to time constraints. Post-emergence herbicides like 2,4-D, atrazine, and pendimethalin are commonly used, though some have been globally prohibited.

6.5.3 Sorghum Cultivation

Sorghum (*Sorghum bicolor*), a member of the Poaceae family, ranks as the second most important cereal crop after maize in sub-Saharan Africa, including Southern Africa (Fig. 6.3, Table 6.3). Its drought tolerance and ability to thrive in areas with moisture deficits, high ambient temperatures, and high soil temperatures make it a resilient crop, outperforming many other cereals under harsh conditions.

Versatility and Uses

Sorghum is cultivated extensively across the region, primarily by resource-poor small-scale farmers. It serves multiple purposes:

- **Grain**: Used for household subsistence as a staple food.
- **Sweet stems**: Chewed for their sweetness.
- **Livestock feed**: Provides fodder for animals.
- **Biomass**: Used as mulch to improve soil health.

Production Areas

In 1992, Mozambique, South Africa, Tanzania, and Zimbabwe had the largest areas under sorghum cultivation (de Milliano 1992). In Zimbabwe, sorghum is a principal source of carbohydrates for most of the population. The crop remains stable in cultivation areas in Lesotho and Malawi, with increasing production in South Africa and the DR Congo (FAO 1986, 1988).

Fig. 6.3 A field of hish green rows of sorghum plants during the growing season in one of smallholder farmers in Copperbelt Province of Zambia

Table 6.3 Area ('000 ha) sown to sorghum and proportion (%) of the total area sown to sorghum of ten southern African countries and Africa, 1969–1971, 1979–1981, and 1984–1986

Country	1969–1971		1979–1981		1984–1986	
	Area ('000 ha)	Proportion (%)	Area ('000 ha)	Proportion (%)	Area ('000 ha)	Proportion (%)
Botswana	133.1	8	98.4	–	55.7	3
Lesotho	75.0	5	57.5	3	60.7	3
Malawi	106.7	6	126.7	7	135.3	7
Mozambique	260.9	16	254.7	14	200.0	10
Swaziland	6.8	1	2.1	1	2.0	1
South Africa	362.8	22	359.8	20	467.0	25
Tanzania	310.0	19	713.3	39	700.0	37
DR Congo	28.7	2	31.1	2	38.0	2
Zambia	75.8	5	38.8	2	34.2	2
Zimbabwe	290.9	17	139.9	8	200.0	11
Total area sown to sorghum						
Ten Southern African countries	1650.7		1822.3		1892.9	
Africa	13,803.9		12,535.1		15,686.0	

Source: FAO tapes, 1986 (1988) for the Republic of South Africa, and for the other countries, FAO 1972, 1981, pp. 106–109, and 1986, pp. 120–121

Ecological Adaptability

Sorghum's ability to produce acceptable yields in a wide range of ecological conditions unsuitable for other cereals highlights its importance in ensuring food security in the region.

The attributes that help sorghum withstand drought conditions are numerous:

- A more extensive root system than other cereal crops.
- A smaller leaf surface area compared to maize.
- Leaves covered with wax which helps reduce water loss through evapotranspiration.
- The ability to stay dormant when conditions are not favorable.

Importance of Sorghum: Human Consumption and Animal Feed

- Sorghum grains—which contain numerous nutrients (Table 6.4)—are eaten as a cooked porridge.
- Pop sorghum, which has small grains with horn endosperm toward a grain's periphery, is prepared and consumed in a similar way to popcorn.

Table 6.4 Chemical composition of sorghum

Nutrient	Unit[a]	Value/100 g	# of data points	Std. error	Cup 192 g
Water	g	12.40	10	0.312	23.81
Energy	kcal	329	–	–	632
Protein	g	10.62	5	0.379	20.39
Total lipid (fat)	g	3.46	5	0.050	6.64
Ash	g	1.43	5	0.072	2.75
Carbohydrate, by difference	g	72.09	–	–	138.41
Fiber, total dietary	g	6.7	6	0.400	12.9
Sugars, total	g	2.53	–	–	4.86
Calcium, Ca	mg	13			
Magnesium, Mg	mg	165	5	0.192	6.45
Phosphorus, P	mg	289	6	24.900	317
Potassium, K	mg	363	5	16.975	555
Sodium, Na	mg	2	5	21.061	697
Zinc, Zn	mg	1.67	5	0.626	4
Copper, Cu	mg	0.284	6	0.125	3.21
Manganese, Mn	mg	1.605	5	–	0.545
Vitamin C, total ascorbic acid			–	–	23.4
Vitamin B-6	mg	0.367	5	0.422	7.081
Vitamin B-12	μg	20	1	–	38
Vitamin A, RAE	μg	0.00	–	–	38
Vitamin D	μg	0.0	–	–	0.96

Reproduced from USDA (2016). Food Composition Databases of Nutrient Database for Standard Reference. Available at: https://ndb.nal.usda.gov/ndb/search/list

[a] g, gram; IU, international unit; kcal, kilocalorie; mg, milligram; μg, microgram

- Immature sorghum varieties with sugary grains are eaten as sweet corn.
- Sorghum is widely used for brewing local beer.
- In stems, sorghum has a large amount of sweet juice that contains up to 10% sucrose. Stems are used for chewing and manufacture of syrup.
- The plant stalks and leaves are used as feed for livestock.

Production Systems

Sorghum is a life-sustaining crop for many people, particularly in the dry areas of Southern Africa. Its ability to thrive under harsh conditions makes it an essential crop for food security and resilience in Southern Africa. These regions often experience marginal rainfall and poor soil fertility, conditions under which maize typically fails. Sorghum is usually grown alongside other crops in mixed cultivation, although it can sometimes depress the yields of these crops by competing for soil nitrate and moisture. Therefore, rotating sorghum with other crops is crucial to prevent the buildup of the parasitic weed striga.

Land Selection and Seedbed Preparation

Sorghum is sensitive to soil acidity and aluminum toxicity, so it should be grown in areas with a pH between 5 and 8. Since birds can severely damage sorghum crops, areas at high risk should be avoided. Sorghum seeds are small and have lower seedling vigor compared to maize, making a fine seedbed essential for good yields. The soil should be pulverized to a depth of 15–20 cm and be clean and firm to ensure good seed-to-soil contact. However, overworking the soil should be avoided to prevent the formation of a hard crust that can hinder seedling emergence.

Planting Time

The planting time for sorghum depends on the rainfall pattern. Local varieties should be planted early when the rains start, and planting should occur when there is adequate soil moisture for good germination and seedling establishment. Planting can be done in rows or in randomly distributed stations, especially for farmers using hand hoes. About two to four plants per station should be retained after thinning, particularly in very dry areas. Some farmers prefer broadcasting seed, which requires practice to distribute the correct amount per hectare. A fine seedbed is critical for broadcasting seeds to ensure even distribution and good coverage to a depth of 3–5 cm.

Climatic Features and Planting Period

The planting period is guided by the climatic features and the length of the growing season in a given area. In low-rainfall areas, planting in the first to third week of December is best for higher yields, with planting up to mid-January being acceptable. In high-rainfall areas, planting should occur from the second to last week of December. Planting should commence when the rains have stabilized to avoid delays that can cause significant yield losses. Planting too early is not advisable for sorghum as late-season rains can damage early maturing grains.

Spacing

Planting is typically done with a spacing of 60–90 cm between rows and 30–50 cm between stations, resulting in a plant population of about 130,000–150,000 plants per hectare (Fig. 6.4). For farmers planting the crop randomly, a spacing of 75 × 50 cm is appropriate.

Number of Seeds per Station

It is difficult to plant seed at a uniform depth with the hand hoe commonly used by small-scale farmers, so germination is seldom complete. Therefore, farmers often plant at a high seed rate (about 10 kg/ha) and then adjust the population to about 120,000–180,000 plants/ha by thinning during the first weeding. Each station may have three to four plants before thinning. When farmers who can afford them use precision planters, a lower seed rate is employed.

Fig. 6.4 A sorghum crop grown at right spacing in Tanzania. (Source: ICRISAT 2006)

Planting Depth

Sorghum seeds are planted into holes about 3 cm deep and 2.5–5 cm in diameter using a hand hoe. Shallow holes are avoided as rains can easily wash the seeds away, while deeper planting may lead to poor germination.

Field Management After Planting Sorghum

To achieve optimal yields at harvest, the condition of the field and the management of the crop throughout the cultivation period are crucial. The following operations are essential for obtaining acceptable yields:

Weeding

Weeding is critical during the early stages of sorghum growth. The first flush of weeds following good rains should be removed to reduce the weed population. Sorghum is particularly susceptible to weeds early on. The first weeding should occur 2–3 weeks after planting, with a second weeding, if necessary, 5–7 weeks later. Once established, sorghum can overcome the adverse effects of weeds.

Thinning

High plant populations can negatively impact crop growth, especially under drought conditions. Excessive plants compete for nutrients, water, and sunlight, slowing growth. Small-scale farmers typically thin their crops about 3 weeks after planting to remove extra and unhealthy plants. Healthy thinned plants can be transplanted to fill gaps in the field, although they are not as productive as directly seeded plants. Farmers should also rogue out off-type plants before they flower to maintain seed purity.

Fertilizer Application

Sorghum requires fertile, well-drained soils. A basal dressing of 200 kg/ha of "D" compound fertilizer ensures a good start for the crop. Additionally, a top dressing of 100 kg/ha of urea can be applied 1–2 weeks before flower initiation. In conditions where water is very limited and field management is poor, sorghum's response to fertilization is minimal. Crop rotation with legumes, which provide organic matter and fix nitrogen, can improve soil fertility.

6.5.4 Cultivation of Millets

Millets, including rice, maize, and sorghum, are among the four most important cereals grown in the tropics, particularly in Southern Africa. While millet production remains largely at the subsistence level, it is a staple food and beverage in many areas.

Nutritional Value of Millets

Millets are nutritionally equivalent or superior to most cereals, containing high levels of essential amino acids like methionine, cystine, and others crucial for human health. They are a unique source of pro-vitamin A and micronutrients such as zinc (Zn), iron (Fe), and copper (Cu), with finger millet being particularly nutrient-dense.

Cultivation and Importance

Pearl millet is the most widely cultivated millet, especially in semi-arid areas. Despite its importance, millet production in the region remains low. Millets are usually grown as part of mixed or intercropping systems, which offer numerous advantages to farmers.

Advantages of Millets

Millets thrive on light-textured and well-drained soils but can grow in almost any type of soil and are drought-resistant. These characteristics make millets vital food staples, particularly in semi-arid areas of Southern Africa where other crops often fail due to inadequate rainfall and poor soil conditions. Together with sorghum, millets provide 75% of the total caloric intake for poor people in drought-prone areas. According to Obilana (2003), millets alone provide 13.4 kg of food per capita annually.

Types of Millets in Southern Africa

While there are many types of millets globally, only two, finger millet and pearl millet, are produced in Southern Africa.

Finger Millet

Finger millet (*Eleusine coracana* (L.) Gaertn.) belongs to the Poaceae family and is a tufted annual crop growing 30–150 cm tall and maturing in 75–160 days. It is indigenous to Africa and well-adapted to Southern African conditions, commonly grown as a subsistence food crop. The grain, which is nutritionally rich, is used as a staple, often consumed as porridge made from flour sometimes blended with cassava flour. It is also used to make local beers. Finger millet is a strategic food security crop due to its good storability and potential as a cash and industrial crop.

Benefits of Finger Millet

Finger millet is especially valuable for its high methionine content, an amino acid lacking in diets based on starchy staples like cassava, plantain, rice, or maize meal.

Nutritional and Industrial Value of Finger Millet

Finger millet holds significant nutritional and industrial importance, particularly in Southern Africa where it is a staple food. Let's delve into the key attributes and uses of this versatile grain.

Nutritional Content

Finger millet is packed with essential nutrients, making it a crucial part of the diet in many regions (Table 6.5):

- **Protein content**: Approximately 7.6% of the grain is protein.
- **Calcium content**: High calcium levels, averaging about 0.34% of the grain.
- **Methionine content**: Methionine, a sulfur-containing amino acid, constitutes between 2.0% and 3.0% of the total protein.

Culinary Uses

- Finger millet grain is incredibly versatile in culinary applications:
- **Flour**: Often ground into flour for various dishes such as gruel, porridges (both fermented and unfermented), and bread. These food products can be cooked, baked, or fried, sometimes mixed with condiments, spices, or other foods.
- **Alcoholic beverages**: Used in brewing local alcoholic drinks.
- **Sweet dishes**: Incorporated into sweet culinary creations.

Malting Characteristics

- One of the standout features of finger millet is its superior malting qualities:
- **Malt production**: The grains are germinated in water, then heated, and dried. During this process, starch in the endosperm converts to sugar.

Table 6.5 Millet nutritional values per 100 g for edible portion of raw material

Nutrient	Unit[a]	Value/100 g	# of data points	Std. error
Water	g	8.67	8	0.584
Energy	kcal	378	–	–
Protein	g	11.02	9	0.685
Total lipid (fat)	g	4.22	9	0.228
Ash	g	3.25	8	0.350
Carbohydrate, by difference	g	72.85	–	–
Fiber, total dietary	g	8.5	–	–
Calcium, Ca	mg	8	19	0.388
Iron, Fe	mg	3.01	19	0.096
Magnesium, Mg	mg	114	19	4.733
Phosphorus, P	mg	285	1	–
Potassium, K	mg	195	19	5.610
Sodium, Na	mg	5	19	0.741
Zinc, Zn	mg	1.68	19	0.114
Copper, Cu	mg	0.750	19	0.028
Manganese, Mn	mg	1.632	19	0.076
Vitamin B-6	mg	0.384	1	–
Vitamin K (phylloquinone)	µg	0.9	–	–
Fatty acids, total saturated	g	0.723	–	–

Reproduced from USDA (2016). Food Composition Databases of Nutrient Database for Standard Reference. Available at: https://ndb.nal.usda.gov/ndb/search/list

[a] g, gram; IU, international unit; kcal, kilocalorie; mg, milligram; µg, microgram

- **Milk-based drinks**: White finger millet is ideal for making malt used in milk-based drinks for young children, thanks to its pleasant flavor and starch-hydrolyzing enzymes.
- **Weaning food**: The malt is rich in calcium and methionine, making it an excellent base for weaning foods.

Industrial Applications

Beyond its nutritional benefits, finger millet has several industrial uses:

- **Malting and brewing**: Used extensively in these processes.
- **Starch making**: An important raw material.
- **Animal feed**: Utilized in the production of livestock feed.

Finger millet is not only a staple food but also possesses numerous medicinal and nutritional benefits. It's one of the most nutritious cereals, making it highly beneficial for maintaining good health.

Medicinal Benefits

Finger millet is recommended for various health conditions:

- **Blood pressure and heart health**: Its food products are beneficial in managing blood pressure and heart conditions.
- **Liver disorders and asthma**: Finger millet is recommended for liver disorders and asthma.
- **Cholesterol reduction**: Contains lecithin and methionine, which help reduce cholesterol levels by eliminating excess fat from the liver.
- **Blood sugar control**: Phytochemicals in finger millet help control blood sugar levels, making it beneficial for diabetics.
- **Weight loss**: High fiber content aids in weight loss.
- **Iron source**: Important for the diets of anemic children and nursing mothers.

Nutritional Profile

Finger millet is rich in several key nutrients:

- **Protein content**: About 7.6% of the grain.
- **Calcium content**: High, averaging about 0.34%.
- **Methionine content**: Between 2.0% and 3.0% of the total protein.

Production Systems

Land Selection and Preparation

Finger millet thrives best in well-drained, fertile sand to sandy loams with good water-holding capacity. It prefers a pH between 5 and 7 but can tolerate highly alkaline soils up to pH 11. However, it performs poorly on infertile soils and cannot withstand waterlogging.

Optimal fields for finger millet production should avoid previous season's plantings to reduce soil-borne diseases and volunteer plants. Weed-free land is also desirable.

Land Preparation Practices

Small-scale farmers employ various practices, ranging from slash-and-burn to multiple cultivations before sowing. Farmers typically apply farmyard manure or plough in a green manure crop to sustain soil fertility. One ploughing to a depth of 15–20 cm is often done off-season to reduce pest and disease buildup.

Planting

Due to the small size of finger millet seeds, the crop is planted in well-prepared seedbeds, often by broadcasting followed by subsequent weedings. Planting typically occurs at the onset of the rainy season. Mid-November to the second week of December is considered ideal for planting the rainfed crop.

Spacing and Plant Depth

Two planting methods are common: direct seeding and transplanting nursery seedlings. In direct seeding, seeds are broadcast and covered by a hoe or animal-drawn equipment. Seed rates range from 10 to 30 kg/ha, with thinning to leave three to six plants per hole. Transplanting involves raising seedlings in nurseries and planting them in the field 3–5 weeks later.

Weed Control

Weeding finger millet is labor-intensive, especially in fields with predominant grasses like *E. indica* and *E. africana* (Fig. 6.5). Effective weed control requires planting in rows and manual weeding 2–3 weeks after emergence.

Fertilization

Fertility is boosted through crop rotation practices. Finger millet responds well to nitrogen (N) and phosphorus (P) applications but not to potassium (K). N should be applied in split doses, with 50% as basal fertilizer and the remainder 35 days after emergence. Manures such as farmyard manure, silt, or earth from old village sites are also often applied by farmers at rates about 2–10 tons/ha.

Fig. 6.5 *Eleusine indica* weed commonly found in millet farmers' fields in Southern Africa

Rotation

Rotating finger millet fields with crops such as legumes, cotton, tobacco, or other well-fertilized and weeded crops can help maintain soil fertility. The most common rotations involve soybeans or groundnuts, which are well-suited to millet growing areas and leave behind residual fertility that supports finger millet growth. This practice can eliminate the need for basal fertilizer applications.

Irrigation

Finger millet is typically grown as a rainfed crop, but irrigation may be necessary during severe dry spells. For the irrigated crop, watering should occur at intervals of 8 days, requiring about 10–15 irrigations from planting to maturity to ensure optimal growth.

Pearl Millet Cultivation

Pearl millet (*Pennisetum glaucum* (L.) *R. Br.*), commonly known as bulrush millet, is one of the most drought-tolerant cereals (Fig. 6.6). As a member of the grass family Gramineae, pearl millet has the ability to regenerate and produce new basal tillers to compensate for losses caused by drought or other unfavorable conditions. It performs exceptionally well on poor soils under moisture stress and high heat, showing a remarkable response to favorable environmental conditions and good field management practices. Pearl millet is predominantly grown as a rainfed crop

Fig. 6.6 A field of lush green pearl millet plants during the growing season in one of smallholder farmers in Copperbelt Province of Zambia

in the semi-arid regions of Southern Africa, where the mean annual rainfall ranges from 200 to 800 mm.

Benefits of Pearl Millet

- **Drought tolerance**: Pearl millet's ability to withstand drought makes it ideal for semi-arid regions.
- **Regeneration capacity**: The crop can regenerate and produce new tillers in response to drought.
- **Soil adaptability**: Performs well on poor soils with moisture stress and high temperatures.

Importance of Pearl Millet

Pearl millet, widely known as bulrush millet, is among the most nutritious cereals, boasting high protein content and exceptional quality. Let's explore its nutritional profile, uses, and best practices for cultivation.

Nutritional Profile

Pearl millet stands out for its impressive nutritional content:

- **Protein content**: 11%, with high-quality lysine at 3.68 mg/g protein, compared to 2.24 for wheat, 3.36 for rice, 3.0 for maize, and 3.2 for sorghum.
- **Carbohydrate content**: 67–72%, with starch as the major component.
- **Fat content**: 4–8%, contributing to its palatability.
- **Ash content**: Varies from 1.5% to 2.7%.
- **Phosphorous content**: High, ranging from 270 to 390 ppm.
- **Vitamins**: Well-represented, particularly thiamine, riboflavin, and nicotinic acid.

Uses of Pearl Millet

- **Human consumption**: Consumed in various forms including sweet and alcoholic beverages.
- **Grain decortication**: Decorticated before grinding for different products. However, traditional hand-pounding produces low yields (around 75%) and poor storage quality.
- **Animal feed**: Stiff, strong straw used for building material and cattle feed during the dry season.
- **Forage**: Specific cultivars bred for forage with numerous vigorous and leafy tillers.

Cultivation Practices

Land Selection and Preparation

Fields selected for pearl millet should be flat and have a history of successful millet cultivation. Avoid growing millet on the same land consecutively to prevent volunteer plants. Though pearl millet can thrive on poor, acidic soils, fields should be reasonably fertile and free from waterlogging. A fine seedbed is preferred due to the small seed size.

Planting

Traditionally, pearl millet is broadcast or sown in hills in dry soils before the rainy season starts. This practice benefits late-maturing local landraces but can pose germination and growth challenges in inadequate early rains. Planting is best done when the soil is moist, typically between mid-November and mid-January.

Spacing and Planting Depth

Seeds are generally broadcast, but hill planting involves placing four to eight seeds per station at a depth of 2–3 cm. Row planting is recommended for better management, with 75 cm between rows and 20 cm between plants within a row.

Number of Seeds per Station

The sowing rate is about 4–6 kg of seed/ha, with thinning to three plants per station during the first weeding to establish a plant population of 60,000–80,000 plants per hectare. Traditional practices result in lower plant populations.

Cropping Systems and Management

Pearl millet is often intercropped with cereals, legumes, and root crops. Sowing forage legumes into a pure stand of pearl millet provides nitrogen and high-quality forage. Agroforestry components, such as *Faidherbia albida*, improve soil organic matter and nitrogen content while serving as windbreaks.

Weed Control and Thinning

Weeding is crucial to prevent competition for soil moisture and nutrients. One weeding is typically sufficient, with thinning to two to three plants per station. Chemical weed control is an option if affordable.

Fertilization

In traditional systems, inorganic fertilizers are not commonly used. Animal manure can sustain fertility initially. For improved yields, inorganic fertilizers like compound "D" or "X" at 100–200 kg/ha for basal dressing and urea or ammonium nitrate at 50–100 kg/ha for top dressing are recommended.

6.5.5 *Rice Cultivation*

Rice (*Oryza sativa*), a member of the Gramineae family, is a versatile, tropical, annual plant characterized by round, hollow, and jointed culms (Fig. 6.7). This crop thrives under a wide range of environmental conditions, from hot, dry climates to the cooler climates of high altitudes. In Southern Africa, rice is a crucial cereal crop, primarily cultivated in flood plains along rivers, river valleys, and dambos (shallow wetlands). It is also successfully grown under rainfed, upland conditions in regions with high rainfall, often supplemented with irrigation schemes to ensure optimal growth and yield.

Importance and Nutritional Value

The rice grain contains a high percentage of protein that compares favorably with that of maize and finger millet. Fats, carbohydrates, fiber, minerals, and vitamins are also available in rice (Table 6.6). The fibrous bran of brown rice is rich in oil, protein, some B vitamins (thiamin, riboflavin, and niacin), and the minerals iron, phosphorus, and potassium. To make white rice, the bran is removed; however, it is in the bran that most of the vitamins and minerals are located. Therefore, white rice is less nutritious than brown rice and must be enriched with the addition of vitamins and minerals to increase its nutritive value.

Fig. 6.7 A field of lush green rice crop during the growing season in one of smallholder farmers in the Copperbelt of the DR Congo

Table 6.6 Nutritional value per 100 g of raw long-grain white rice

Element	Quantity[a]	%	Element	Quantity	%
Energy	365 kcal	–	Vitamin B6	0.164 mg	13.0
Carbohydrates	80 g	–	Calcium	28 mg	3.0
Sugars	0.12 g	–			
Dietary fiber	1.30 g	–			
Fat	0.66 g	–	Iron	0.80 mg	6.0
Protein	7.13 g	–	Magnesium	25 mg	7.0
Water	11.61 g	–	Manganese	1.088 mg	52.0
Thiamine (Vit B1)	0.0701 mg	6.0	Phosphorus	115 mg	16.0
Riboflavin (Vit B2)	0.0149 mg	1.0	Potassium	115 mg	2.0
Niacin (Vit B3)	1.62 mg	11.0	Zinc	1.09 mg	11
Pantothenic acid (Vit B5)	1.014 mg	20.0	–	–	–

Percentages are relative to dietary reference intakes for adults
Reproduced from USDA (2016). Food Composition Databases of Nutrient Database for Standard Reference. Available at: https://ndb.nal.usda.gov/ndb/search/list
[a] g, gram; IU, international unit; kcal, kilocalorie; mg, milligram; µg, microgram

Human Consumption

Rice is consumed as a cereal after it is boiled or fried, and it can be eaten with milk, legumes, fish and meat, etc.

Industrial Uses

Rice is cultivated worldwide and serves multiple purposes beyond being a staple food. It's a raw material for producing starch, rice flour, glucose, acetic acid, and fermented and unfermented alcohol like rice milk and sake.

Medicinal Uses

Rice offers numerous health benefits:

- **Hypertension**: Low sodium content makes it beneficial for managing blood pressure.
- **Diarrhea treatment**: Rice water is effective as an oral rehydration fluid for infants and one of the recommended foods for adults with diarrhea.

Cultivation Practices

Farmers employ different cultivation methods based on environmental conditions:

Rainfed or Upland Rice

- Depending entirely on rainfall.
- Grown in plains and sloping lands at various altitudes.
- Land preparation ensures roots can access the water table.

Irrigated or Lowland Rice

- Requires plots bounded by bunds with irrigation systems such as canals and drains.
- Depends on manual watering.

Incomplete Water Control System

- Relies on both rainfall and manual watering.
- Often grown on raised beds separated by drains.
- Rotated with crops like beans, maize, or tuber crops.

Time of Planting

Planting times depend on the maturity period of the rice variety:

- **Late maturing varieties** (160–180 days): Planted during the first 2 weeks of rainfall.
- **Medium maturing varieties** (140–160 days): Planted during the last 2 weeks of rain onset.
- **Early maturing varieties** (120–140 days): Planted from mid-December to mid-January.

Planting Methods

Transplanting

- Efficient but labor-intensive.
- Seedlings grown in nursery beds for 30 days are transplanted to the main field.
- Typically done in rows about 25 cm apart with 10–15 cm spacing within rows.

Direct Sowing

Two methods are commonly used:

- **Drilling**: Furrows 2–3 cm deep and 25 cm apart are made, with 40–50 seeds per meter sprinkled into the furrow.
- **Broadcasting**: Popular among farmers for its speed and ease.

Growth and Stages of Development

The life cycle of rice is divided into three main stages: vegetative, reproductive, and ripening.

Vegetative Stage

- **Germination to panicle formation**: This stage begins with the germination of seedlings and continues until panicle initiation. It is marked by active tillering, increased plant growth, and leaf emergence. Secondary adventitious roots develop after transplanting, absorbing nutrients from the soil to sustain growth.

Reproductive Stage

- **Panicle formation to heading**: This stage occurs about 20–30 days from panicle initiation. The booting stage involves the swelling of the flag leaf sheath due to the increasing size of the panicle. During this stage, meiosis occurs. The heading phase begins when panicles extend out of the boots, and flowers develop at the tips, leading to fertilization and grain filling.

Ripening Stage

- **Heading to grain filling and maturation**: This final stage starts after ovary fertilization. Grains increase in size and weight due to the translocation of starch and sugars from the culm and leaf sheaths. They transition from green to gold and straw colors, and the leaves senesce as the plants mature.

Integrated Rice-Cum-Fish Farming

In some Southern African countries, such as Zambia, rice-cum-fish farming has been incorporated into agricultural practices. This method involves stocking rice fields with fingerlings to obtain both rice and fish crops, enhancing land utilization efficiency.

Advantages of Rice-Cum-Fish Farming

- **Economic utilization of land**: Produces both rice and fish.
- **Labor efficiency**: Minimal extra labor is required for fish care.
- **Reduced weeding**: Fish feed on weeds.
- **Increased rice yield**: Fewer insects, better soil mineralization, improved soil aeration, and increased organic fertilization.
- **Resource utilization**: Waste products from one system, such as organic fertilizer or weeds, are used as inputs for another.
- **Improved cash flow**: Provides higher income per land area.
- **Diet improvement**: Offers a better diet for the farmer and family.
- **Risk management**: Spreads biological and economic risks.

Challenges

Despite its potential, rice-cum-fish farming has not been widely adopted in Southern Africa due to:

- Adverse environmental conditions for fish growth (aridity, high altitude/low temperature).
- Shifting agriculture and seasonal labor shortages.
- Lack of interest in fish husbandry and competition from capture fisheries.
- Health implications from water-borne diseases such as bilharzia or malaria.

Weeding

Weed control is a major challenge in rice production, contributing significantly to low yields. Effective methods are crucial for both upland and irrigated rice fields.

Upland Rice

In upland rice fields, which are typically smallholder plots, weeding is performed at least two to three times:

1. **First weeding**: Within 4 weeks of germination, when weeds are about 4 cm tall.
2. **Second weeding**: About 6 weeks later, before tillering.
3. **Third weeding**: If necessary, at 8 weeks.

Weeds are usually removed manually using a hoe. In larger fields, draft animals may be used for the first weeding, or chemical weed killers are applied by those who can afford them. A post-planting and pre-emergence application of oxadiazon is often recommended for controlling grasses. Herbicides like Basagran, Basagran PL2, MCPA, and Butachlor are used by farmers at recommended rates.

Irrigated Rice

In irrigated or flooded rice fields, maintaining a water depth of about 10–25 cm during cultivation effectively controls weeds. This depth must not exceed 5 cm during tillering to avoid hindering plant development. One or two weedings are needed, the first early and the second before tillering. Chemical weed control can be achieved by spreading herbicide emulsions or granules under water or spraying after drying out the rice field.

Fertilization

Rice responds well to fertile soils, and the type and timing of fertilizer application depend on the cultivation system.

Upland Rice

In upland rice, fertilizer effectiveness largely depends on water supply. Nitrogen (N) and phosphorus (P) are applied on acidic soils, while potassium (K) is used for maintenance dressing. In semi-intensive cultivation, applying 50 kg of N, 25 kg of P_2O_5, and 30 kg of K_2O per hectare is advisable. These amounts can be doubled in intensive cultivation.

Lowland Rice

For lowland rice, N fertilization is crucial. If water depth is maintained from the start, half or one-third of N is incorporated with P and K at transplanting. The remaining N is applied at tillering or split between tillering and panicle initiation.

Swamp Rice

Swamp rice cultivation uses ammonium sulfate (20% N), calcium cyanamide (22% N), or urea (48% N) as the primary N sources. Nitrates are not recommended due to their rapid release and leaching.

Irrigation Management

Water management aims to maintain desirable irrigation levels during growth, serving both plant hydration and weed control. Flooding irrigation involves leveling fields and maintaining water flow from ditches. After transplanting, a water level of 5–10 cm is maintained until physiological maturity, when fields are drained to leave soil wet but not submerged.

Challenges with Irrigation

Salt accumulation from irrigation water is a common issue. As water evaporates, salts concentrate and can crust on the soil surface or affect root levels, hindering plant growth.

6.5.6 Root and Tuber Crop Production

Introduction

In terms of calories, maize is Africa's most crucial food crop. It's a model crop for meeting the growing urban demand for convenient food products. However, maize production in Southern Africa is often risky and inadequate due to unreliable rainfall. As agriculture in the region is predominantly rainfed, maize yields have been highly variable, falling short due to harsh environmental conditions and limited access to essential inputs like fertilizers.

Benefits of Root and Tuber Crops

On the other hand, root and tuber crops such as cassava (*Manihot esculenta*), sweetpotato (*Ipomoea batatas*), and yams (*Dioscorea* spp.) offer a more reliable food source under marginal agroecological conditions. Compared to other staples, these crops:

- Provide larger amounts of energy and nutrition per unit area and time.
- Require less intensive management, even in challenging climatic conditions.
- Can be processed into various food, feed, and industrial products by both community-based units and sophisticated industries.

These crops are essential for the livelihood and food security of numerous households in Southern Africa, often grown on marginal land (Fig. 6.8). They are particularly important in combating hunger caused by droughts, floods, climatic disasters, pests, diseases, or civil strife.

Nutritional Value

For approximately 30% of the population in some parts of the region, cassava and sweetpotato are fundamental sources of energy and vital nutrients. Certain cultivars contribute significant amounts of iron, zinc, calcium, and vitamins A and C to diets (Latham 1979). The carbohydrate yield from cassava and sweetpotato per unit of land is higher than other major staples. Some sweetpotato varieties are rich in beta-carotene, a precursor of vitamin A (Faber et al. 2010), which is crucial for

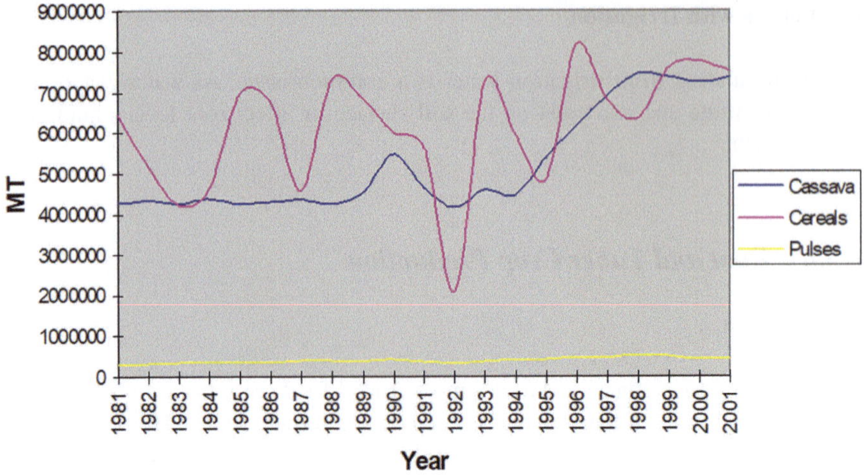

Fig. 6.8 Production of cassava, cereals, and pulses in five different drought-affected countries in Southern Africa (FAO 2002)

improving the nutritional status of children, lactating mothers, and people living with HIV/AIDS.

Economic and Social Impact

Cassava and sweetpotato also play crucial roles in Southern African development:

- **Famine reserve food**: Major food staples during periods of food scarcity.
- **Cash crops**: Serve as important industrial raw materials and livestock feed.
- **Wheat flour substitution**: Partial substitution with cassava or sweetpotato flour reduces wheat imports, saving foreign exchange.

Root and tuber crops are vital as a safety net in households affected by the HIV/AIDS epidemic. Southern Africa has some of the highest HIV/AIDS infection rates globally, with national infection rates ranging between 14% and 42% (Mukelabai et al. 2006). Approximately 10 million people in the region are infected, and 5 million children have been impacted, with 2.7 million orphaned. These crops provide food security and help sustain households with weakened labor forces.

Production and Marketing Challenges

Cassava and sweetpotato are typically grown as subsistence crops and marketed through traditional systems with low levels of technology and poor organizational, institutional, and policy support. Production and use are concentrated in countries with lower per capita incomes, playing a significant role in food systems in remote, marginal areas.

Root and tuber crops such as cassava, sweetpotato, and yams are invaluable for food security in Southern Africa due to their resilience. These crops thrive despite drought, poor soil, and minimal inputs like fertilizers. They provide year-round food, reducing recurrent shortages.

In 1998, a FAO/World Food Program (WFP) Crop and Food Supply Assessment Mission in Zambia concluded that cassava-producing provinces had greater food security than cereal-producing areas (FAO/WFP 1998). The contributions of cassava and sweetpotato to food security in Zambia are well-recognized (FAO/WFP 1998; MFNP 2002). However, a significant challenge remains of expanding the utilization and distribution of these crops to food-insecure regions (Soroko 1998).

Economic and Nutritional Potential

Root and tuber crops offer immense economic potential due to their unique characteristics:

- **Higher yields**: These crops yield an average of 7.7 tons/ha, compared to less than 1.0 tons/ha for cereals.
- **Nutritional value**: They provide larger amounts of nutrition per unit area and time.
- **Management**: Require less intensive management, even under marginal and risky climatic conditions.
- **Processing**: Suitable for processing at various levels, from community-based units to sophisticated industrial applications for food, feed, and bio-fuels (SARRNET 2008).

Cassava, in particular, remains green during dry seasons when other crops fail, making it a reliable food source.

Cultivation and Utilization

Approximately 20 million farm families in Southern Africa cultivate around 5 million hectares of root and tuber crops (Minde and Jumbe 1997). Cassava and sweetpotato are the most important, alongside Irish potato (*Solanum tuberosum*), yams (known as *Busela* in Zambia), and cocoyams—*Colocasia esculenta* (*Nduma* in Tanzania) and *Xanthosoma sagittifolium*.

Challenges and Solutions

To transform the status of these crops in the region, comprehensive sector information is needed to identify opportunities, constraints, and needs. Priority areas for technological, institutional, organizational, and policy interventions should be aimed at:

- Increasing production
- Enhancing utilization
- Promoting commercialization
- Meeting the demand created by cereal deficits

Constraints to Production

Several factors contribute to the low production of root and tuber crops:

- **Inadequate farmer training**
- **Low yielding varieties**
- **Scarcity of quality planting material**
- **Inappropriate production practices**
- **High post-harvest losses**
- **Biological constraints**: Pests, diseases, and weeds

Addressing these constraints is crucial for improving agricultural production and ensuring food security in the region.

6.5.7 Cassava Cultivation

Cassava is a perennial woody shrub belonging to the Euphorbiaceae family, commonly grown as an annual crop (Fig. 6.9). Known by various names such as yuca, tapioca, and manioc, cassava is a crucial source of low-cost carbohydrates, especially in the humid tropics. Brazil tops the list as the world's largest producer of cassava, followed by Thailand, Nigeria, the DR Congo, and Indonesia. Notably, about half of the world's cassava production comes from Africa. The crop is cultivated in around 40 African countries, playing an essential role in food security, particularly in Nigeria, the DR Congo, Tanzania, Ghana, Mozambique, Uganda, Madagascar, Angola, Côte d'Ivoire, Cameroon, Benin, and Kenyaé.

Economic and Nutritional Significance

Cassava is a vital food crop, providing essential calories and nutrients to millions. Its adaptability to poor soils and resistance to drought make it a reliable crop in regions prone to food insecurity. Cassava's versatility extends beyond food; it is also used in producing starch, animal feed, and biofuels, adding economic value to its cultivation.

Cassava is a crucial crop for many African countries, providing essential food and income. Around 70% of Africa's cassava output is harvested in Nigeria, the DR Congo, and Tanzania (IFAD and FAO 2000). In the DR Congo and Angola's Kwanza Norte province, cassava is fundamental to the diet and economy, with the tubers serving as food for both humans and livestock.

Fig. 6.9 A field of cassava crop during the growing season in a multiplication plot of a research program in the Copperbelt of the DR Congo

Economic and Nutritional Value

Cassava's importance extends beyond its role as a staple food. According to Bokanga (1995), cassava can significantly boost rural incomes by providing starch for agro-industries producing paper, boards, adhesives, alcohol, and more nutritive elements (Table 6.7). Cassava leaves are also valuable, accounting for about 68% of all vegetable output in some regions (Tshibaka and Lumpungu 1989).

Expansion Efforts and Challenges

Although already a staple throughout Southern Africa, efforts are ongoing to expand cassava cultivation to enhance food security amid recurrent droughts. Population growth and external pressures are straining production systems, making sustainability a pressing concern.

Significance in Southern Africa

In Southern Africa, cassava is integral to traditional cropping systems, meeting the caloric needs of the rapidly growing population. It is particularly vital for about 84 million of the poorest people in the region (Minde and Jumbe 1997; Kambewa 2010). The crop's resilience and adaptability to harsh conditions make it a dependable source of carbohydrates and a safeguard against food insecurity, especially in drought-prone areas.

Table 6.7 Cassava nutritional values per 100 g for edible portion of raw material

Nutrient	Unit[a]	Values/100 g	Cup 206 g	Root 408 g
Water	g	59.68	122.94	243.49
Energy	kcal	160	330	653
Protein	g	1.36	2.80	5.55
Total lipid (fat)	g	0.28	0.58	1.14
Carbohydrate, by difference	g	38.06	78.40	155.28
Fiber, total dietary	g	1.8	3.7	7.3
Sugars, total	g	1.70	3.50	6.94
Calcium, Ca	mg	16	33	65
Iron, Fe	mg	0.27	0.56	1.10
Magnesium, Mg	mg	21	43	86
Phosphorus, P	mg	27	56	110
Potassium, K	mg	271	558	1106
Sodium, Na	mg	14	29	57
Zinc, Zn	mg	0.34	0.70	1.39
Vitamin C, total ascorbic acid	mg	20.6	42.4	84.0
Vitamin B-6	mg	0.088	0.181	0.359
Vitamin A, IU	IU	13	27	53
Vitamin K (phylloquinone)	µg	1.9	3.9	7.8
Fatty acids, total saturated	g	0.074	0.152	0.302

Reproduced from USDA (2016). Food Composition Databases of Nutrient Database for Standard Reference. Available at: https://ndb.nal.usda.gov/ndb/search/list

[a] g, gram; IU, international unit; kcal, kilocalorie; mg, milligram; µg, microgram

Advantages over Other Crops

Cassava offers several advantages compared to other carbohydrate sources:

- **High productivity**: Thrives under marginal climatic conditions and on fertility-depleted lands.
- **Drought tolerance**: Remains productive even during dry periods.
- **Low input requirement**: Does not need expensive fertilizers or purchased seeds as it is vegetatively propagated.
- **Longevity**: Can remain in the soil and be harvested later without significant quality deterioration during periods of civil strife or drought.

Cassava is a perennial woody shrub widely grown as an annual crop. Known as yuca, tapioca, and manioc, cassava is essential for food security in Africa. Even during temporary displacement, cassava often remains undisturbed in fields due to the labor-intensive process of harvesting and transporting heavy roots.

Nutritional and Industrial Value

Cassava roots have a high dry matter content (35–40%), with over 85% consisting of highly digestible starch (Bradbury and Holloway 1988). Cassava starch's excellent agglutinant properties make it suitable for shrimp and fish feed, replacing costly artificial agglutinants. It is also used in textile and pharmaceutical industries in various forms such as glucose, maltose, sucrose, and syrup (Nweke 2005).

Potential Disadvantages

Cassava has some drawbacks, including its bulk, rapid perishability, and low protein content. Additionally, some cassava varieties contain cyanogenic glucosides, which can convert to toxic hydrogen cyanide (HCN) when the cells are ruptured. Simple processing, like natural drying of tubers, can eliminate most HCN (Mahungu et al. 1987). Despite these challenges, dried cassava products overcome bulk and perishability issues, leaving low protein content as the primary disadvantage compared to other carbohydrate sources.

Uses of Roots and Leaves

Both cassava roots and leaves are economically significant. In some regions like DR Congo, leaves are more valued than roots, accounting for about 68% of all vegetables consumed (Tshibaka and Lumpungu 1989). Lukombo et al. (2022) reported that in the DR Congo, leaves of both common cassava (*Manihot esculenta* Crantz) and the wild (*Manihot glaziovii*) are consumed as vegetables, and that those of *M. glaziovii* although being less abundant are highly appreciated for their sweet taste. While cassava is already a staple food in the humid and sub-humid, cassava roots are a key carbohydrate source, rich in dietary fiber, calcium, iron, magnesium, sodium, riboflavin, thiamine, nicotinic acid, and citrate (Bradbury and Holloway 1988). Cassava leaves, containing up to 7% protein per fresh weight, provide essential vitamins and minerals, with an amino acid composition comparable to chicken eggs (Leung et al. 1972; Lancaster and Brooks 1983). In contrast to cassava roots, the leaves (Fig. 6.10) provide a valuable supplement to predominantly starchy diets.

Shifts in Staple Food Production

In Zambia, for instance, two significant factors led farmers to shift from maize to cassava:

1. **Removal of maize subsidies**: Government withdrawal of maize market guarantees reduced incentives for maize cultivation.
2. **Introduction of improved cassava varieties**: Disease-resistant and early-maturing varieties significantly outyielded conventional ones, increasing cassava production by about 3.4% annually for a decade (Haggblade and Nyembe 2007; Soroko 1998).

In addition to all the above qualities of cassava, it can be summarized that the crop has the following advantages.

Fig. 6.10 A trader taking cassava leaves on his bicycle for sale at marketplace (top left), a hired trolley carrying cassava leaves to a marketplace (top right), women traders marketing cassava leaves in the market (bottom left), and (bottom right) a wifehouse carrying her cassava leaf product at home to make relish. (Source: Courtesy S. Lukombo, IITA DR Congo)

- Cassava responds well to improvement of agronomic practices.
- It responds well to genetic improvement.
- It yields well under good environmental conditions.
- It performs better than other crops under suboptimal environmental conditions.
- Its starch, in the tuberous roots, is a valuable source of cheap calories.
- Its leaves can add well-balanced protein and vitamins to starchy diets.
- Cassava can be harvested at different times, thus providing food at any time of the year.

In summary, cassava is an essential crop that significantly impacts the livelihood of populations across Zambia and the DR Congo. As a drought-resistant crop, cassava thrives in varied climates and poor soils, making it ideal for regions where other crops struggle to grow. It serves as a dependable source of food for millions, providing sustenance during times of scarcity and ensuring food security. Cassava's ability to produce high yields on small plots of land also supports smallholder farmers, who rely on it as both a source of nourishment and income.

The nutritional value of cassava cannot be overlooked. Its roots are rich in carbohydrates, making it an excellent energy source for populations who depend on calorie-dense foods. Additionally, cassava leaves are a valuable source of vitamins and minerals, further contributing to balanced diets. With many communities processing cassava into products like flour, chips, and starch, the crop fosters local entrepreneurship and economic growth.

Moreover, cassava plays a role in poverty reduction across Africa. Its versatility enables farmers to sell both raw and processed cassava products in local and international markets, generating income and improving livelihoods. The crop's resilience to climate change also makes it a sustainable choice for farmers, ensuring long-term economic stability.

Beyond food, cassava has industrial applications that benefit the region's economy. Cassava starch is used in textile production, paper manufacturing, and even biofuel generation. This adds another dimension to its usefulness, creating opportunities for industry development and employment.

Production Practices: Field Selection and Preparation

Cassava thrives in light sandy loams or loamy sands that are moist, fertile, and deep. However, it is also adaptable to soils ranging from sands to clays and those with relatively low fertility. Even soils depleted by repeated cultivation can yield cassava economically. In very rich soils, the plant may produce more stems and leaves at the expense of roots. Cassava grows well across a pH range of 4.0–8.0, provided the soil texture is friable enough for good root development.

Land Preparation for Cassava

First Crop in Cleared Forest

When cassava is the first crop in newly cleared forest, no additional soil preparation is necessary. Clearing the land allows more sunlight to reach the ground and removes weeds and undergrowth that might compete with the crop.

Following Other Crops

Cassava can often be planted without additional soil preparation once the preceding crop has been harvested. If the previous crop left significant debris or weeds, the field should be ploughed two or three times to ensure it is free from grass and unwanted plants.

Land Clearing Practices

Farmers often do not cut large trees but instead burn plant debris covering the land. This burning removes small branches and underbrush, creates a layer of ash that increases the availability of potassium salts, and destroys surface soil parasites. However, complete clearing of the soil can lead to degradation due to nutrient leaching from sunlight and rainfall.

Preparation of Planting Material

The propagation of cassava using vegetative material (e.g., cuttings) (Fig. 6.11) makes the crop more vulnerable to adverse conditions, and to diseases and pests. When exposed to the sun after cutting, cuttings can lose their viability and quality very quickly through dehydration. Excessive moisture is also not good as it may cause bud sprouting before planting. Sprouting is best if stem cuttings harvested shortly before planting are used. For the best crop yields, fresh cuttings from mature plants are ideal. Cuttings that are stored in a dry, well-ventilated, shaded area where direct sunlight and dampness are avoided maintain their viability longer.

The quality of cassava planting material is influenced by several factors, including the age of stems and plants, the thickness of cuttings, the number of nodes per cutting, and the overall health of the cuttings.

Age of Stems and Plants

Cuttings from older, more mature parts of the stem make better planting material compared to those from younger sections. While younger stem cuttings will sprout, they are highly susceptible to soil-borne pathogens and sucking insects. Ideal cuttings should be taken from plants that have been growing for at least 8–15 months. Cuttings from plants older than 18 months tend to be highly lignified with fewer food reserves, reducing bud viability and hampering root and shoot formation.

Fig. 6.11 A farmer busy preparing cassava cuttings for planting (left) and farmers in happy mood after receiving IITA (right) improved cassava stems for multiplication. (Source: Reproduced from Businessdiary Ph. Available at http://businessdiary.com.ph/3546/cassava-production-guide/)

Thickness of Cuttings

Thin stems or branches with poor nutrient reserves are unsuitable as propagating material because they produce weak shoots and small tuberous roots. Farmers should avoid using thin, still-developing plant tips.

Number of Nodes per Cutting

Nodes are crucial as they house the buds where shoots and roots originate. Given that some buds may lose viability during preparation, transportation, and planting, longer cuttings with more nodes increase the chances of successful growth. Cuttings should have at least five to seven nodes and be a minimum of 20 cm long. In sandy soils, farmers often use longer cuttings (60–100 cm) to mitigate termite damage.

Health of Cuttings

Propagation material should be taken from disease-free and insect-free plants (Fig. 6.12). It is essential to avoid wounding the stems during selection, harvesting, and transportation to prevent microbial infections. Selecting high-quality cassava planting material is vital for ensuring robust plant growth and high yields. Following these best practices leads to successful cassava cultivation and contribute to food security in the region.

Fig. 6.12 Good quality cassava planting material is essential for high crop yields. (Source: IITA)

Planting Cassava

Small-scale farmers plant cassava either as a single crop or intercropped with other crops like maize, legumes, or vegetables. Mixed planting is a common practice to maximize land use efficiency and crop diversity.

Planting cassava involves strategic decisions to mitigate the risk of crop loss due to unfavorable weather and pests. By intercropping cassava with other crops, farmers spread the risk over plants with different levels of susceptibility.

Planting Methods

Cassava cuttings can be planted vertically, horizontally, or diagonally on flat land, ridges, or mounds (Fig. 6.13).

Time of Planting

Farmers typically plant cassava at the beginning of the rainy season, although planting can occur months before or after depending on weather conditions and the availability of planting material. In regions with bimodal rainfall patterns, planting is done during both rainy seasons. In areas with year-round rainfall, planting can occur throughout the year. Water is crucial for cassava growth until the plants are well established. Sprouting generally occurs within the first week, but about 5% of cuttings may fail, necessitating surplus cuttings to fill gaps.

Spacing

Optimal spacing between rows is about 80–100 cm, with plant spacing within rows adjusted to local conditions. The number of plants per hectare varies by country, ranging between 10,000 and 15,000 or more.

Fertilization

Cassava's widespread cultivation is partly due to its ability to grow in impoverished soils that cannot support other staple crops. Its extensive root system allows it to access nutrients that other crops cannot. Cassava is often grown without fertilization and can still yield modest fresh tuberous roots when the land is freshly cleared.

Like all rapidly growing carbohydrate crops, cassava also has high nutrient requirements and can quickly deplete soil nutrients. Sustaining cassava production over several years necessitates replenishing the soil with essential nutrients through organic or chemical fertilization.

Fig. 6.13 Woman farmers planting cassava in a diagonally slanted position (above) and cassava grown vertically and diagonally slanted on mounds (below) in Mwinilunga District, Northwestern Province of Zambia (Reproduced from IITA)

Nutrient Requirements

Cassava requires nitrogen (N), phosphorus (P), and large quantities of potassium (K). Potassium is crucial for starch formation, while N and P are essential for overall growth. Excess nitrogen can lead to heavy vegetative growth without a corresponding increase in root production. A chemical fertilizer with an NPK ratio of

Fig. 6.14 Workers and community members weeding a cassava field. (Source: IITA)

12–12–18 or 15–15–15 is recommended, with application rates ranging from 300 to 650 kg/ha, depending on soil type and conditions. Fertilizer should be applied in two stages:

1. **First application**: Immediately after planting.
2. **Second application**: When the crop canopy covers the soil.

Weeding Practices

Cassava is particularly susceptible to weed competition during the first 3–4 months after planting, when the sprouts are developing. Weeds can harbor insects that transport and transmit plant pathogens. Therefore, weeding is crucial during this early stage (Fig. 6.14). It is recommended to weed the crop two or three times until the plants are well established and their shade suppresses further weed growth. This task is often carried out by women in farming communities. Once the canopy is dense, further weeding becomes unnecessary as the cassava plants themselves prevent weed growth.

Multiplication of Planting Material

Cassava cuttings have a notably low multiplication ratio. For instance, a cassava stem cutting (approximately 25 cm long) planted today produces about 10 new stem cuttings after 12 months, resulting in a multiplication ratio of 1:10. This is significantly lower compared to a maize plant which, through true seeds, yields a cob with about 300 seeds, boasting a multiplication ratio of 1:300.

To address this, the International Institute of Tropical Agriculture (IITA) in Ibadan, Nigeria, developed a rapid multiplication technique, now widely used in Southern Africa and other regions to boost the availability of cassava planting material. This technique is based on several key principles:

1. **Utilization of entire mother stem**: The whole stem of the mother plant is used for cuttings, with the primary goal of producing daughter stems.
2. **Health and quality**: Only healthy, disease-free, and pest-free stems are selected for multiplication, ensuring high-quality planting materials (Fig. 6.15).
3. **Axillary bud development**: By removing apical dominance, each axillary bud on a stem can develop into a new shoot.

Preparation of Ministem Cuttings

The stem is cut into several small pieces of at least one or more nodes each, depending on the portion of the stem from which it was cut. The cuttings are prepared using farm equipment such as shears, secateurs, machetes, sharp knives, or hand saws (Fig. 6.16).

Planting Ministem Cuttings in the Nursery

Hardwood Meristem Cuttings

These cuttings are also planted in either nursery beds or polyethylene bags:

1. **Nursery beds**: Space the cuttings at 10 × 10 cm.
2. **Polyethylene bags**: Plant at a depth of about 4–5 cm to avoid exposing the cuttings after watering cycles. Ensure that two opposite nodes are on the right and left sides when buried to prevent one node from being planted deeper than the other. Shoots from deeply planted nodes often struggle to emerge and can be weak and fragile.

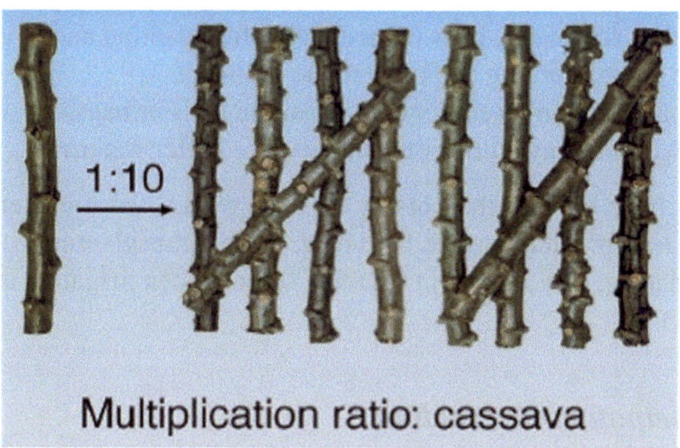

Fig. 6.15 A 1:10 cassava stem multiplication ratio. (Source: Otoo (IITA) 1996)

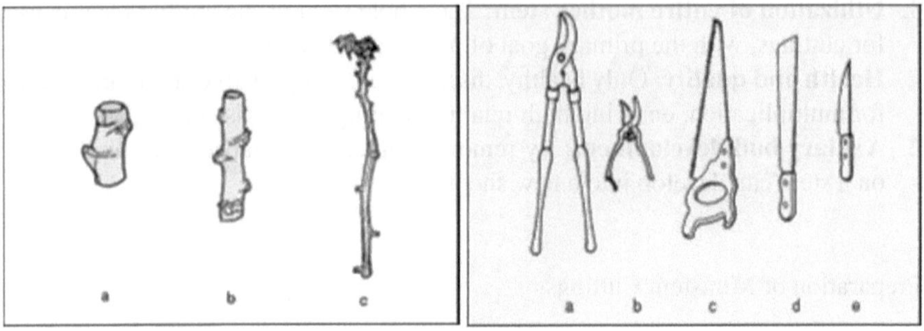

Fig. 6.16 Typical ministem cuttings (left): hardwood (a), semi-mature (b), tip shoot (c). The tools used for their preparation (right) including shear (a), secateur (b), hand saw (c), machete (d), and knife (e). (Source: Otoo–IITA 1996)

Planting Orientation

Proper orientation is crucial for the healthy development of the cuttings. Ensuring that nodes are correctly positioned helps avoid the emergence of weak sprouts, which are prone to breakage during transplanting.

Semi-matured Meristem Cuttings Such cuttings are usually 7–10 cm long and are also planted at spacing of 10 × 10 cm (Fig. 6.17).

Sprouting and Nursery Maintenance

Cuttings often shoot out 7–10 days after planting with profuse, fibrous roots on the buried nodes. To obtain good sprouting and plant growth, maintenance of nursery beds should include:

- Watering beds and plastic bags immediately after planting and twice a day thereafter (once in the morning and once in the evening).
- Removing by hand any weeds which appear in beds or bags.
- Covering with soil any cuttings that are exposed after watering.

After they have been in the nursery for 4–6 weeks, the ministem cuttings are uprooted and transplanted into the field (Fig. 6.18). Transplanting is carried during the dry season using irrigation or in the rainy season when irrigation is not available.

6.5.8 Sweetpotato Cultivation

Sweetpotato (*Ipomoea batatas* (L.) Lam.) (Fig. 6.19) belongs to the Convolvulaceae family and is a dicotyledon widely cultivated for its starchy tubers in tropical, subtropical, and some temperate areas around the globe (Purseglove 1976). In regions like Southern Africa, sweetpotato leaves are also harvested and cooked as a vegetable similar to spinach.

Fig. 6.17 Semi-mature meristem cuttings planted in a nursery bed. (Source: IITA 1990)

Fig. 6.18 Several shoots sprout from each stump, but leave only 2 or 3 which will mature into stems (left) and sprouting in the nursery bed (right). (Source: IITA 1990)

Characteristics and Propagation

Sweetpotato is a herbaceous perennial, capable of regrowing from its rootstock even after the leaves and vines die back. The crop is commonly propagated using stem vines from the previous season's plants. Unlike Irish potatoes, planting sweetpotato root pieces directly in the field does not yield uniformly sized storage roots. Propagation through true seeds is not viable for sweetpotato as the crop is genetically complex and plants growing from true seed are extremely variable.

Fig. 6.19 Sweetpotato crop plants in field of a smallholder farmer (left) and harvested tubers packed in sacks ready for transport to market for sale (right) in Northwesten Province of Zambia

Sweetpotato is an ideal crop for most smallholder farming systems as it requires few agricultural inputs. It is more drought tolerant than other vegetable crops, crowds out weeds, suffers from relatively few pests and diseases, and can produce acceptable yields in soils with low fertility levels. Orange-fleshed cultivars are used to mitigate vitamin A deficiency particularly in children, which is widespread in various countries in the region.

Importance of Sweetpotato

In Southern Africa, sweetpotato is a vital root crop due to its adaptability to diverse edapho-climatic conditions, including drought, and its high return on minimal investment. It plays a crucial role in farming systems and household food security across the region. The rising costs of agricultural inputs for other crops, such as maize, and increasing drought occurrences have amplified the importance of sweetpotato. Policymakers and the general population now emphasize sweetpotato for food security, income generation, and as a drought relief measure.

Nutritional Value

Sweetpotato is valued not only for its calorific content (Table 6.8) but also for its rich vitamin content, particularly vitamin A. The water content of sweetpotato tubers ranges from 57% to 78% of their fresh weight. The remaining composition includes the following:

- **Starch**: 13–33%
- **Sucrose**: 2.6–6.0%
- **Reducing sugars**: 0.3–0.8%
- **Minerals**: 0.8–2.2%

Table 6.8 Sweetpotato nutritional values per 100 g for edible portion of raw material

Nutrient	Unit[a]	Value/100 g	# of data points	Std. error	Cup, cubes 133 g	Sweetpotato 130 g
Water	g	77.28	7	0.650	102.78	100.46
Energy	kcal	86	–	–	114	112
Protein	g	1.57	4	0.145	2.09	2.04
Total lipid (fat)	g	0.05	4	0.033	0.07	0.07
Ash	g	0.99	4	0.056	1.32	1.29
Carbohydrate, by difference	g	20.12	–	–	26.76	26.16
Fiber, total dietary	g	3.0	1	–	4.0	3.9
Sugars, total	g	4.18	–	–	5.56	5.43
Starch	g	12.65	4	2.709	16.82	16.45
Calcium, Ca	mg	30	3	4.632	40	39
Iron, Fe	mg	0.61	3	0.070	0.81	0.79
Magnesium, Mg	mg	25	3	3.180	33	32
Phosphorus, P	mg	47	3	7.413	63	61
Potassium, K	mg	337	3	27.552	448	438
Sodium, Na	mg	55	3	26.394	73	72
Zinc, Zn	mg	0.30	3	0.033	0.40	0.39
Copper, Cu	mg	0.151	3	0.015	0.201	0.196
Manganese, Mn	mg	0.258	3	0.052	0.343	0.335
Vitamin C, total ascorbic acid	mg	2.4	4	1.038	3.2	3.1
Vitamin B-6	mg	0.209	4	0.014	0.278	0.272
Carotene, beta	μg	8509	43	224.125	11,317	11,062
Vitamin A, IU	IU	14,187	–	–	18,869	18,443
Vitamin K (phylloquinone)	μg	1.8	2	–	2.4	2.3
Fatty acids, total saturated	g	0.018	–	–	0.024	0.023

Reproduced from USDA (2016). Food Composition Databases of Nutrient Database for Standard Reference. Available at: https://ndb.nal.usda.gov/ndb/search/list
[a] g, gram; IU, international unit; kcal, kilocalorie; mg, milligram; μg, microgram

- **Protein**: 0.8–2.2%
- **Cellulose**: 0.9–1.2%
- **Carotene**: 0–24 mg per 100 g of fresh tuber
- **Vitamin C (ascorbic acid)**: 23–43 mg/100 g

The high carbohydrate content gives sweetpotato its sweet flavor, and its energy value surpasses that of cassava and other roots and tubers. Additionally, the leaves serve as a nutritious relish and livestock feed, while the roots can be used for various industrial purposes. FAO Agrostat data shows an increasing trend in sweetpotato production in the region, particularly after 1990 (Minde and Jumbe 1997).

Agricultural and Health Benefits

Sweetpotato's significance is further highlighted by its diverse roles in agriculture and health:

- **Food security**: Essential for sustenance during droughts.
- **Income generation**: Provides economic benefits to farmers.
- **Industrial uses**: Serves as raw material for various industries.

As Raemaekers (2001) and Faber et al. (2016) explain, orange-fleshed sweetpotato is an excellent source of bio-available beta-carotene, which the body converts to vitamin A. The flesh color, ranging from cream to dark orange, correlates with beta-carotene content. Thus, the darker the flesh, the higher the vitamin A value. Faber et al. (2016) recommend using orange to dark-orange sweetpotato varieties to combat vitamin A deficiency.

Food Security for the Population

In rural areas, sweetpotato is usually eaten with groundnuts either as breakfast, lunch, or supper, providing food for households particularly during periods of famine (Fig. 6.20). The roots can also be eaten raw, roasted, peeled, or fried in cooking oil as sweetpotato chips much like Irish potatochips. They are also used as pie filling, puree, in candies, and in other foods like cookies/crackers. Sweetpotato leaves—with acceptable level of vitamins A and C and iron (Laurie 2010)—are edible and can improve the vision and immune systems particularly for young

Fig. 6.20 Boiled sweetpotato tubers on a dinner plate served for breakfast together with a cup of tea

children and infants. Furthermore, the peels are used to produce a local beverage consumed by the local people.

Animal Feed

Sweetpotato is a highly valued crop in Southern Africa due to its adaptability and multiple uses. It serves as a low-cost, nutritious feed for livestock such as cattle, pigs, goats, and domesticated rabbits. According to Peters (2008), sweetpotato feed is 90% as efficient as corn meal feed for lactating dairy cows and serves effectively as a primary carbohydrate source for pig diets.

Industrial Use

A significant portion of the sweetpotato harvest is used industrially, particularly in biscuit production in countries like Botswana and South Africa.

Production Systems: Land Selection and Preparation

Sweetpotato is typically grown as a pure stand but can also be part of various cultivation systems, including relay cropping, intercropping, and rotations with other crops such as cassava, rice, maize, and beans.

Soil Requirements

Sweetpotato thrives in a wide range of soils, with the best being light, friable, and well-drained fine sandy loams. Poor drainage can decrease yields, and heavy clay soils can lead to rough and irregular root formation. Soils with more than 2% organic matter can also reduce production. Optimal pH for sweetpotato is between 5.8 and 6.2, but it can tolerate pH levels from 4.5 to 7.5 and grow in aluminum-rich and phosphorus-poor acid soils.

Land Preparation

Land preparation for sweetpotato cultivation typically occurs between November and February. The soil is turned upside down using hoes to invert weeds and grasses, which are then covered with more soil. Ridges are commonly used to allow for deep root development and easy tuber formation.

Planting

The planting period for sweetpotato for tuberous root production is generally from early November to late January. To obtain planting material, the crop is planted during the dry season in lowland areas or areas irrigated using nearby water sources, between August and September. Planting is done on a slant with the basal portion of a vine cutting inserted into the soil at a spacing of 25–30 cm within rows on ridges 60–100 cm apart. Planting on mounds is also common and supports good tuber formation (Fig. 6.21).

Planting Material

Sweetpotato is typically propagated using cuttings (vines) harvested from the previous crop or nursery plants. To obtain vine cuttings, nurseries are either raised from healthy tubers or selected vines. Terminal vine cuttings, about 20–40 cm long, are stored with intact leaves under shade for 2 days before planting, which improves sprouting and tuber yield (Nair et al. 1989). Directly planting sweetpotato root pieces in the field, as done with Irish potatoes, does not yield uniform storage roots. Most farmers produce their own planting material in locations where water allows off-season growth, while others obtain cuttings from friends, relatives, or producers.

Weeding

Feeder roots develop rapidly after planting, allowing the crop to compete effectively with weeds once the vegetation covers the ground. Weed control is necessary before this coverage, with the first weeding done 4 weeks after planting and the second

Fig. 6.21 Sweetpotato newly planted on large ridges to facilitate good developmental production of the tubers

7–9 weeks later. Weeding is performed using a hand hoe, followed by earthing up (re-ridging) to improve soil conditions. Higher tuber yields are obtained when weeding and earthing up are done 15 or 30 days after planting (Nair et al. 1989). However, many growers do not weed because sweetpotato vines spread quickly, covering the ground and suppressing weeds.

Fertilization

Most small-scale farmers do not use chemical fertilizers for sweetpotato. The crop relies on decomposing organic matter from weeds and grasses incorporated into ridges. Excessive organic matter, however, can lead to tuber rot. Farmers often use farmyard manure or compost. Intercropping with leguminous plants like *Sesbania sesban* improves yield (Chalwe et al. 1998).

Though chemical fertilizers are generally avoided, sweetpotato can respond well to them. The type and dosage of fertilizers depend on soil types, but excessive nitrogen should be avoided as it promotes excessive shoot growth at the expense of storage roots. A split application is recommended in Zambia (Muliokela 1997):

- **First dose**: Applied immediately after planting, as a band or spot beside the plant.
- **Second dose**: Applied 5 or 6 weeks after planting.

6.5.9 Cultivation of Yams

Yams (family Dioscoreaceae, genus Dioscorea) are climbing perennial monocots with underground, and sometimes aerial, starchy tubers grown as staple food (Fig. 6.22). Widely consumed in Africa, especially in West Africa, yams are a dietary mainstay for millions.

Fig. 6.22 Yam crop plants in field of a smallholder farmer (left) and harvested tuberous roots in Luapula Province of Zambia

Importance and Uses

According to Coursey (1967), yams serve as the staple food for millions in tropical and subtropical regions, with significant secondary food value for millions more. Besides cultivated yams, wild yams are collected and consumed during food shortages, providing a crucial nutritional backup.

Cultivation Challenges

Yam cultivation is labor-intensive, requiring significant effort for land preparation, planting, staking, weeding, harvesting, and post-harvest handling. Major constraints in yam production include pests and diseases, both in the field and during storage, leading to high production costs and lower tuber yields.

Yam Species

Yams are a staple food across Africa, particularly in West Africa. They are climbing perennial monocots with starchy tubers, and several species are cultivated extensively in the humid tropics, including Southern Africa.

Dioscorea rotundata and *Dioscorea cayenensis*

The most common yam species in Southern Africa are distinguished by the color of their tubers:

- ***Dioscorea rotundata*** (white yam): The tubers are roughly cylindrical, with smooth brown skin and white, firm flesh. This species is widely cultivated by small-scale farmers in Botswana, Malawi, Mozambique, Zambia, and Zimbabwe.
- ***Dioscorea cayenensis*** (yellow yam): Named for its yellow flesh, which is due to the presence of carotenoids. The tubers have less extensively grooved skin compared to white yams. Yellow yam has a longer vegetative period and shorter dormancy. It is a marginal crop in many regions due to its need for high and consistent rainfall levels (Wilkin 2001).

Dioscorea alata

Known as "water yam," "winged yam," and "purple yam," *Dioscorea alata* originated in Southeast Asia and is now globally distributed. It is the second most popular yam in Africa after the white yam. Found in Angola, Malawi, Mozambique, and Zambia, this yam has square, winged vines and generally cylindrical tubers with white, watery flesh.

Dioscorea bulbifera

Common in farmers' fields and home gardens, *Dioscorea bulbifera* produces long vines and underground tubers, but its aerial bulbils at the leaf base are more significant as food (Fig. 6.23). It produces a crop after only 4 months and continues for up to 2 years. The bulbils are easily harvested for boiling and consumption at any time.

Dioscorea dumetorum

The "bitter yam," *Dioscorea dumetorum* (Kunth) Pax, is popular as a vegetable. The wild forms of this species are very toxic and are sometimes used to poison animals when mixed with bait. According to Wilkin (2001), *D. dumetorum* is found in Zambia, Zimbabwe, and Mozambique with considerable number of vernacular names, ranging from "idiya" in Mozambique to "chinkolongo/inkolongo," "nkansa," and "ndiya" in Zambia. The tubers are used as a famine food after washing and drying to remove toxins.

Importance of Yams

Originally, yams (*Dioscorea* spp.) were primarily consumed by rural households with limited resources. However, the crop has gradually made its way into urban areas across sub-Saharan Africa, thanks to innovative and appealing methods of

Fig. 6.23 Different types of bulbils that have developed on the aerial part of *Dioscorea bulbifera* yam plants. Bulbils on a mature plant (left) and a developing bulbil on a young plant (right)

preparation. The traditional methods of boiling and roasting have expanded to include frying yam chips for both household consumption and street food sales, which has become a burgeoning business and a significant source of income for many.

Culinary Uses and Nutritional Value

After peeling, yam tubers can be boiled, pounded into exquisite pounded yam, or dried and ground into flour for preparing thick porridge (Nshima) served with stew and vegetables, especially in West Africa. Yams are an essential source of starch due to their high carbohydrate content, providing about 110 calories per 100 g of cooked yam (Coursey 1967). Yams also contain appreciable amounts of protein, with *D. dumetorum* having the highest crude protein content among yam species. While rich in phenylalanine and threonine, yams are limited in tryptophan and sulfur-containing amino acids (cystine and methionine). They are also high in vitamins C and B6, potassium, manganese, and dietary fiber, while low in saturated fat and sodium (Table 6.9).

Table 6.9 Nutritional value of yam per 100 g of raw material (percentages are relative to US recommendations for adults)

Nutrient	Unit[a]	Value/100 g	Cup, cubes 150 g
Water	g	69.60	104.40
Energy	kcal	118	177
Protein	g	1.53	2.29
Total lipid (fat)	g	0.17	0.26
Carbohydrate, by difference	g	27.88	41.82
Fiber, total dietary	g	4.1	6.2
Sugars, total	g	0.50	0.75
Calcium, Ca	mg	17	26
Iron, Fe	mg	0.54	0.81
Magnesium, Mg	mg	21	32
Phosphorus, P	mg	55	82
Potassium, K	mg	816	1224
Sodium, Na	mg	9	14
Zinc, Zn	mg	0.24	0.36
Vitamin C, total ascorbic acid	mg	17.1	25.6
Vitamin B-6	mg	0.293	0.439
Vitamin A, IU	IU	138	207
Vitamin E (alpha-tocopherol)	mg	0.35	0.53
Vitamin K (phylloquinone)	µg	2.3	3.5
Fatty acids, total saturated	g	0.037	0.056

Source: USDA. Agricultural Research Service. (2018). USDA National Nutrient Database for Standard Reference, Legacy (2018). Retrieved from https://ndb.nal.usda.gov/ndb/ at 2018-12-18

Year-Round Availability and Food Security

Yam tubers are available year-round, making them more reliable than seasonal crops. Their availability and versatility make yams a preferred food and a culturally significant food security crop in Southern Africa. Key attributes distinguishing yams from other root and tuber crops like cassava and sweetpotato include:

1. Highly palatable and high cash value.
2. Can be harvested and stored for 4–6 months, extending food availability.
3. Drought-hardy propagules (seeds and tuber pieces) that can be grown before the onset of rains.
4. Multi-species crop with broad ecological adaptation and high yield potential.
5. Tolerance to drought combined with natural tuber dormancy, allowing flexibility in planting and labor use.
6. Different species and varieties provide a continuous food supply throughout the year.

6.5.10 Production Practices: Land Selection and Preparation

Yam cultivation is influenced by several factors, including the length of fallow periods, previous crops, soil moisture, and the intensity of previous cropping systems. Yams require deep, loose, well-drained soils with good water-holding capacity and a pH range of 4.2–6.7. To enhance soil fertility, selected land is often hilled into mounds or ridges, and manure from various grasses is incorporated at the beginning of the rainy season.

Planting Practices

Farmers typically set aside tubers from the previous harvest for planting the next year, rather than purchasing new seed yams. Some farmers replant yams immediately after harvest, leaving some tubers in the ground to sprout. However, yam tubers have long dormancy periods, so tubers harvested early and replanted may take 3–5 months to sprout. Environmental factors such as sunlight and soil temperature significantly affect sprouting.

Traditionally, whole tubers or large tuber pieces are used for planting, but small tubers, known as minisetts, are increasingly used as planting material (Fig. 6.24). The tubers are first hung in an open place to reduce moisture content before being placed in the mounds. Planting is done at the beginning of the rainy season, with whole seed tubers or tuber portions planted into mounds or ridges.

Fig. 6.24 Yam planting materials: vine cuttings (left), whole tubers and minisetts (middle), and tuber fragments (right). (Reproduced from IITA)

Intercropping

In the DR Congo and Angola, small-scale farmers often intercrop yams with cereals and vegetables. In Zambia, yams are also cultivated alongside cassava. This intercropping practice helps maximize land use and improve crop yields. In Zambia, the crop is also cultivated alongside with cassava. After planting, the crop grows for 6–10 months and becomes dormant for 2–4 months, depending on the species. The growth and dormant phases correspond, respectively, to the wet and the dry seasons in a given location.

Growth Stages of Yams

The growth rate of tubers in most *Dioscorea* species starts slow immediately after tuber initiation, accelerates rapidly during full canopy formation, and then slows as the plant approaches maturity. Yams survive the dry period and sprout after a dormancy of 2–4 months, initially drawing on moisture and nutrients from the planted parent tuber. An extensive root system quickly establishes, and the sprouted vine, covered in a waxy bloom to reduce moisture loss, elongates before developing leaves.

1. **Phase 1 (sprouting to 6 weeks):** Planted whole tubers or pieces sprout, forming a primary nodal complex from which roots and vines develop. The waxy vine emerges from the soil about 6 weeks after planting, initially without leaves and incapable of photosynthesis.
2. **Phase 2 (6–13 weeks):** Rapid cell division and expansion occur, leading to vine elongation and leaf production. Full leaf canopy is achieved by 10 weeks after sprouting, root growth peaks, and tubers begin to grow, influenced by plant population, nutrient application, organic matter, and soil moisture.

3. **Phase 3 (14–18 weeks):** Maximum canopy development and significant tuber bulking occur due to rapid dry matter accumulation.
4. **Phase 4 (20–32 weeks):** Leaf canopy senesces, and tuber dry matter accumulation declines. Tubers enter dormancy, either harvested or left underground, respiring at a higher rate while losing dry weight post-maturity.

Weeding

Initial field preparation is crucial to minimize weed challenges. Farmers typically weed three to four times at intervals of 3, 8, 12, and 16 weeks after planting using hoes.

Fertilization

Both farm manure and chemical fertilizers can be used to improve soil fertility for yams. Various organic sources such as composts, farmyard manures, animal manures, green manures, organic mulch, and crop residues are incorporated into the soil after harvest. Leguminous fallow crops also help improve soil fertility. The slow nutrient release from organic fertilizers provides longer residual effects than mineral fertilizers.

Mineral fertilizers are less common in Southern Africa's yam production but are used in West Africa for higher yields. Fertilizer application should be split: the first around a month after emergence and the second during tuber bulking, approximately 3 months after emergence, placed 10–15 cm away from the plant.

Staking

Stakes support yam stems, allowing them to climb and access sufficient sunlight (Fig. 6.25). Staking improves crop yields, with bamboo or termite-resistant poles preferred. Common staking methods include individual staking and pyramid staking, where stakes from adjacent stands form a pyramid. Trellises, with a network of ropes, provide additional climbing space.

6.5.11 Wild Yams

In addition to the cultivated varieties, many wild species of yams thrive in the natural forest habitats of Zambia (Fig. 6.26, Table 6.10). Wild yams represent a fascinating array of species with unique characteristics and traditional uses. These plants thrive in a range of environments, from tropical forests to temperate areas, showcasing their remarkable adaptability. Historically, wild yams have played a significant

Fig. 6.25 Yam crop in a smallholder farmer's field stacked with bamboo sticks to prevent plants from bending or breaking under their own weight and allow better airflow

role in the diets and medicinal practices of indigenous cultures, particularly in regions where they are native. They are not only valued for their nutritional properties but also for their potential health benefits, including anti-inflammatory and hormonal-regulating effects. These wild yams are annually harvested for both household consumption and trading. The planting material for yams primarily comes from their edible parts, which ensures a regular food supply, especially during emergency periods.

Their significance can be observed in various aspects, including culinary practices, traditional medicine, and cultural symbolism.

6.5.12 *Culinary Uses*

In Southern African communities, wild yams are utilized as a food source, particularly in regions where cultivated crops may be less accessible. The tubers of wild yams are often harvested during the rainy season when they are most abundant. They are typically boiled, roasted, or mashed and can be eaten alone or combined with other ingredients to enhance their nutritional profile and flavor. Used primarily as a starch source, the tubers provide carbohydrates and are occasionally incorporated into traditional dishes.

Different species of wild yams are recognized for their distinct flavors and textures, allowing them to be featured in diverse culinary applications. In addition to their nutritional value, they serve as a food supplement during times of scarcity, playing a crucial role in food security in rural communities.

Fig. 6.26 Survey for wild yam occurrence in Zambia in 2005. A team of experts from the Copperbelt University prospecting the village nearby forest (above left), a village woman holding a wild yam plant (above right), and a well-developed yam plant in a village yard in Southern Province of Zambia

Table 6.10 Yams (*Dioscorea* spp.) and their distribution in different provinces of Zambia

Yam species	Distribution by province					
Dioscorea asterricus	W	N	–	C	S	B
Dioscorea baya	W	–	–	–	–	–
Dioscorea cochleari-apiculata	W	N	E	C	S	B
Dioscorea dumetorum	W	N	E	C	S	B
Dioscorea hirtiflora	W	N	E	C	S	B
Dioscorea hylophila	–	N	–	–	–	–
Dioscorea mundtii	–	N	–	C	–	–
Dioscorea praehensilis	W	N	E	C	S	B
Dioscorea quartiniana	–	N	–	–	–	–
Dioscorea sansibarensis	W	N	E	–	–	–
Dioscorea sylvatica	–	–	–	C	S	B

Source: Fanshawe (1973). Key: W, Copperbelt & Northwestern Provinces; N, Northern & Luapula Provinces; E, Eastern Province; C, Central Province; S, Southern Province; B, Western Province

6.5.13 Traditional Medicine

Beyond their dietary applications, wild yams are widely acknowledged in Southern African traditional medicine. Various parts of the yam plant, including the tubers and leaves, are used to prepare traditional remedies for a range of ailments. The tubers are often believed to have anti-inflammatory properties and are used to alleviate symptoms of conditions such as arthritis and digestive disorders.

Furthermore, some communities utilize wild yams as herbal treatments for women's health issues, particularly in relation to hormonal balance and reproductive health. The presence of steroidal saponins and phytoestrogens in certain yam species has led to research into their potential health benefits, bolstering their traditional uses with emerging scientific inquiry.

6.5.14 Cultural Significance

Wild yams are not only an essential food source but also hold cultural significance in many Southern African societies. They often feature in folklore, rituals, and festivals, symbolizing fertility and prosperity. During traditional ceremonies, wild yams may be offered to signify the harvest's abundance and as a means of honoring ancestors.

In some cultures, the practice of harvesting wild yams is embedded within communal rites, strengthening social bonds among community members and roles in cultural identity. The gathering and sharing of wild yams can serve as a social activity that reinforces communal ties, particularly among women and families.

Ultimately, the utilization of wild yams in Southern Africa reflects a deep connection to the land and a rich heritage of knowledge. As communities continue to adapt to changing ecological and economic conditions, wild yams stand as a testament to the resilience of traditional practices and the potential of indigenous plants in modern dietary and medicinal contexts. Their multifaceted roles in culinary, medicinal, and cultural domains highlight the need for continued research and preservation efforts, ensuring that these valuable resources remain integral to the lives of Southern African communities.

6.5.15 Harvesting Practices and Challenges

The current standard method of harvesting yams involves removing all vines and tubers without leaving any portion in the ground for regeneration. This practice significantly threatens the crop's sustainability. To ensure the long-term availability of wild yams, more sustainable harvesting techniques are needed.

6.5.16 *Smallholder Contribution*

Smallholder rural households predominantly harvest and domesticate yams, adapting the crop to local agro-ecological conditions. This adaptation helps integrate yams into general cultivation practices, contributing to agricultural development and food security in Zambia.

6.5.17 *Tube Characteristics of Species of Wild Yams Occurring in Natural Forest Habitats in Zambia*

Previous research on yams in Zambia has primarily focused on taxonomic and eco-geographical distribution. However, despite the vital role plant genetic resources play in our daily lives, and the threat of extinction facing some valuable plants like yams due to unsustainable harvesting practices, there has been little effort to conserve these resources.

6.5.18 *Conservation Approaches*

To ensure the survival of yam species, several conservation strategies can be implemented:

1. **In situ conservation**: Protecting yams within their natural habitats.
2. **Ex situ conservation**: Preserving yams in controlled environments outside their natural habitats.
3. **On-farm clonal domestication**: Encouraging the cultivation and propagation of yams on farms, maintaining genetic diversity, and adapting to local agro-ecological conditions.

6.5.19 *Morphological Variability*

Yam tubers exhibit significant morphological diversity, with various shapes including:

- Roughly cylindrical
- Regular-irregular rounded
- Flattened
- Oval
- Oblong to cylindrical
- Rounded and oval-oblong

- Oval-cylindrical
- Elongated (Fig. 6.27)

Ashley Steel et al. (2022) conducted a focus group study to identify the most commonly collected species within nine categories of wild food products such as mushrooms, insects, green leafy vegetables, tubers, fruits, nuts, wild meat, wild fish, and aquatic plants. Their findings revealed the followings:

- **Wild food collection**: Almost all surveyed households collected wild foods, with an average of five different types per household. The most commonly collected were mushrooms, fruits, and green leafy vegetables.
- **Volume collected**: The quantity of wild foods collected varied significantly by household, product type, and study area. Rural households in Zambia are estimated to collect over 380 million liters (380,000 m^3) of wild foods annually, with 88% sourced directly from forests. Considering uncertainties, at least 238,000 m^3 of wild foods are collected from forests each year, equating to approximately 12 million large (20 L) collecting buckets.
- **Correlation with wealth**: The volume of wild foods collected was not strongly correlated with wealth, indicating widespread consumption of wild foods. However, the most food-insecure households collected particularly high volumes.
- **Economic potential**: Pilot market surveys revealed local value chains for wild foods, suggesting economic benefits. The findings highlight the importance of locally disaggregated data and the need for national estimates of wild food collection to inform forest policy and management strategies.

In the DR Congo, Adejumobi et al. (2022) evaluated the potential of wild yam cultivation and highlighted the activities necessary for its improvement. They found

Fig. 6.27 (Top) Round and yellow flesh tubers, bottle-like tubers, finger-like tubers; (bottom) round and black-skin tubers with burbils, elongated tubers, and small shaped tubers of wild yams

that despite several challenges, yam diversity has been primarily preserved through traditional farming methods in house gardens and small farmlands. Reversing the current stigma associated with yam is a significant challenge for both the scientific community and the general population. Overcoming this stigma involves:

- Increasing awareness of yam's nutritional benefits.
- Enhancing genetic diversity through scientific research.
- Implementing policies to support yam cultivation and consumption.
- Educating farmers on modern yam farming techniques.

By adopting these strategies, the DR Congo can promote wild yams as a vital crop for food security and nutrition.

6.5.20 Leguminous Crop Cultivation

Introduction

For most small-scale farmers in Southern Africa, the use of fertilizers to replenish their fields has become economically unfeasible due to rising costs and climatic risks. This has led to degraded soil fertility, resulting in low crop yields and widespread malnutrition, particularly among children. Waddington et al. (2004) highlighted that mixed maize, legume, and livestock cropping systems dominate sub-humid and some semi-arid zones of the region where soil fertility improvement is needed. However, nutrient balance studies indicate that soil depletion rates far exceed replenishment in these systems.

Challenges in Soil Fertility

Most smallholder farms receive little or no mineral fertilizer, relying on small amounts of organic and mineral nutrients from farm-sourced materials. Despite this, nutrient off-take in various crops is often high due to intensive continuous cropping, and losses from sources such as leaching can also be significant. Farmers are forced to feed both themselves and the soil with limited resources, leading to continuous cropping without inputs, which depletes nutrients and results in soil erosion and degradation. The major challenge remains finding ways for smallholders to maintain soil fertility and crop productivity under constraints of low income, limited land, and insufficient labor.

Sustainable Solutions

A range of organic and inorganic (mineral) soil fertility technologies, such as green manure crops and cropping system options, have been developed for smallholders in Southern Africa. These technologies were created through consultative and participatory on-farm research and testing, particularly in Malawi and Zimbabwe. Most of these technologies provide short-term soil fertility and crop productivity benefits, making them attractive to farmers due to their multiple uses. They are compatible with various farming circumstances and are effective within resource constraints (e.g., cash, labor, and land).

Role of Legumes

For the poorest families in developing countries, dietary protein primarily comes from vegetable legumes. These plants are harvested for their leaves, tender shoots, and roots, which are used as vegetables. Leguminous crops are rich in phosphorus, calcium, iron, and essential vitamins. Crops like soybeans, groundnut, and Bambara groundnut are also rich in fats and vitamins A and E. Although leguminous vegetables lack some sulfur-containing amino acids, consuming cereals with legumes compensates for this deficiency.

Nitrogen Fixation

Legumes are unique among vegetables because they can obtain free atmospheric nitrogen through their symbiotic association with nitrogen-fixing bacteria (Rhizobia) found in root nodules. Nitrogen gas from the atmosphere is converted into ammonia within the nodules, which is then assimilated into amino acids, nucleotides, and other cellular constituents. This nitrogen is available to the plant, improving yields in nodulated plants and enriching the soil when the plant matter decomposes. Legumes can grow well in poor soils where there is insufficient fixed nitrogen to support other plants.

After harvest, decomposing legume roots release organic nitrogen compounds for uptake by the next season's plants. Farmers take advantage of this natural fertilization by rotating leguminous crops with non-leguminous ones. Natural nitrogen fixation reduces the need for artificial fertilizers, saving money and preventing issues like eutrophication, acid rain, and the overgrowth of non-food crops.

6.5.21 Cultivation of Common Beans

Common beans (*Phaseolus vulgaris* L.) belong to the family Leguminosae (Fabaceae), subfamily Papilionoideae (Faboideae), and tribe Phaseoleae. These herbaceous annual dicotyledonous plants are among the preferred leguminous crops grown by small-scale farmers in Southern Africa (Fig. 6.28).

Fig. 6.28 Common bean plants in field. Rows of growing plants at vegetative stage (left) and fully grown plants bearing flowers and green pods (right) in Copperbelt Province of Zambia

Importance and Cultivation

Estimates indicate that bean consumption in East and Southern Africa exceeds 50 kg per person per year (Mauyo et al. 2007). Beans are cost-effective to cultivate and require minimal land, making them ideal for intercropping with other food crops. Although classified as a single species, common beans include bush (dwarf), climbing, and semi-climbing types.

Climate and Soil Requirements

Beans thrive at higher altitudes and cooler temperatures, tolerating low temperatures and surviving mild frost. They grow more vigorously and yield higher in the cool, dry season, making them unsuitable for lowlands or arid areas with prolonged high temperatures.

The crop is harvested both as dry beans and fresh legumes, providing a primary source of protein for many low-income families with limited access to animal proteins. Beans offer a balanced diet when consumed with staple foods and significantly contribute to soil fertility when grown in intercrops or rotations, enriching the soil with nitrogen needed by other crops.

6.5.22 Importance of Common Beans

Common beans, belonging to the family Leguminosae (Fabaceae), subfamily Papilionoideae (Faboideae), and tribe Phaseoleae, are a vital leguminous crop for small-scale farmers in Zambia and the DR Congo.

Benefits in Crop Rotation

Common beans play a crucial role in crop rotations due to their ability to fix atmospheric nitrogen through symbiosis with *Rhizobium leguminosarum* biovar phaseoli (Ramírez-Bahena et al. 2008). This symbiotic relationship enriches the soil, enhancing fertility for subsequent crops.

Nutritional Value

Beans boast a high protein content in both seeds and leaves, surpassing that of maize, rice, and cassava (Table 6.11). When consumed with these carbohydrate staples, beans provide a balanced diet for low-income households (CIAT 2004).

Table 6.11 Beans' nutritional values per 100 g for edible portion of raw material

Nutrient	Unit[a]	Value/100 g	# of data points	Std. error	Cup 100 g	Bean 55 g
Water	g	90.32	165	0.273	90.32	49.68
Energy	kcal	31	–	–	31	17
Protein	g	1.83	104	0.048	1.83	1.01
Total lipid (fat)	g	0.22	14	0.031	0.22	0.12
Ash	g	0.66	144	0.018	0.66	0.36
Carbohydrate, by difference	g	6.97	–	–	6.97	3.83
Fiber, total dietary	g	2.7	4	0.050	2.7	1.5
Sugars, total	g	3.26	4	0.235	3.26	1.79
Starch	g	0.88	4	0.141	0.88	0.48
Calcium, Ca	mg	37	153	1.369	37	20
Iron, Fe	mg	1.03	155	0.077	1.03	0.57
Magnesium, Mg	mg	25	151	0.633	25	14
Phosphorus, P	mg	38	140	0.786	38	21
Potassium, K	mg	211	154	4.448	211	116
Sodium, Na	mg	6	154	0.176	6	3
Zinc, Zn	mg	0.24	152	0.021	0.24	0.13
Copper, Cu	mg	0.069	161	0.004	0.069	0.038
Manganese, Mn	mg	0.216	150	0.008	0.216	0.119
Vitamin C, total ascorbic acid	mg	12.2	9	0.752	12.2	6.7
Vitamin B-6	mg	0.141	4	0.002	0.141	0.078
Vitamin A, RAE	μg	35	–	–	35	19
Carotene, beta	μg	379	77	48.979	379	208
Vitamin A, IU	IU	690	–	–	690	380
Fatty acids, total saturated	g	0.050	–	–	0.050	0.028

Reproduced from USDA (2016). Food Composition Databases of Nutrient Database for Standard Reference. Available at https://ndb.nal.usda.gov/ndb/search/list

[a] g, gram; IU, international unit; kcal, kilocalorie; mg, milligram; μg, microgram

Additionally, beans have been used as substitutes for other plant and animal protein sources in industrial food preparations such as ice cream and regional dishes like miso in Japan and kishk in Eastern Europe (Nestares et al. 2001).

Agricultural and Economic Importance

Common beans can be cultivated twice a season in high rainfall areas, ensuring year-round food supply and fodder for livestock. The dry legume grain is relatively easy to store at the small-scale farm level. Beyond their nutritional value, beans are an important source of income due to increasing demand in domestic and export markets (Mauyo et al. 2007). Some farmers also attribute medicinal properties to beans, believing they help treat conditions like rheumatism, acne, eczema, diarrhea, and dysentery.

Growth and Stages of Development

The hypocotyl is the first part of a bean embryo to emerge from a planted seed (Fig. 6.29). It promptly turns downward, regardless of the seed's orientation. By the time the epicotyl begins rapid development, the radicle at the lower end of the hypocotyl has already formed a young root system capable of anchoring the plant in the soil and absorbing water and minerals. Root nodules also start forming to enhance nitrogen fixation.

Root System Development

As the taproot elongates, lateral roots multiply and elongate, with both root types bearing root hairs near the tips. Initially, seedlings grow slowly in length. However, they enter a period of rapid growth, producing more leaves and branches and

Fig. 6.29 Three bean developmental stages: common beans hypocotyl emerging from the seed embryo (left) followed by the growth of the seedling that comes out of the soil (middle) and foliage development

elongating the climbing point. This stage is accompanied by flowering, which begins on the lower branches and continues as the plant grows.

Maturity and Harvest

Being an annual plant, a bean's rapid vegetative growth slows or stops as it approaches maturity, with growth becoming more evident in the pods. After about 90–120 days, pod production ceases, and pods at the base of the plant turn yellow, indicating readiness for harvest. The ripening process follows an upward sequence, with the topmost pods being the last to ripen. Leaves turn yellow and fall off, followed by the drying of the entire plant.

6.5.23 Production Systems: Land Selection and Preparation

Common beans are produced in agroecological zones where altitude and rainfall patterns are favorable. In smallholder farms, beans are grown either as a sole crop or in various crop associations such as mixed cropping, intercropping, alternate strip cropping, and relay cropping. Farmers typically pair common beans with crops like maize, cassava, banana, and groundnuts. Common cropping systems include combinations like bean-maize, bean-cassava, bean-maize-cassava, and bean-banana.

Soil Requirements and Preparation

Beans thrive in well-drained loamy clay to light sandy loam soils. Highly compacted clay soils can hinder germination and lead to stunted plants with restricted root development and few nodules. If the soil pH is below 5, liming may be necessary. Small-scale farmers prepare the land well before the rainy season, using hand tools or animal traction to achieve a loose, deep soil with a fine tilth for even seeding and uniform crop emergence. Beans can be planted on level ground, mounds, or ridges. In high rainfall areas, planting on ridges helps avoid waterlogged fields.

Preferred Soil Conditions

Like many legumes, common beans prefer well-aerated, well-drained soils with a pH of 6–7.5. Waterlogging can acidify soils, inhibiting the development of nodule-forming, nitrogen-fixing bacteria essential for bean growth. Acidic soils with a pH below 5 or above 8.1 can negatively affect the crop.

Seed Selection

Selecting high-quality seed is crucial before planting. Seeds can be rounded and elliptical, rounded and elongated, or somewhat flattened in shape, with a rich assortment of coat colors and patterns (Fig. 6.30). They typically weigh between 20 and 60 g per 100 seeds. Farmers are aware that seeds can be damaged by storage pests and seed-borne pathogens and have an impact on the yield at harvest. Bean seed certified free from pests and pathogens is expensive and difficult to get, so small-scale farmers often use seeds of local varieties, which they grow and clean themselves and then select those of the best quality for sowing.

Planting

Smallholder farmers cultivate common beans under various conditions, with specific practices to optimize their growth and yield.

Planting Densities and Spacing

In pure culture and under optimal soil conditions:

- **Dwarf (bush) beans**: Planting densities range from 150,000 to 200,000 plants/ha.
- **Twining (pole) beans**: Planting densities range from 75,000 to 100,000 plants/ha.

Recommended spacing is 60–75 cm × 10 cm or 45–60 cm × 10 cm, using 35 kg/ha for small-seeded varieties and 50 kg/ha for larger-seeded varieties. The sowing depth ranges from 3 to 6 cm, depending on moisture availability and soil type. Dwarf beans require 90–110 days to mature, while some climbers need up to 150 days. Farmers often prefer early maturing and large-seeded varieties for their higher market value.

Fig. 6.30 Diversity of seeds of common beans characterized by their various colors and sizes

Planting Time and Seed Rate

For rainfed crops, the planting date should be adjusted to allow harvest after the main rains. Planting from late November to early January generally yields good results. Dry season planting is done around July, often in lowland dambos or under irrigation.

The seed rate depends on seed size:

- Small seeds: 30–35 kg/ha.
- Large seeds: 50–80 kg/ha.

Weed Control

Beans are planted in weed-free fields as they do not compete well with weeds for nutrients, water, and light, especially early on. Fields are typically weeded using traditional hoes or by hand-pulling weeds twice:

- First weeding: 2 weeks after germination.
- Second weeding: 3–4 weeks later.

Weeding is crucial during the first 30 days before the crop canopy covers the soil. Earthing up soil around the base of plants during weeding encourages adventitious root growth, minimizes erosion, and helps plants survive pests like bean flies.

Fertilization

In traditional farming, manure, compost, crop residues, and ashes are used to amend the soil. Crops benefit from residual fertility left from the previous season. Despite beans primarily obtaining nitrogen from nitrogen-fixing rhizobia, lime and phosphate are vital in acidic soils (pH 4–5) to promote rhizobia activity. A basal dressing of 100–300 kg/ha compound "D" (50 kg/ha N, 75 kg/ha P, and 50 kg/ha K) before planting and a top dressing of ammonium nitrate or urea (100–150 kg/ha) at early flowering are beneficial in sandy soils prone to nutrient leaching.

Crop Rotation

Crop rotation is critical in bean production to avoid diseases. Beans should be grown on land where no beans have been grown for the previous 2 years and no other legumes have been grown for at least a season. This practice minimizes the buildup of soil-borne pathogens and reduces the risk of seed-borne pathogens spreading from volunteer plants of the previous crop.

6.5.24 Groundnut Cultivation

Groundnuts (*Arachis hypogaea* L.) are a vital crop for smallholder farmers in Southern Africa, grown either once or twice a year depending on rainfall patterns (Fig. 6.31). Groundnuts are primarily cultivated as a subsistence food with minimal inputs. They are a principal source of protein (25–34%), cooking oil (44–56%), and essential vitamins like thiamine, riboflavin, and niacin. The straw and their stems serve as livestock feed. The crop is also a significant source of cash income, contributing to household food security and poverty alleviation. Groundnuts are an essential legume for smallholder farmers across large areas of the region particularly in Angola, DR Congo, Malawi, Mozambique, South Africa, Tanzania, Zambia, and Zimbabwe. Malawi and Zimbabwe are the largest producers, and the crop is not grown in Lesotho (Nigam et al. 1983). According to FAO (2003), groundnut is grown on 26.4 million ha worldwide with a total production of 37.1 million metric tons and an average productivity of 1.4 metric tons/ha. In 1969, approximately 155,000 ha of groundnut were planted in Mozambique representing 3.5% of the total area under cultivation and making groundnuts the fifth most important crop in terms of area (Missao de Inquerito Agricola-MIA 1969). Groundnut production suffers from various constraints in SADC member states, but diseases are generally regarded as the most damaging threat to its production throughout the region.

Types of Groundnuts

Groundnuts are a self-pollinated tropical annual legume belonging to the genus *Arachis* in the subtribe Stylosanthinae of the tribe Aeschynomeneae and the family Fabaceae (Leguminosae). It includes two subspecies: hypogaea, which has alternate

Fig. 6.31 Groundnut field visit at the Zambia experimental research station in Northern Province of Zambia

branching, and fastigiata, which develops sequential branching. According to ICRISAT, these subspecies are further divided into two botanical varieties: var. hypogaea and var. aequatoriana (ICRISAT 2006). Each variety exhibits different plant, pod, and seed characteristics. Most commercially cultivated groundnuts belong to the hypogaea botanical variety, including the Virginia or runner type, Fastigiata (Valencia) type, and Vulgaris (Spanish) type.

Characteristics of Groundnut Varieties

Arachis hypogaea ssp. Hypogaea (Virginia Variety)

- Long vegetative period
- Dark green leaves
- Many branches with a spreading form
- Generally, two seeds per pod
- Distinct dormancy period of about 30–180 days

Arachis hypogaea ssp. Fastigiata (Spanish and Valencia Varieties)

- Short vegetative period
- Light green leaves
- Upright growth
- Pods with two to six seeds concentrated around the main stem
- Spanish varieties have two seeds per pod
- Valencia varieties have pods with three to six seeds, thick stalks, and fewer secondary or tertiary branches than Spanish varieties
- Plants do not undergo dormancy

Importance of Groundnuts

Groundnuts are grown in Southern Africa and other parts of sub-Saharan Africa as a primary protein source for households with limited income. They also provide valuable carbohydrates and fats. Groundnuts are consumed raw, cooked, or roasted and processed into peanut butter, cooking oil, sweets, snacks, and pastes for soups, porridges, and sauces. Groundnut cake, with 40–50% easily digestible protein, enriches other foodstuffs like cassava flour. The crop's foliage serves as protein-rich fodder, while the pods provide fiber for fodder and cellulose for industries, fuel, and composting.

Groundnuts are a significant export commodity for many SADC member states, offering high nutritional value (Table 6.12) and serving as a major source of foreign exchange. For example, groundnuts are grown in all three agroecological zones of Zambia, with the eastern region being the most productive.

Table 6.12 Groundnut nutritional values per 100 g for edible portion of raw material

Nutrient	Unit[a]	Value/100 g	Cup 146 g	Oz 28.35 g
Water	g	4.26	6.22	1.21
Energy	kcal	570	832	162
Protein	g	25.09	36.63	7.11
Total lipid (fat)	g	47.58	69.47	13.49
Carbohydrate, by difference	g	20.91	30.53	5.93
Fiber, total dietary	g	8.7	12.7	2.5
Calcium, Ca	mg	62	91	18
Iron, Fe	mg	2.09	3.05	0.59
Magnesium, Mg	mg	184	269	52
Phosphorus, P	mg	336	491	95
Potassium, K	mg	332	485	94
Sodium, Na	mg	1	1	0
Zinc, Zn	mg	3.34	4.88	0.95
Vitamin C, total ascorbic acid	mg	0.0	0.0	0.0
Vitamin B-6	mg	0.340	0.496	0.096
Fatty acids, total saturated	g	7.329	10.700	2.078

Reproduced from USDA (2016). Food Composition Databases of Nutrient Database for Standard Reference. Available at https://ndb.nal.usda.gov/ndb/search/list

[a] g, gram; IU, international unit; kcal, kilocalorie; mg, milligram; μg, microgram

Human and Industrial Uses

Human Uses

- Edible oil for cooking, salads, canning, deep-frying, margarines, and shortening in pastry and bread.
- Whole seeds eaten raw, boiled, or roasted and salted.
- Chopped in confectioneries or ground into peanut butter.
- Young pods consumed as a vegetable.
- Young leaves and shoot tips used as cooked green vegetables.
- Products like ice cream, massage oil, and peanut milk.

In Malawi, groundnuts are mainly used in confectioneries (largely for export), oil (with residual cake used as cattle feed), local consumption (boiled, roasted, etc.), and hay fed to livestock.

Industrial Uses

- Groundnut oil used in soap fabrication, cold creams, cosmetics, dyes, paints, and pomades.
- Manufacturing of lubricants, emulsions for insect control, and diesel engine fuel.
- Peanut hulls used as filler for fertilizers.

6.5.25 Production Systems: Land Selection and Preparation

Groundnut production in the SADC region is carried out under various cropping systems based on environmental, social, and economic conditions. Factors influencing the choice of system include seed size, mechanization capabilities, and traditional landrace patterns, which reduce the effects of biotic and abiotic stresses.

Cultivation technologies (including the use of some "improved" varieties) have reduced the diversity formerly present in the groundnut landraces.

Planting groundnuts in mixed crop systems is a common practice on small farms across Southern Africa. Groundnuts are often intercropped with cereals like maize or sorghum, and cassava is almost always present in these fields. Thanks to its tolerance for shading, groundnut pairs well with tall-growing crops.

Intercropping Practices

In Mozambique, groundnuts are typically intercropped with maize, cassava, sorghum, millet, and plantation crops such as coconut palm and cashew. The success of mixed cropping systems depends on selecting site-appropriate crops, understanding crop interactions, and correctly integrating specific groundnut varieties with other crops.

Rotation of Crops

Groundnuts should be planted in a 3-year crop rotation to prevent the buildup of soil-borne pathogens and maintain soil humus. Despite the crop's importance, its yield in the region is often low due to unreliable rainfall, lack of improved production technologies, pest and disease prevalence, poor seed quality, and cultivation on marginal land. Political instability and non-supportive smallholder farm policies also negatively impact groundnut production.

Soil Requirements

Groundnuts thrive on well-drained, sandy to sandy loam soils with a pH of 5.0–7.0 and an organic matter content of 2.5% or higher. The ideal soil is rich in calcium for pod filling. Acidic soils with low calcium levels can result in empty pods ("pops"). Liming may be necessary to correct soil acidity before planting.

Seedbed Preparation

A groundnut seedbed should be deep, loose, and not too fine, preventing it from becoming muddy during rainy spells. The top 10 cm of soil should be kept loose during flowering and pod formation to allow the pegs to penetrate and the pods to develop. Compacted soils inhibit taproot penetration, and clay soils that form hard pans are unsuitable for groundnut production.

Primary Soil Cultivation

Primary soil cultivation involves removing debris and breaking the plough layer. Crop residues that spread diseases and harbor pests should be removed or incorporated into the soil. Light soils require shallow raking after the first light rains to eliminate early weeds and break up the soil surface. In wetter areas with heavier soils, a second hoeing or harrowing refines the seedbed.

Land Configuration

Groundnuts can be grown on flat or ridged fields, with ridges made during soil preparation, sowing, or weeding. Raised beds are suitable for heavy soils to limit runoff or waterlogging. Contour ridges help protect against erosion and improve water retention. Groundnuts tolerate flooding for up to a week, provided water drains away completely. In high-rainfall areas, well-drained fields or planting on mounds or ridges is necessary. Moist soil allows pegs to penetrate and develop pods, while drought stress during seed ripening can lead to mold infections, such as *Aspergillus flavus*.

Time of Planting

The optimum time to sow groundnut depends largely on the rain, as yields can be considerably reduced when the crop is sown outside the rainy season when soil conditions and weather are not favorable. The time of planting also varies depending on the maturity period of the variety to be sown. Early maturing varieties are sown from early to mid-January so that they mature in the dry season. Planting much earlier should be avoided as the crop would mature during the latter part of the rainy season, resulting in considerable seed sprouting. Late maturing varieties are always planted at the onset of the rains from late August to early or mid-November. In areas of erratic rainfall such as those in Mozambique, groundnut planting is done from late August to early September. Often farmers plant the crop very late after the onset of rains, mostly because the land is not ready for sowing at the beginning of the rains.

Spacing

Groundnut cultivation practices, including spacing, planting depth, and weeding, play a critical role in optimizing yield and maintaining crop health (Fig. 6.32). The space requirements for groundnut depend on the cultivar:

Fig. 6.32 Recently planted groundnut seeds in a standard spacing and which are emerging from the soil following the germination

- **Small-seeded, early-maturing varieties**: 60 cm between rows and 10 cm between plants, resulting in an optimal population of 130,000 plants/ha.
- **Late-maturing varieties**: 75 cm between rows and 15 cm between plants, resulting in a population of 90,000 plants/ha.
- **Branching Virginia types with long vegetative periods**: 40 × 20 cm spacing, resulting in 100,000–125,000 plants/ha.
- **Bush varieties of Spanish type**: 30 × 15 cm spacing, leading to 200,000–250,000 plants/ha.
- **Spanish runner types under rainfed conditions**: Ideal row width of 90–100 cm between rows.

In areas with erratic rainfall, a preferred population density is about 150,000 plants/ha. In mixed cropping systems, the best approach is based on traditional knowledge and site-specific conditions. For example, groundnuts and maize may be sown together on the same ridges, or rows of another crop could alternate with one to two rows of groundnut. Spanish types typically require about 60 kg/ha of seed, while Virginia types require about 80 kg/ha.

Planting Depth

Groundnut seeds should be sown at a uniform depth of 3–5 cm to ensure equal plant emergence. Pressing the soil down with a hoe after planting maximizes soil contact with seeds and improves germination. For untreated seeds, it is advisable to dress them with a pesticide or plant extracts to minimize damage by soil-borne pests. Seedlings typically emerge in 5–10 days, depending on sowing depth and soil moisture.

Weeding

Groundnuts compete poorly with weeds, especially during early development stages. The slow early growth and lack of ground coverage often lead to significant weed presence. Timely weeding is crucial to prevent nutrient depletion by weeds and ensure optimal groundnut growth.

Common problematic weeds include the following:

- **Grasses**: *Rottboellia exaltata* (itchgrass), *Eleusine indica* (goose grass).
- **Broadleaf weeds**: *Acanthospermum hispidum* (bristly starbur), *Bidens pilosa* (black jack), *Nicandra physalodes* (apple of Peru), *Commelina benghalensis* (tropical spiderwort).
- **Parasitic weeds**: *Alectra* spp. in Malawi.

Timely weeding minimizes damage to the root system. Before groundnut shoots emerge, the soil is tilled with light hoeing. After root development, slight tilling during the hottest part of the day helps kill weeds. Cultivation after emergence keeps soil loose, aiding pegging. Typically, hoeing twice—on the 14th day and before the 60th day after sowing—is sufficient.

Hand weeding with a traditional hoe is effective but demanding (Fig. 6.33). After flowering, minimal cultivation and earthing up are necessary to avoid disturbing pegs and developing pods. Hand pulling weeds is also labor-intensive but effective for broadleaf weeds and *Alectra* spp. Chemical weed control with herbicides has been successful against grasses, but herbicides are often too expensive for smallholder farmers. Gradually forming ridges helps reduce weed growth and prevent damage to the main stem.

Fig. 6.33 Farmers weeding groundnut plots in a field at Wamba in Kenge District in Kwango Province of the DR Congo. (Courtesy Frangoie, Antoine at IITA DR Congo)

Fertilization

Groundnut does not benefit from the direct application of nitrogen (N) fertilizers, as it is a leguminous crop capable of fixing atmospheric nitrogen through symbiosis with bacteria such as *Bradyrhizobium* spp. The efficiency of these nitrogen-fixing bacteria is enhanced by the presence of sulfur and calcium and is diminished by nitrogen-rich fertilizers.

Nitrogen Fixation and Soil Nutrients

Groundnuts are adept at utilizing residual soil nutrients and performs well in rotation with well-fertilized crops like maize (CIAT 2004). In high pH soils, certain elements such as iron and zinc become unavailable, making direct crop fertilization beneficial. This is particularly true for light sandy soils often used for groundnut cultivation in the region.

Fertilizer Application.

For good plant establishment, applying D-Compound fertilizer at a rate of 150 kg/ha at planting time is recommended (Mauyo et al. 2007). The availability of calcium in the upper soil layers is critical for groundnuts, as it allows roots to take up nutrients and helps pods fill as they grow. The presence of empty pods ("popping") is an indicator of calcium deficiency, which can be corrected by adding lime to the soil.

6.5.26 Bambara Groundnut Cultivation

Bambara groundnuts (*Vigna subterranea* (L.) Verdc.) are annual legumes closely related to cowpeas, belonging to the family Fabaceae and sub-family Papilionoideae (Fig. 6.34). According to Swanevelder (1998), the crop has two botanical varieties, namely, ***Vigna subterranea* var. *spontanea*** (wild varieties) and ***Vigna subterranea* var. *subterranea*** (cultivated varieties). The crop produces small (1–5 cm long), roundish or slightly oval-shaped, wrinkled pods and seeds on or just below the soil surface. The pods usually contain one to two seeds, which can be creamy, brownish, reddish, or purplish. In many African countries, Bambara groundnuts have become less important due to the expansion of groundnut production. However, there has been renewed interest in the crop in arid and semi-arid savannah zones due to its drought resilience and ability to yield reasonably well on poor soils.

The renewed interest in Bambara groundnuts is driven by their resilience to drought and capacity to produce a reasonable yield on poor soils. This makes them an ideal crop for arid and semi-arid regions, contributing to food security and

Fig. 6.34 Cultivation of bambara groundnut: a row of plants in the field (left) and harvested pods (right) still attached to a plant. (Source: https://en.wikipedia.org/wiki/Vigna_subterranea#:~:text=The%20origin%20of%20the%20Bambara%20groundnut%20is%20West,the%20northern%20parts%20of%20Nigeria%20and%20Northern%20Ghana)

agricultural sustainability. Bambara groundnuts are known under different names depending on the country in which they are grown.

Cultivation and Importance

In Botswana, Bambara groundnuts play a significant role in agricultural production systems. They are the third most important grain legume grown in intercropping systems after cowpea and groundnuts (Karikari et al. 1997). In Mpumalanga, South Africa, Bambara groundnuts are considered the second most important food legume and the third most important food crop after maize and groundnut. Almost all small-scale farmers cultivate them. In Zambia, however, they are among the minor grain legumes along with cowpea, pigeon pea, chickpea, mungbean, lentils, and adzuki beans. Despite their nutritional benefits and adaptability to marginal lands, Bambara groundnuts are cultivated on a small scale in Western, Southern, and Eastern Provinces (Mbewe et al. 1995).

In Zimbabwe, Bambara groundnuts are one of the most important local crops used for both food and fodder. In 1981, resettlement farms cultivated around 4000 ha of Bambara groundnuts, producing approximately 2200 tons annually, while small-scale farmers cultivated about 1300 ha, yielding about 750 tons of unshelled nuts.

Nutritional Value and Uses

Bambara nuts have the potential to improve malnutrition and boost food availability in the SADC region (Fig. 6.35). The seeds contain the following:

- **Protein**: 14–24%

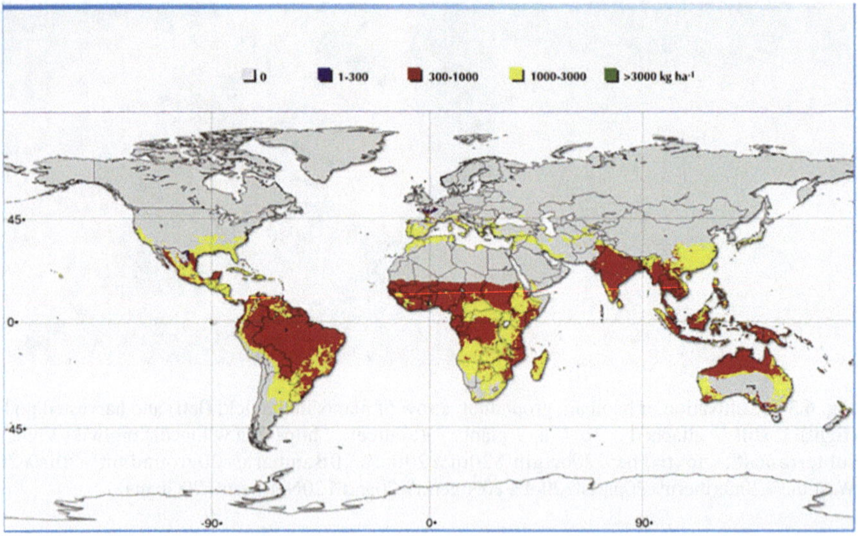

Fig. 6.35 Predicted pod yield of Bambara groundnut (kg/ha) across the world including Southern Africa. (Reproduced from Sayed Azam-Ali, José Aguilar-Manjarrez and Mohammad Bannayan-Avval (2001); Courtesy FAO Available at http://www.fao.org/docrep/003/Y0494E/Y0494e06.htm)

- **Carbohydrates**: 65%
- **Oil**: 6–12% (less than half the quantity found in groundnuts)

Mature seeds are rich in protein (17.5–21.1%), carbohydrate (53–60%), and crude fat (2.3–8.5%) (Ominawo et al. 1999). The protein content is higher in the essential amino acid methionine compared to other grain legumes.

Consumption

- Immature seeds: More palatable, boiled or grilled.
- Mature seeds: Cooked, ground into flour, mixed with oil or butter to form porridge, roasted in oil.

Animal Feed

- Leaves: Used as livestock feed.
- Seeds: Fed to pigs and poultry, often after being soaked in water.

Culinary Uses

- Fresh or grilled seeds consumed as a snack or meal.

Fig. 6.36 The collection of Bambara nuts morphologically characterized to unlock their genetic diversity and potential for improving global food security and poverty eradication (left) and seeds of the crop at IITA Genebank (right). (Reproduced from IITA)

- Boiled (shelled or unshelled) and eaten alone or mixed with immature groundnuts or maize.
- Flour used to make cakes or balls, mixed with other foods like meat stew, rice, spinach, vegetables, maize, and sorghum.

In some restaurants in Angola and Mozambique, boiled and salted Bambara groundnut seeds are served as appetizers. In Botswana, Bambara groundnuts are primarily consumed by humans. Consumers particularly enjoy the immature seeds, which are boiled in the pod, salted, and eaten either alone or mixed with maize seeds. When the seeds are ripe and dry, they are pounded into flour to make a variety of cakes or mixed with cereals to prepare different types of porridge.

The seeds come in various sizes and colors (Fig. 6.36), and they offer versatile uses in cooking including boiled in the pods, salted, and consumed as a snack or mixed with maize or pounded into flour for cakes or mixed with cereals for porridge. The leaves of Bambara groundnuts can be used as livestock feed and are readily consumed by cattle and game. Livestock, especially goats, are particularly fond of the haulm (the plant's stems and leaves), which they graze on after the pods have been harvested.

In semi-arid parts of Tanzania, the crop is considered as an insurance or a security crop. It is therefore eaten in a variety of ways, ranging from the freshly cooked pods to the cooked dry grains. It is always incorporated into main dishes, such as cooked plantains and cereals. The vegetative parts of the plant, and sometimes the haulms, are used to feed livestock. The crop is also useful for rotation cultivation systems, as it contributes a large amount of N to the soils which benefits subsequent crops.

Bambara groundnut is used medicinally and for pest and disease control. Water from the boiled maize and pulse mixture is drunk to treat diarrhea. Bambara groundnuts can also be used to cure nausea suffered by pregnant women. In this case, the raw nut has to be chewed and swallowed. Leaves are pounded with those of *Lantana trifolia* L. or Mexican marigold, and water is then added to make a solution that is used to wash livestock as a preventative insecticide against ticks. When applying the same solution to vegetables, care has to be taken to apply it to the ground level and not to pour it on the leaves as it can burn leaves.

Bambara groundnuts are not only vital for human and animal nutrition but also beneficial for crop rotation systems due to their ability to contribute significant nitrogen (N) to the soil, enhancing the growth of subsequent crops.

Developmental Growth Stages

Bambara groundnuts take about 3–6 months to mature, depending on local weather conditions and the cultivar grown. The seeds germinate within 7–15 days after planting. Flowering begins 30–35 days after sowing and may continue until the plant's life ends. Under long-day conditions, flowering might not occur. The vegetative growth phase occurs between November/February and early March/late April, with pod formation starting in late April/early May. Pods develop within 30 days after fertilization, and seeds set about 10 days later. Seeds are mature when the parenchymatous layer surrounding the embryo disappears, and the pods turn light brown.

6.5.27 Production Systems: Land Selection and Preparation

Bambara groundnut thrives in well-drained soils, with light, sandy loams with a pH of 5.0–6.5 being most suitable. The crop also performs well in nutrient-poor soils. High soil nitrogen levels favor vegetative growth. In Zambia, Bambara groundnut is planted on both flat land and ridges. It yields best on deeply ploughed fields with a fine, well-leveled seedbed but can also be planted on ridges in very wet conditions. Bambara groundnut is cultivated both as an intercrop with maize, cowpeas, and melons and as a sole crop. In Tanzania, it is typically intercropped with cereals such as millet, sorghum, or maize, root and tuber crops, and other legumes (Smyth 1968). Similarly, in Zimbabwe, Bambara groundnut is grown on sandy soils, either as a single crop or in mixed stands with other crops such as maize, groundnuts, cotton, or sorghum.

Planting Time

Nearly all seeds used for planting vary in size and color, composed of local landraces usually kept by farmers. The optimal planting period for Bambara groundnut is from late October to November through December to early January after sufficient rains have moistened the soil. Later plantings in January and February result in lower yields.

Planting Space

Bambara seed size varies considerably, and this determines to a large extent its plant spacing (Fig. 6.37). Usually, planting is done at 10 × 15 cm in single rows spaced 45 × 90 cm apart. The seeding rate can vary from 25 to 75 kg/ha with an average weight of 1000 seeds ranging from about 500 to 750 g.

Planting Depth

Bambara groundnuts are resilient legumes cultivated in various soil types. Proper planting depth, fertilization, and weed control are essential for optimal growth. The planting depth for Bambara groundnuts depends on soil type:

- **Heavy soils or high moisture conditions**: Seeds should be planted 2.5–3.0 cm deep (high moisture conditions are generally not recommended).
- **Sandy soils**: Depth varies from 5.0 to 7.5 cm.

Traditionally, a hoe is used to sow seeds and cover them with soil.

Fertilization

Bambara groundnuts enhance soil fertility, benefiting surrounding crops without requiring fertilizer application. The plant obtains its nitrogen (N) supply from atmospheric N fixation by rhizobia in its root nodules. High soil N content usually results in few pods and seeds above the soil surface. Fertilization is generally unnecessary for Bambara groundnuts due to their nitrogen-fixing capabilities.

Fig. 6.37 A field of Bambara groundnuts planted at recommended spacing in the Eastern Province of Zambia

Weed Control

Weeding in Bambara groundnut fields is primarily done by hand on smallholder farms. Careful weeding is crucial, especially during the flowering stage, as the flower stalks are fragile and can break easily.

6.5.28 Vegetable Production

Introduction

Increased awareness of the health benefits of non-nutrient bio-active compounds in fruits and vegetables has brought significant attention to their cultivation as vital components of daily diets (Fig. 6.38). For sub-Saharan African populations, this focus on vegetables is particularly crucial, as leafy vegetables are essential ingredients in traditional relishes that accompany carbohydrate staples. African indigenous and traditional leafy vegetables play a pivotal role in the success of the World Health Organization's (WHO) global initiative on fruit and vegetable consumption in the sub-continent.

Nutritional Importance

Producing more food alone is not the solution to malnutrition in Africa; the quality and nutrient content of the food are equally important (AVRDC 2003). For instance, cassava and rice are valuable staples widely consumed in Africa, but it takes about 4 kg of cassava and even more rice daily to meet one person's vitamin A and iron

Fig. 6.38 A farmer watering his vegetable nursery (left) and a vegetable trader at marketplace in Southern Africa (right). (Reproduced from Dhaliwal, M.S., Wanduky, N., 2009. Vegetable Cultivation: Seedling Nurseries for Better Vegetable Production. AVRDC, Regional Center for Africa, Yaounde, Cameroon at World Vegetable Center, 2003. Vegetables for Life: Confronting the Crisis in Africa. World Vegetable Center Publication No. 03–564, Shanhua, 28 p)

requirements—an impractical solution. Inadequate micronutrient intake has profoundly harmful effects.

Policy and Diversification

Governments of SADC Member States have adopted crop diversification policies to ensure constant food availability, considering the region's changing weather patterns, which often result in prolonged droughts or constant flooding. The FANR at SADC indicates that the heavy reliance on maize is a food insecurity problem, and diversification should be seen as an improved risk management element (http://www1.sadc.int/content/english/key_documents/risdp/pics/risdp02.jpg). Angola, the DR Congo, Malawi, Mozambique, Tanzania, and Zambia are major cassava-growing countries in the region, collectively producing an estimated 30 million tons of fresh cassava annually, equivalent to about 12.5 million tons of maize in nutritional value. Cassava's importance to food security in these countries has led to its inclusion in the food balance sheet. Other significant root and tuber crops produced in the region include potatoes (2.3 million tons), sweetpotato (1.0 million tons), and yams, though production statistics for yams are yet to be established. Agricultural diversification is prioritized as it can mitigate commodity dependence in adverse climatic conditions. The vegetable production sub-sector, in particular, offers real opportunities to enhance smallholder farm incomes and reduce poverty in Southern Africa.

Nutritional Benefits of Vegetables

Vegetables are vital in diets for their vitamin content, especially vitamin A, vitamin C, and folate (folic acid, folacin) (http://www.naturalhub.com/natural_food_guide_vegetables.htm, Table 6.13). Besides these, vegetables are also a good source of thiamine (B1). Potatoes, broccoli, cauliflower, and tomatoes are primary sources of pantothenic acid (B5). Additionally, all brassicas, potatoes, spinach, peas, carrots, watercress, and onions provide pyridoxine (B6), essential for brain function, immune system function, and as a precursor to several important hormones.

Many vegetables encompass also small but useful amounts of vitamin E in addition to other several minerals except the iron.

FANR points out that carotenes, especially beta-carotene, are only obtainable from plant foods. Beta-carotene, often referred to as "pro-vitamin A," is converted by the body into vitamin A and is generally non-toxic. For example, carrots account for about 30% of the vitamin A in the US diet, highlighting not only the limited variety of vegetables consumed but also how cultural practices can significantly impact health.

The way food is prepared and cooked also greatly affects its vitamin A content. The conversion rate of beta-carotene to vitamin A increases when the food is finely chewed or grated. Moderate cooking improves availability by breaking down the

Table 6.13 Vegetable production in SADC countries (in 1000 tons and as percentage of world vegetable production)

Countries	Area harvest (1000 ha)					Production (1000 tons)				
	1994–1996	1999–2001	2004	2005	2006	1994–1996	1999–2001	2004	2005	2006
Angola	46	47	48	48	48	256	259	271	271	271
Botswana	3	3	3	3	3	16	17	16	16	16
DR Congo	74	72	87	88	73	523	440	525	527	445
Lesotho	3	3	4	4	3	23	20	29	32	18
Madagascar	43	43	43	43	43	341	348	344	344	344
Malawi	33	30	31	31	31	258	250	313	313	310
Mauritius	6	6	7	6	6	77	85	93	79	86
Mozambique	25	23	21	21	21	147	135	117	117	117
Namibia	2	3	3	3	3	9	15	15	15	15
South Africa	122	127	122	122	112	2089	2195	2248	2243	2006
Swaziland	1	1	1	1	1	12	11	11	11	11
Tanzania	176	183	188	188	187	1146	1218	1246	1246	1242
Zambia	37	38	41	42	38	266	266	287	298	287
Zimbabwe	21	22	26	26	24	150	154	180	181	161

Reproduced from FAO Statistical Yearbook 2007–08. Available at: http://www.fao.org/economic/ess/ess-publications/ess-yearbook/fao-statistical-yearbook-2007-2008/b-agricultural-production/en/

vegetable's cell walls. However, repeated cooking at high temperatures can destroy some of the vitamin A content.

Faber et al. (2010) observed that the deficiency in vitamin A can bring about several deterrent effects mainly in children. Among these the following are the most important:

1. **Poor child growth and development**: Children lacking in vitamin A often experience a loss of appetite, leading to weight loss and malnutrition. Malnourished children have a weaker immune system, making them more susceptible to infections. Serious infections, such as measles and diarrheal diseases, can cause significant weight loss, often hindering proper growth and development.
2. **Increased risk of infection**: A deficiency in vitamin A makes children more prone to infections, particularly gastrointestinal (causing diarrhea) and respiratory infections.
3. **Increased severity of infection**: The impact of infections, especially measles, is more severe in children with vitamin A deficiency.
4. **Higher mortality rates**: Vitamin A deficiency increases the likelihood of death in children compared to their well-nourished counterparts.
5. **Eye-related problems**: One of the first signs of vitamin A deficiency is night blindness, characterized by difficulty or inability to see in dim light. This condition can progress to more severe structural damage in the eyes, such as Bitot's spots (foamy white patches on the white part of the eye) and, in extreme cases, irreversible blindness.

Ensuring adequate intake of vitamin A is crucial for preventing these severe health issues in children.

According to Nyirenda et al. (2011), vegetable production in Southern Africa is transitioning from traditional backyard gardens and small plots (150–300 m^2) to more intensive systems, especially in areas with supplementary irrigation. Despite this shift, cultivation remains dominated by the same fields year after year, relying heavily on family labor and locally available inputs for soil nutrient replenishment and pest control.

Importance and Economic Value

In Malawi and Zambia, vegetables are crucial as they provide dietary vitamins and minerals in the predominantly maize-based diet of rural households. Their high market value generates income for poor households year-round. However, the smallholder horticulture sector in Zambia is relatively underdeveloped, with few farmers specializing in fruit or vegetable production (Nenguwo 2004). Horticultural production is typically secondary, producing food for home consumption with surpluses sold locally.

In Farming Systems

As noted by Okigbo (1990), vegetables in Africa are produced within various farming systems, including traditional shifting cultivation, home gardens, upland and lowland fields, market gardening, and specialized horticultural systems. Farmers can be categorized as small-scale (below 5 ha), medium-scale emergent (5–20 ha), and large-scale commercial producers. Small-scale farmers usually have limited access to credit and face challenges in transporting produce to markets due to poor infrastructure. They often use family labor and simple tools, and their limited capital constrains their use of high-quality seeds and fertilizers, resulting in lower productivity.

Nutritional and Economic Role

In countries like Zambia and the DR Congo, vegetables provide essential nutrients and help improve small farmers' incomes. Local horticultural production meets only about 20% of the national demand, with the rest imported. This highlights the potential to increase local production. Vegetable production in these countries is seasonal, influenced by varying climatic conditions. Madisa et al. (2010) reported that the 2008/2009 national demand for vegetables was estimated at 50,000 metric tons, with local production at 31,150 metric tons.

Regional Production

The SADC region produces 5.7 million metric tons of vegetables annually, up from 4.4 million tons in 1980, with an average annual growth rate of 1% over the last 25 years. South Africa leads with 40% of the production, followed by Tanzania (22%), DR Congo (8.4%), Madagascar (6.4%), and Angola and Zambia (5% each). Significant long-term growth trends have been recorded in Namibia, Mauritius, South Africa, and Malawi.

Challenges

Vegetable production in Southern Africa faces several challenges:

- Recurrent drought and water shortages
- Poor soils
- Insufficient infrastructure
- Pest and disease outbreaks
- Poor farming practices
- Low adoption of improved cultivation technologies
- Poor organization of markets

- Lack of security of tenure

The promotion of vegetable production and addressing the numerous challenges faced by smallholder farmers would significantly enhance food security, nutrition, and economic stability in Southern Africa.

6.5.29 Cabbage Cultivation

Cabbage (*Brassica oleracea* L. var. capitata) is a vegetable that belongs to the group of the Brassicaceae (mustard) family (Fig. 6.39). It is a cole crop that is widely grown in many areas of the world including the SADC region as an irrigated crop. Cabbage is a unique type of cole vegetable crop whose leaves form a head-like structure, and it does not grow much in length like rape, and at maturity, the crop is closer to the ground surface than other vegetable crops. It performs well in cool-dry areas; hence its growth in Southern Africa region is during the cool-dry season. Cabbage is of great importance for the population due to its vitamin A and vitamin C contents which are peculiar for human sight and disease resistance (Table 6.14). It is usually eaten boiled, although it can also be consumed raw as a salad.

Cabbage has become the third most important crop in Tanzania, particularly in the cooler highland areas of Arusha, Tanga, Iringa, Mbeya, and Morogoro (Massomo et al. 2005). Its cultivation began in the early 1970s, with a notable increase in growers from 1997 to 2001. Cabbage is primarily grown as a monocrop or intercropped with bananas and coffee. Production occurs year-round, peaking from July to February.

Fig. 6.39 A well-developed and healthy cabbage head tready for harvest in a farmer's garden in Copperbelt Province of Zambia

Table 6.14 Cabbage nutritional values per 100 g for edible portion of raw material

Nutrient	Unit[a]	Value/100 g	# data points	Std. error	Cup, chopped 89 g	Leaf large 33 g	Head 908 g	Leaf 15 g
Water	g	92.18	20	0.128	1150.41	30.42	836.99	13.83
Energy	kcal	25	–	–	312	8	227	4
Protein	g	1.28	4	0.097	15.97	0.42	11.62	0.19
Total lipid (fat)	g	0.10	13	0.015	1.25	0.03	0.91	0.01
Ash	g	0.64	4	0.039	7.99	0.21	5.81	0.10
Carbohydrate, by difference	g	5.80	–	–	72.38	1.91	52.66	0.87
Fiber, total dietary	g	2.5	2	–	31.2	0.8	22.7	0.4
Sugars, total	g	3.20	–	–	39.94	1.06	29.06	0.48
Starch	g	0.00	1	–	0.00	0.00	0.00	0.00
Calcium, Ca	mg	40	30	1.307	499	13	363	6
Iron, Fe	mg	0.47	31	0.035	5.87	0.16	4.27	0.07
Magnesium, Mg	mg	12	31	0.325	150	4	109	2
Phosphorus, P	mg	26	30	0.657	324	9	236	4
Potassium, K	mg	170	31	4.811	2122	56	1544	26
Sodium, Na	mg	18	46	1.651	225	6	163	3
Zinc, Zn	mg	0.18	31	0.009	2.25	0.06	1.63	0.03
Copper, Cu	mg	0.019	31	0.002	0.237	0.006	0.173	0.003
Manganese, Mn	mg	0.160	31	0.012	1.997	0.053	1.453	0.024
Vitamin C, total ascorbic acid	mg	36.6	4	3.388	456.8	12.1	332.3	5.5
Vitamin B-6	mg	0.124	4	0.020	1.548	0.041	1.126	0.019
Vitamin A, RAE	µg	5	–	–	62	2	45	1
Carotene, beta	µg	42	25	0.860	524	14	381	6
Vitamin A, IU	IU	98	–	–	1223	32	890	15
Fatty acids, total saturated	g	0.034	–	–	0.424	0.011	0.309	0.005

Reproduced from USDA (2016). Food Composition Databases of Nutrient Database for Standard Reference. Available at https://ndb.nal.usda.gov/ndb/search/list

[a] g, gram; IU, international unit; kcal, kilocalorie; mg, milligram; µg, microgram

Labor and Production

Most farmers invest 5–65 man-days per hectare for cabbage cultivation (Mtaita and Msuya 1994), growing two to four crops per season. About 80% of the harvested cabbage is sold to wholesale or retail traders who collect it on the farm, while only 18% is sold directly by farmers at local markets. Based on commercial seed sales, at least 40,000 ha of white-headed cabbage are grown in Tanzania, and 10,000 ha in Malawi, Zambia, and Zimbabwe. Nearly all white-headed cabbage is produced for local urban markets. Mozambique imports significant quantities of headed cabbage from South Africa and, until recently, from Zimbabwe.

Nutritional Value and Utilization

Cabbage is low in calories and protein but rich in minerals like potassium and vitamins A and C. Green cultivars generally contain more vitamin A than red cultivars, and savoy types have more vitamin A than smooth types. Additionally, cabbage is believed to have medicinal properties, potentially treating ailments like gout, headaches, and deafness.

Production Systems

Cabbage can easily be produced using very cheap organic fertilizers such as cattle and chicken manure. Therefore, any smallholder farmer is able to effectively produce the crop.

Propagation

Crops of the Brassicaceae family, including cabbage, are typically propagated by seeds. However, some cabbage varieties can also be propagated using cuttings when seeds are scarce. In this method, topped stems are left in the ground, and the shoots that develop from the cut surface are used as propagation material. This technique eliminates the need for nursery beds and maintains the vegetable's quality and purity. However, plants grown from cuttings tend to be smaller than those grown from seeds. This propagation method is useful when seeds are in short supply or a particular cabbage variety sells well.

Land Selection and Preparation

Cabbage thrives in a wide range of soils, but well-drained loam or sandy loam soils with good organic content and moisture retention are ideal (Fig. 6.40). Always avoid soils that dry quickly. The soil pH should be neutral (6.5–7.5), and lime may be added to acidic soils to maintain the preferred pH level. The land should be well ploughed and followed up with hoeing to achieve a fine tilth. Rows should be prepared a week before planting. Thorough soil pulverization ensures good root development and plant establishment. To prevent soil-borne diseases and nematode problems, avoid planting cabbage in soils that have previously grown the same crop or other cole crops in the past 2–3 years. Crop rotation is an effective way to reduce disease severity.

Planting Time

Cabbage can be planted directly or through nursery establishments before transplanting to the main field. In nurseries, seeds are sown in seedbeds prepared to a fine tilth, with about 300 g of seeds adequate for planting 1 ha of land. Compost or other

Fig. 6.40 Cabbage seedlings emerging from a well-prepared nursery bed

suitable fertilizing material should be mixed well with the soil before sowing. For direct planting, the same method is used.

The optimal planting time depends on the cabbage variety and the climate. The best time is during the cool-dry season, typically between May and September when the temperature and soil moisture are suitable. When using seedlings, transplanting should occur in the evening or on cloudy days if the temperature is high. Transplanting should be done when plantlets have at least two pairs of true leaves and are about 15 cm high, usually 4–5 weeks after seeding. Seedlings should be hardened off by exposing them to the growing environment before transplanting. Water immediately after transplanting and regularly for 5–6 days until seedlings regain turgidity. Cabbage prefers firm soil around the seedlings. Early varieties require 85–90 days from seeding to harvest, while main-season crops require 110–115 days.

Plant Spacing

Spacing within the growing of any crop is very important as it determines the level of competition between plants. For cabbage, the spacing of about 45–60 cm by 30–45 cm depending on the variety is always observed by farmers, making sure that late maturing varieties need larger spacing than early maturing varieties. Generally, a spacing of 50 × 40 cm or 40 × 40 cm is recommended for early maturing cultivars, while all others can be planted 50 × 50 cm apart. This spacing is the same for both direct planting and when seedlings are used. However in a nursery, seeds are sown into the seedbeds at 0.5 × 1.0 cm spacing deep below the surface although spacing can also be varied depending on the head size required with narrow spacing for small heads and wider spacing for bigger heads.

Planting Depth

Seeding depth during planting is often about 0.5 cm below the ground surface of a well-prepared seedbed. However, the exact planting depth is always adjusted based on the size of the crop seeds, sowing larger seeds deeper.

6.5.30 Field Maintenance After Planting

Cabbage is a heavy feeder of nutrients from soil, and as such, soil nutrition management is crucial for its growth. Effective cabbage cultivation requires meticulous attention to weed management, fertilization, irrigation, and crop rotation.

Weed Management

Weed control in cabbage fields starts with tillage before seeding or transplanting to destroy any emerged weeds. Manual weeding using hoes and other tools is common among farmers. Some smallholder farmers also use selective herbicides to prevent crop damage. For instance:

- **Trifluralin**: Applied to control annual grasses and small-seeded broadleaf weeds.
- **Devrinol**: Used pre-emergence after seeding or transplanting to control similar weeds as trifluralin.

Fertilization

Fertilizers are applied before and after planting cabbage to ensure optimal growth. Common fertilizers used for basal dressing include:

- **D Compound** (800 kg/ha)
- **Compost**
- **Chicken manure**
- **Tea leaves (tea manure)**

When using animal manure or compost, it is added to the soil at least 2 weeks before planting. The organic matter is thoroughly incorporated into the soil to a depth of 30–50 cm. For top dressing, ammonium **nitrate** (200 kg/ha) is used to improve the quality and flavor of cabbage heads. The crop responds well to nitrogen and should receive regular top-dressings. Additionally, cabbage benefits from sulfur and is sensitive to magnesium and boron deficiencies.

Irrigation

Irrigation is critical for consistent vegetable production and quality. Cabbage requires a reliable water source, as rainwater alone is insufficient. The soil must be kept moist to a minimum depth of 45 cm. Although cabbage is relatively drought-tolerant, maintaining adequate soil moisture levels is essential to maximize yields. The most critical irrigation periods are:

- Post-seeding or transplanting
- During head development

Stress from a lack of water during these periods can lead to smaller heads, growth cracks, or tip-burn, reducing marketability. Even water supply throughout the growing season prevents head cracking.

Roguing

Roguing involves removing offset crops from the field after regular inspections. Plants are removed based on factors like poor heading, leaf discoloration, and cabbage cracking.

Crop Rotation

Crop rotation, the practice of growing different crops in succession on the same land, is crucial. It prevents the buildup of soil-borne diseases and nematodes. Avoid planting cabbage in fields that previously grew cabbage or other cole crops in the past 2–3 years. Crop rotation effectively reduces disease severity and incidence.

6.5.31 Okra Cultivation

Okra (*Abelmoschus esculentus* (L.) Moench) is an annual plant from the Malvaceae family, known for its pod-forming characteristics and impressive height, reaching up to about 2 m (Fig. 6.41). Cultivated for its green seed pods, okra is harvested when the pods are still immature and used in various soups. The young fruit is highly valued as a vegetable and is particularly noted for its mucus-like juice, which thickens stews such as gumbo. It is widely cultivated throughout the year in tropical, subtropical, and warm temperate regions. The plant thrives in warm weather and struggles in frost and cold temperatures. Known for its resilience, okra is among the most heat- and drought-tolerant vegetables. It can grow with minimal water and withstand extremely hot conditions, making it an ideal crop for challenging climates. The word "okra" originates from West Africa, and in various Bantu

Fig. 6.41 Okra crop plants in field of a smallholder farmer in Kalulushi District of the Copperbelt Province of Zambia

languages, it is referred to as "kingombo." This highlights its cultural importance and historical roots in the region.

Importance of Okra

Although okra is mainly cultivated for its pod's consumption as salads or in various soups, it can also be used for income generation by selling the immature pods. Okra is mainly used in traditional cooking, especially in West Africa where immature fruits are cut into pieces and cooked with meats, and because of their high mucilaginous content, the pods are ideal for both thickening and flavoring stews and soups. Chopped pods are stewed to soften the mucilaginous content and then mixed with other vegetables, rice, or meat. Okra can be served raw, marinated in salads, or cooked on its own and goes well with tomatoes, onions, corn, peppers, and eggplant. Whole, fresh okra pods also make excellent pickles. Its mild flavor can be compared to eggplant, though the texture is somewhat unusual. The leaves are also eaten in certain areas of the Southern Africa region. The crop is rich in various nutrients and necessary elements that offer a variety of body health benefits (Table 6.15).

Health Benefits of Okra

- The pods are among the very low calorie vegetables. They provide just 30 calories per 100 g besides containing no saturated fats or cholesterol. Nonetheless, they are rich sources of dietary fiber, minerals, and vitamins; often recommended by nutritionists in cholesterol controlling and weight reduction programs.
- The rich fiber and *mucilaginous* content in okra pods help in smooth peristalsis of digested food particles and relieve constipation condition.
- The pods contain healthy amounts of vitamin A, and flavonoid anti-oxidants such as beta carotenes, xanthin and lutein. It is one of the *green* vegetables with highest levels of these anti-oxidants. These compounds are known to have antioxidant properties and are essential for vision. Vitamin A is also required for maintaining healthy mucus membranes and skin. Consumption of natural vegetables and fruits rich in flavonoids helps to protect from lung and oral cavity cancers.
- Fresh pods are the good source of folates; provide about 22% of RDA per 100 g. Consumption of foods rich in folates, especially during the pre-conception period helps decrease the incidence of neural tube defects in the offspring.
- The gumbo pods are also an excellent source of anti-oxidant vitamin, **vitamin-C**, providing about 36% of daily-recommended levels. Research suggests that consumption of foods rich in vitamin-C helps the body develop immunity against infectious agents, reduce episodes of cold and cough and protect the body from harmful free radicals.
- The veggies are rich in B-complex group of vitamins like niacin, vitamin B-6 (pyridoxine), thiamin and pantothenic acid. The pods also contain good amounts of vitamin K. Vitamin K is a co-factor for blood clotting enzymes and is required for strengthening of bones.
- The pods are an also good source of many important minerals such as iron, calcium, manganese and magnesium.

Reproduced with permission from http://www.nutrition-and-you.com/okra.html

Production Systems: Land Selection and Preparation

Okra thrives in a variety of soils with good drainage, though sandy loam or loam soils high in organic matter are ideal. To bring soil pH to a neutral range (6.0–8), lime or dolomite may be added during preparation. To avoid contamination and soil-borne pathogens, avoid planting okra on land that had okra crops in the previous season for certified seeds and the previous two seasons for basic seeds.

Soil Preparation

Primary tillage and planting groundwork involve:

- Chopping prior crop residues
- Removing the hardest residues

Table 6.15 Okra nutritional values per 100 g for edible portion of raw material

Nutrient	Unit[a]	Value/100 g	# of data points	Std. error	Cup 100 g	Pod 95 g
Water	g	89.58	19	0.224	89.58	85.10
Energy	kcal	33	–	–	33	31
Protein	g	1.93	4	0.033	1.93	1.83
Total lipid (fat)	g	0.19	5	0.014	0.19	0.18
Ash	g	0.86	2	–	0.86	0.82
Carbohydrate, by difference	g	7.45	–	–	7.45	7.08
Fiber, total dietary	g	3.2	–	–	3.2	3.0
Sugars, total	g	1.48	–	–	1.48	1.41
Starch	g	0.34	1	–	0.34	0.32
Calcium, Ca	mg	82	4	4.500	82	78
Iron, Fe	mg	0.62	4	0.088	0.62	0.59
Magnesium, Mg	mg	57	6	0.726	57	54
Phosphorus, P	mg	61	4	1.825	61	58
Potassium, K	mg	299	6	7.245	299	284
Sodium, Na	mg	7	6	0.705	7	7
Zinc, Zn	mg	0.58	1	–	0.58	0.55
Copper, Cu	mg	0.109	1	–	0.109	0.104
Manganese, Mn	mg	0.788	1	–	0.788	0.749
Vitamin C, total ascorbic acid	mg	23.0	12	2.129	23.0	21.9
Vitamin B-6	mg	0.215	2	–	0.215	0.204
Carotene, beta	μg	416	22	–	416	395
Vitamin A, IU	IU	716	–	–	716	680
Vitamin K (phylloquinone)	μg	31.3	1	–	31.3	29.7
Fatty acids, total saturated	g	0.026	–	–	0.026	0.025

Reproduced from USDA (2016). Food Composition Databases of Nutrient Database for Standard Reference. Available at https://ndb.nal.usda.gov/ndb/search/list

[a] g, gram; IU, international unit; kcal, kilocalorie; mg, milligram; μg, microgram

- Listing seedbeds

These tasks are typically performed from September through January using hoes, spades, rakes, and forks. The chopped residues are ploughed into the soil while it's still moist to avoid hard clods on heavier soils. The ground is then leveled using a hoe to ensure efficient water supply and drainage. Chicken manure may be broadcasted and incorporated into the soil.

Planting

Okra planting is done using a hoe, with about 8 kg/ha of seeds. Seeds are soaked for 4–6 h before planting. Planting usually occurs with the first rain from November to March. In well-prepared, moist soil, planting holes are spaced at 90 cm × 60 cm for monocrops and 80 × 40 cm or 60 × 30 cm for intercropped okra. Typically, three to four seeds are dibbled into each hole.

Planting Depth

Seeds are sown to a depth of 2–2.5 cm. Deeper sowing can delay germination and result in weaker shoots or failed germination, leading to lower plant populations.

Fertilization

Before planting, a basal dressing of 300–500 kg/ha of Compound-D or Compound-C is applied and cultivated into the soil. Starting 2–3 weeks after seedling emergence, three top dressings of 50 kg/ha ammonium nitrate are applied at 2- or 3-week intervals. Sometimes, chicken manure (4–5 tons/ha) is used instead of chemical fertilizers, broadcasted, and ploughed into the soil.

Weeding

Weeding is performed 2 and 4 weeks after emergence and during the early flowering period. Weeding 4 weeks after planting coincides with fertilizer application, and inter-row cultivation is necessary after each top dressing to destroy weeds and slightly earth up the plants.

Pruning

Okra plants do not develop branches until they reach about 30–40 cm in height. Typically, 8–14 fruit-bearing shoots develop. If further offshoots grow, thinning may be necessary. Two weeks after sowing, thinning is performed to leave only one to two plants per station, with uprooted shoots filling in vacant holes within a week. Ratooning, or cutting back old plants to allow them to re-shoot, is also practiced rejuvenating plants and producing new crops when prices are higher.

Irrigation

Okra, being a heavily foliaged crop, requires a significant amount of water for growth and pod development. The exact amount and frequency of watering depend on the site, time of year, soil type, and water availability. Regular irrigation is essential after planting if rain is insufficient until germination and plant establishment.

6.5.32 Onion Cultivation

Onion (*Allium cepa* L.) (Fig. 6.42) belongs to the genus *Allium*, of the family Lilliaceae, also classified by others as Alliacea (Raemaekers 2001). The true onion genus comprises over 600 species, but only 6–7 are cultivated, incredibly versatile and adapted to various climates, ranging from tropical to sub-arctic regions. This adaptability is largely due to their unique response to day length, which affects bulbing rather than flowering.

- **Nodding wild onion**: Known scientifically as *Allium cernuum.*
- **Shallot onion**: Referred to as *Allium ascalonicum.*
- **Chive onion**: Known as *Allium schoenoprasum.*
- **Green onion or common leek**: Scientifically named *Allium ampeloprasum.*

Onions are a significant crop across the SADC region, providing an essential source of income for many smallholder farmers. Their ability to adapt to different day lengths and climates makes them a reliable and valuable vegetable crop.

Onion Shapes

The shape of onion bulbs varies significantly depending on the variety, ranging from spherical, globular, or flattened to conical or elongated. Years of domestication and breeding have resulted in cultivars where the bulb swells considerably during the first year of growth (Raemaekers 2001). Onion varieties also differ in color—white, yellow, brown, or red—and in taste and storage qualities.

Common Onion Cultivars

Farmers grow a wide variety of onion cultivars, including:

Fig. 6.42 Onion cultivation: profusely growing leafy plants in field of a famer (left) and a bumpy bulb harvest (right) in the Copperbelt Province of Zambia

- **Storage onions**: Superba (F1 Hybrid), Hyduro.
- **Early onions**: Early Vanguards, Spanish Yellow, Yellow Valencia.
- **White onions**: Barletta White, Texas Early Grano 502, Yellow Granex.
- **Egyptian onions**: Rocambole.

In Africa, popular cultivars include Red Creole, Red Kano, and Violette de Galmi. Small white cultivars are among the most productive. Egyptian onions develop bulblets on the flowering spike, which can be planted to produce relatively large bulbs. Each farmer, however, has his/her own preference for any of the cultivars based on specific household criteria.

Importance of Onions

Onion has been known since ancient times by the Egyptians, Greeks, and Romans who initially used it as a medicinal plant before it became a consumable vegetable that became widely used in making raw salads and also cooked in a variety of stews and soups. In addition to healthful fiber, onions serve up several vitamins and minerals useful for meeting human body nutritional needs. The nutrients found in onions are numerous (Table 6.16), making the crop among the most cultivated worldwide. Wolf (2011), available at http://www.livestrong.com/article/392981-what-nutritional-value-do-onions-have/#ixzz2CIm27gT3) acknowledges that a one-cup serving of chopped onions contains 51 calories and 0.13 g of fat. Each cup of onions also provides 11.6 g of carbohydrates, while the crop is also a secondary source of about 1.3 g protein, which is needed for body energy; onions are a healthy choice for getting nearly 2.1 g fiber that contributes toward soluble-fiber intake. The fiber in onions lowers the chances of developing cardiovascular diseases, such as high blood pressure, stroke, heart failure, and heart attacks, and may also reduce cholesterol. Moreover, one serving of onions provides 11.5% of vitamin B6 needed

Table 6.16 Onion nutritional values per 100 g of edible portion of raw material

Nutrient	Content[a]	Nutrient	Content		Nutrient	Content	
Energy	40 kcal	Water	89.1 g		Vit E	0.02 mg	0%
Carbohydrates	9.34 g	Vit A	Equiv 0 μg	0%	Vit K	0.4 μg	0%
• Sugar	4.24 g	Vit B1	0.046 mg	4%	Ca	23 mg	2%
• Fiber	1.70 g	Vit B2	0.027 mg	2%	Iron	0.21 mg	2%
Fat	0.10 g	Vit B3	0.116 mg	1%	Mg	0.13 mg	0%
• Saturate	0.042 g	Vit B6	0.12 mg	9%	P	29 mg	4%
• Monounsaturated	0.013 g	Folate	19.0 μg	5%	K	146 mg	3%
• Polyunsaturated	0.017 g	Vit C	7.40 mg	12%	Na	4.0 mg	0%
–	–	–	–	–	Zinc	0.17 mg	2%

Percentages are relative to US recommendations for adults
Reproduced from USDA Nutrient database
[a] g, gram; IU, international unit; kcal, kilocalorie; mg, milligram; μg, microgram

for cognitive and nerve functions as it helps in the manufacturing of serotonin and making of myelin, the layer of fats and proteins covering the nerves. Onions also are a source of about 5.9% daily folate intake, 3.9% vitamin C, and 8.5% calcium. The calcium from onions lends critical strength and density to bones and teeth. Additionally, each serving of onions provides a small amount of manganese as well.

6.5.33 Production Systems: Land Selection and Preparation

Small-scale farmers typically grow onions mainly as a monocrop for household consumption, with any surplus sold at markets for additional income. When selecting land for onion cultivation, factors such as soil type and temperature, slope, environmental humidity, and water availability are considered. Soils, being the most crucial factor, must provide the nutrients required for the crop's growth and support plant roots. Onions grow best in fertile sandy loams and silt loams that are well-drained, rich in organic matter, and have a pH range of 5.8–6.8. While onions can tolerate alkaline conditions, they are very sensitive to high soil acidity and salinity. Heavy clay soils and coarse sandy soils are often avoided because clay soils can lead to pungent onions due to their higher sulfur content, and sandy soils require more fertilizers and water supply.

Land Preparation

The first step in preparing the land involves clearing bushes, burning them, and uprooting stumps, followed by raking to leave a fine tilth. Removing overgrown grasses and burying weeds using hoes ensure proper weed control.

Planting Methods

There are three primary methods for growing onions:

1. Direct drilling
2. Sowing in the nursery followed by transplanting
3. Planting bulblets

Small-scale farmers prefer sowing seeds in the nursery due to their small size and the need for adequate and even moisture for germination. In Tanzania, for instance, seeds are often imported or sourced from the farmers' own grown onions. They are sown first in nursery beds and protected under shade. Transplanting in the field occurs 5–8 weeks after sowing, once seedlings have about five to six leaves and are at least 15 cm tall.

Transplant Production

On-farm transplant production for dry bulb onions involves seeding in high-density plantings. Common practice includes using at least 3–4 kg of seeds per linear 30 cm in 1 ha, with four to five rows planted 10–15 cm apart on well-prepared beds. Hand planting sets are suitable for smaller operations.

Planting Time

To capitalize on market prices, planting is done as early as possible. Early sowing can lead to bolting and thick necks, whereas late sowing reduces these defects but also reduces bulb size and yield. Planting should be done from mid to late February and not beyond this period. Transplanting can start from early March to mid-April depending on different crop varieties with each having distinct species.

Weed Control

Onion seedlings are notoriously poor competitors against weeds due to their erect growth habit, making weed control essential for optimum yield, especially during early development (Fig. 6.43). Smallholder farmers typically manage weeds using manual methods, such as hand weeding and hoes, although those who can afford it may also use chemical herbicides.

Fig. 6.43 A farmer's onion field well maintained and without weeds in Zambia

Manual Weeding

Weeding between plants by hand is common practice. Inter-row weeding is necessary and should be shallow but frequent to prevent the soil surface from becoming caked or cracked. This method not only helps suppress weeds but also ensures that the bulbs are not too deeply covered by soil, which could affect their development.

Chemical Weeding

For those who can afford it, chemical herbicides offer an alternative means of weed control. However, care must be taken to select appropriate herbicides that do not harm the crop.

Earthing up the seed crop, which also helps to control weeds, is also performed, but as ripening approaches, about 80% of the individual bulbs may well stand above the ground level.

Some smallholder farmers use chemicals for weed control, and both pre- and post-emergence herbicides are applied to the soil (Table 6.17).

Table 6.17 Commonly used herbicides for control of weeds in onion

Common name	Chemical name	Rate (kg/ha)	Time of application	Weeds controlled	Precautions
Decthal	Chlorthal	8–12	After sowing before weeds appear	Annual grasses and some broad-leaved weeds	Three months residual
Tortil	Inoxynil	2–2.751	After seedlings reach 3–4 leaf stage	Broad-leaved weeds	No irrigation for 24 h after application
Afalon	Linuron	0.5–1.751	5–20 days from transplanting or within 10 days for sets	Broad leaved	Do not use in light soils. No irrigation for 24 h after application
Perforan	Flourodifen	8–141	Immediately after planting	Broad-leaved and grasses	Irrigate immediately
Ronstar	Oxidiazon	4–51	Immediately after planting	Broad-leaved and grasses	Irrigate immediately
Tribunal	Methanbenz-Thiazuron	2–3	After weed emergence for transplanted crops	Broad-leaved	Do not use in sets. No irrigation for 24 h after application

Reproduced with permission from Zambia Seed Co. for Mathai, P.J., 1988. Vegetable Growing in Zambia. Berlings, Arlov, Sweden

Irrigation

Onions are shallow-rooted crops, requiring frequent watering to maintain a rooting depth of about 30 cm. Soil in this zone must be kept moist to ensure optimal growth. Small-scale onion production in Southern Africa often relies on irrigation, though rainfed production is also common, particularly for household consumption and market sales.

Irrigation Practices

For small-scale irrigation, farmers typically use cans or treadle pumps. Regular watering is especially crucial during the onset of bulb formation. Once the bulbs approach maturity—indicated by one-third of the plant tops falling over—watering is stopped. During the first 6 weeks after direct drilling and the first week after transplanting, watering is frequent (every 1–2 days) to establish the crop. In warmer areas, more watering compensates for higher plant evapotranspiration. Sandy soils require more frequent watering, while heavy soils need less. Mulching with grass and other plant debris helps conserve soil moisture and control weeds. Mulch is regularly removed and covered back to the soil to expose the bulbs to the sun.

Roguing

Roguing is essential in onion production to eliminate infested and off-type plants. At least four roguings are necessary:

1. **Active growing stage**: Remove diseased plants to promote general development and vigor.
2. **Bulb lifting**: Check bulb characteristics such as shape, size, color, and the presence of double bulbs.
3. **Additional roguings**: Further inspection and removal as needed to maintain crop health and quality.

The third roguing is done for the same bulb characteristics just before replanting after storage. Finally, all early or late flowering and diseased plants are removed.

6.5.34 Pepper Cultivation

Pepper (*Capsicum annum*) is one of the well-known and most widely utilized spice all over the world (Fig. 6.44). The crop includes numerous varieties, but the most of economic importance and which are widely grown are species from two families, namely, Solanaceae and Piperaceae, which are *Capsicum* spp. and *Piper* spp., respectively. The first family includes the bell pepper, red pepper, chilli pepper, pimento, and paprika, whereas in the second family, there are the black, green,

Fig. 6.44 Pepper following a bumper harvest by female farmers and a growing plant with pepper fruits. (Reproduced from World Vegetable Center, 2003. Vegetables for Life: Confronting the Crisis in Africa. World Vegetable Center Publication No. 03–564. Shanhua, p. 28)

white, and Ashanty pepper varieties. The popularly grown pepper in Southern Africa belongs to the *Capsicum* spp.

Importance of Pepper

Pepper, a crucial spice, is not only a staple in almost all foods but also a valuable commodity often used as collateral or to earn foreign currency. Its significance extends beyond culinary uses to various medical applications.

Culinary Uses

Pepper is an essential ingredient in preparing a wide range of foods and spice mixtures, including sausage, mincemeat, and other meat products. It is as ubiquitous as salt in many kitchens around the world.

Medical Uses

Pepper, particularly capsaicin found in chili peppers, is used in numerous medical formulations, especially in tropical medications. It acts as a circulatory stimulant and pain reliever and is noted for its potential in treating cancer cells. Additionally, pepper is prescribed for a variety of ailments including:

- Constipation
- Diarrhea
- Earache

- Gangrene
- Heart disease
- Hernia
- Hoarseness
- Indigestion
- Insect bites
- Insomnia
- Joint pain
- Liver problems
- Lung disease
- Oral abscesses
- Sunburn
- Tooth decay
- Toothaches

Nutritional Value

Pepper is rich in vitamins A, C, and E, with sweet peppers particularly high in these nutrients. For instance, a small amount of raw sweet pepper contains about 40 mg of vitamin C, which is two-thirds of the daily recommended intake of the vitamin (Table 6.18).

6.5.35 Types of Cultivars and Varieties

Fruits of most *Capsicum* spp. contain capsaicin (methyl vanillyl nonenamide), a lipophilic chemical that yields a strong burning sensation in the mouth and, if not properly digested, at the anus of unaccustomed eaters. The amount of capsaicin is highly variable and depends on the plant type. The only *Capsicum* pepper without capsaicin is *C. annuum*, a bell pepper which is known to have zero rating on the Scoville scale. Black pepper causes similar burning sensations, but this is caused by another chemical compound called piperine.

Piper nigrum

Only the following selection of *Piper* spp. varieties is said to be of importance as spice among the 700 or so varieties:

- Black pepper (*Piper nigrum*) from India, Malaysia, and Indonesia.
- Bengal pepper (*Piper longum* L.) from the mountains of the lower Himalayas.
- Java pepper (*Piper retrofractum* Vahl) from Malaysia and Indonesia.
- Ashanti pepper (*Piper guineense* Schum. et Thonn) from tropical Africa.

Table 6.18 Pepper (hot chili, red) nutritional values per 100 g for edible portion of raw material

Nutrient	Unit[a]	Value/100 g	Pepper 45 g	Sliced 75 g
Water	g	88.02	39.61	66.02
Energy	kcal	40	18	30
Protein	g	1.87	0.84	1.40
Total lipid (fat)	g	0.44	0.20	0.33
Carbohydrate, by difference	g	8.81	3.96	6.61
Fiber, total dietary	g	1.5	0.7	1.1
Sugars, total	g	5.30	2.38	3.98
Calcium, Ca	mg	14	6	10
Iron, Fe	mg	1.03	0.46	0.77
Magnesium, Mg	mg	23	10	17
Phosphorus, P	mg	43	19	32
Potassium, K	mg	322	145	242
Sodium, Na	mg	9	4	7
Zinc, Zn	mg	0.26	0.12	0.20
Vitamin C, total ascorbic acid	mg	143.7	64.7	107.8
Vitamin B-6	mg	0.506	0.228	0.380
Vitamin A, IU	IU	952	428	714
Vitamin E (alpha-tocopherol)	mg	0.69	0.31	0.52
Vitamin K (phylloquinone)	µg	14.0	6.3	10.5
Fatty acids, total saturated	g	0.042	0.019	0.032

Reproduced from USDA (2016). Food Composition Databases of Nutrient Database for Standard Reference. Available at: https://ndb.nal.usda.gov/ndb/search/list

[a] g, gram; IU, international unit; kcal, kilocalorie; mg, milligram; µg, microgram

- Kubeben pepper (*Piper cubeba* L.f.) which is found in Indonesia and Malaysia.

  There are many varieties of the black pepper, but the two main groups include:

– **Varieties with large leaves**
  They have large leaves and produce small fruits. Varieties in this group include the very productive varieties "Balamacotta" from India, "Kuching" from Malaysia which is very susceptible to stalk rot, and "Belantung" from Indonesia.

– ***Varieties with small leaves***
  These varieties produce smaller leaves but larger single fruits and are more resistant against diseases. The most prominent varieties among them include "Kalluvalli" from India (relatively resistant to drought), "Cheriakaedan" (Highly resistant to stalk rot), "Bangka" from Indonesia, and other many more.

Capsicum annuum

This is by far the largest domesticated species of pepper, both in the number of different varieties and in the area of cultivation worldwide. This species includes a wide range of pepper shapes, sizes, and characteristics. Peppers in this species most often ripen green to red but come in a variety of other colors as well. They include many of the most common and best-known pepper varieties, such as:

- Sweet pepper
  Sweet peppers refer to those without any heat and typically are used for flavoring, cooking, or stuffing. The sweet pepper includes:

1. Bell pepper
   This pepper is mostly blocky in shape with three or four lobes on the bottom of the fruit (Fig. 6.45). It matures into a variety of colors, including red, yellow, orange, lavender, purple, and chocolate. The bell peppers have a crisp, thick flesh and are suitable for eating fresh or for stuffing and baking.
2. Paprika
   Paprika is a sweet pepper that is grown mainly for its color that makes various foods more attractive (Fig. 6.46). The fruits are long, banana shaped, and when mature develop into a dark-red color. When dried and ground, this thin-walled pepper becomes the flavorful condiment paprika.
3. Sweet cherry
   This pepper is harvested when mature as green to deep-red pepper and is generally used in processing as pickled.

Fig. 6.45 Some of most common colors of the bell pepper found at marketplaces in Southern Africa

Fig. 6.46 A female farmer with a rainfed paprika crop in Tanzania. (Reproduced with permission from USAID 2005)

- Hot peppers

  Hot peppers refer to those that taste hot and can range from the relatively mild Anaheim to the blazingly hot pepper. Among the known hot peppers, the following are the most important:

1. Red chili

   These small cone-shaped peppers are often 1–4 mm long and have medium-thick flesh. They are usually used dried and ground in chili powder.
2. Green chili

   These are the long, green two-celled mildly pungent Anaheim-type peppers that are so flavorful in chili spicing. They turn red at maturity but are nearly always harvested when they are still green for roasting.
3. Red hot peppers

   These peppers vary in plant and fruit size and shape. These scorchers, such as chiltepin, chili pequin, Tabasco, and Thai, mature red and zest up foods.

6.5.36 Production Systems: Land Selection and Preparation

Peppers thrive in various soil types but prefer fertile loam soils rich in organic matter. Well-drained soils are essential as waterlogging can cause leaf drop. While green peppers can grow in a wide range of soils, they perform best in sandy-loam and loamy soils with good water-holding capacity, like alluvial or volcanic soils, and a pH range between 5.5 and 6.8. They also grow well in soils within young secondary forests and on forest edges.

Land Preparation

Before growing peppers, farmers prepare the land to ensure easy water and nutrient movement. The soil is finely tilled, firm, and free from weeds to allow easy root penetration. Manure or compost is added before transplanting seedlings to improve moisture retention, especially during hot, dry months.

Planting Methods and Timing

Pepper planting depends on the species or variety. While pepper tolerates various climates, it thrives in warm, dry conditions. Planting is done year-round except during the dry period from May through September, which impairs fruit setting. Optimal planting times are during the rainy season for seed production. Seeds are sown in shallow furrows in seedbeds, and about ½ kg of seeds suffices for 1 ha. Mulch may be used under dry conditions to retain moisture and removed once seedlings appear. In frost-prone areas, sowing begins after the frost danger has passed and temperatures are favorable.

Transplanting seedlings from nurseries to fields occurs 4–5 weeks after planting, preferably in the evening or on cool days to prevent wilting. Plants are watered immediately and regularly until harvest, with tree sticks provided as supports.

Spacing and Seed Rate

Optimal spacing is 50–70 cm between single rows and 40–60 cm between plants within rows. Double rows may also be planted 60–80 cm apart with 50–70 cm between plants to facilitate inspection and harvesting. The recommended seed rate in nurseries is 3 g/m^2, yielding about 1.2 kg seed/ha.

Propagation Methods

- **Seeds**: Fully ripened berries are soaked in water for 2–3 days, the flesh removed, and seeds dried in the shade. Seeds are then sown in moist, shaded beds.

- **Cuttings**: Preferred for retaining genotype and shortening juvenile periods. Cuttings are selected from the main shoot's terminal part of a healthy parent plant, planted at an angle of 45° with three to four nodes.

Field Maintenance

Peppers require well-structured, friable sandy loam soils, warm temperatures, and weed-free fields.

6.5.37 *Pumpkin Cultivation*

Pumpkins, belonging to the Cucurbitaceae family, are a vital crop for smallholder farmers in Southern Africa. They are cultivated for both their leaves, used as vegetables, and their fruits, grown in home gardens and farms (Fig. 6.47). Often intercropped with maize, pumpkins benefit from twining around maize stalks. They are highly nutritious vegetables rich in vitamins and minerals. These crops are mainly grown during the rainy season, though the area planted with pumpkins at the farm level in the region is relatively small, and there are three distinct cultivated pumpkin families, namely, *C. maxima, C. moschata*, and *C. mixta*, and the three most important genera whose species are cultivated for consumption of the fruits in the topics are *Cucurbita* (pumpkins), *Cucumis* (cucumber), and *Citrullus* (watermelon).

Fig. 6.47 Profusely developed pumpkin leaves commonly consumed by households in Copperbelt region of Zambia and the DR Congo

In Southern Africa, the most commonly cultivated pumpkin varieties are *Cucurbita pepo* and *Cucurbita maxima*. The fruits of *C. maxima* are generally known for their superior flavor and popularity compared to the lantern types of *C. pepo*.

Characteristics of Pumpkin Varieties

- ***Cucurbita maxima***: This variety has long, trailing, soft stems that are round in cross section. Its leaves are flaccid, dark green, and kidney-shaped with large basal nodes and entire margins. It is primarily grown for its flavorful fruits.
- ***Cucurbita pepo***: Known for its immature fruits used as squashes, *C. pepo* plants are trailing annuals with an extensive root system. They feature angular, prickly stems and large, roughly triangular leaves divided into five palmately arranged lobes. This variety is cultivated for both vegetables and fruits in Southern Africa.

Nutritional and Economic Importance

Pumpkins hold significant nutritional and economic value:

- **Edible leaves and fruits**: Pumpkin leaves are consumed as vegetables, while the fruits are enjoyed in desserts.
- **Seeds**: Edible roasted seeds are a common delicacy, and in some regions, pumpkin fruits are used as livestock fodder.
- **Oil extraction**: Seed extracts provide vegetable oil in certain cases.
- **Nutritional value**: Pumpkins are rich in iron, proteins, and vitamins, making them a valuable crop for improving nutrition (Table 6.19).

Health Benefits

Beyond their nutritional content, young pumpkin leaves and shoots offer numerous health benefits, including:

- Boosting the immune system.
- Improving eyesight.
- Reducing malaria, thyroid issues, high blood pressure, stomachache, edema, and constipation (Kimiywe et al. 2008).

6.5.38 Production Systems: Land Selection and Preparation

Pumpkins flourish in fertile, well-drained soils with adequate organic matter and a pH range of 5.0–7.0. Sandy loam and loam soils are ideal. Organic mulching is often used to preserve soil moisture. Farmers typically avoid fields that have recently cultivated other vine crops to prevent disease and pest buildup.

Table 6.19 Pumpkin nutritional values per 100 g for edible portion of raw material

Nutrient	Unit[a]	Leaves, raw		Fruits	
		Value/100 g	Cup 39 g	Value/100 g	Cup 116 g
Water	g	92.88	36.22	91.60	106.26
Energy	kcal	19	7	26	30
Protein	g	3.15	1.23	1.00	1.16
Total lipid (fat)	g	0.40	0.16	0.10	0.12
Carbohydrate, by difference	g	2.33	0.91	6.50	7.54
Fiber, total dietary	g	–	–	0.5	0.6
Calcium, Ca	mg	39	15	21	24
Iron, Fe	mg	2.22	0.87	0.80	0.93
Magnesium, Mg	mg	38	15	12	14
Phosphorus, P	mg	104	41	44	51
Potassium, K	mg	436	170	340	394
Sodium, Na	mg	11	4	1	1
Zinc, Zn	mg	0.20	0.08	0.32	0.37
Vitamin C, total ascorbic acid	mg	11.0	4.3	9.0	10.4
Vitamin B-6	mg	0.207	0.081	0.061	0.071
Vitamin A, IU	IU	1942	757	8513	9875
Fatty acids, total saturated	g	0.207	0.081	0.052	0.060

Reproduced from USDA (2016). Food Composition Databases of Nutrient Database for Standard Reference. Available at https://ndb.nal.usda.gov/ndb/search/list

[a] g, gram; IU, international unit; kcal, kilocalorie; mg, milligram; μg, microgram

Land Preparation

Land preparation involves clearing fields and using simple tools like hoes and rakes to ensure the soil is fine, firm, and free of weeds. It's crucial to select land carefully, avoiding areas previously planted with the same crop to minimize disease spread.

Planting Techniques

In smallholder farms, pumpkins are often planted on farmer-made mounds or at the base of termite mounds. Planting is done just before or at the onset of rains for optimal establishment. Later-planted pumpkins face higher disease and insect risks. Pumpkins are intercropped with maize and groundnuts on flat land, ridges, or mounds. They are grown either by direct sowing of seeds or by transplanting seedlings. Seeds are planted 3–5 cm deep in well-prepared holes spaced 2–3 m apart. About three to five seeds are sown per station and later thinned to two vigorous plants.

Farmers usually use their own saved seeds or obtain them from neighbors and relatives. In areas like Arusha, Tanzania, farmers can also purchase seeds from seed companies or agro-dealers.

Fertilization

To improve soil fertility, farmers apply organic fertilizers like mulch, compost, and farmyard manure. Manure is supplied at a rate of 30–50 tons/ha and ploughed into the soil before planting. Top-dressing with 35 kg/ha N is done when plants start producing runners.

Weeding

Weed-free soil is essential for pumpkin production. Farmers conduct hand weeding at least three times before the crop matures. Shallow cultivation around the root area is performed to avoid root damage.

6.5.39 Rape Cultivation

Rape (*Brassica napus* L.) is a member of the Brassicaceae (Cruciferae) family and the *Brassica* genus. Widely grown worldwide, rape is cultivated for both forage and as an oilseed crop (Fig. 6.48). In Southern Africa, it is an important vegetable crop, grown throughout many member countries, often thriving in the cooler seasons. The leaves of rape are highly nutritious and also serve as excellent green fodder for livestock. Generally, rape leaves are auricled and stem-clasping, characterized by their lyrate pinnatipartite form. There is considerable variation in leaf shape among rape cultivars, influenced by factors such as the depth of incision in the margin, number of lobes, and the shape and size of the terminal lobe. Some varieties feature spatula-type leaves with entire margins, while others have glabrous or hairy leaves, with hairiness varying between cultivars. The crop is versatile and can be cultivated during any season, though it is most commonly grown in cooler seasons in Southern Africa. This

Fig. 6.48 Plants of rape in vegetative stage (left) and grown plants in home garden in the Copperbelt Province of Zambia

adaptability makes it a reliable crop for both nutritional and agricultural purposes. The crop, commonly referred to as canola or rapeseed, is an important vegetable and oilseed crop in the southern Africa region. Its significance is multifaceted, encompassing nutritional, economic, agricultural, and environmental aspects. Rape greens are a rich source of vitamins A, C, and K, as well as essential minerals such as calcium, iron, and magnesium. The leaves and stems can be consumed in various culinary forms, including salads, stir-fries, and soups. This nutritional profile makes rape an essential part of a balanced diet, contributing to better health outcomes in communities.

6.5.40 Importance of Rape

Rape, widely grown as a forage and oilseed crop in regions like Europe, New Zealand, and Russia, has versatile applications. In countries like Tanzania, it was originally grown in India for its edible oil, but it gained importance as an industrial and lamp oil in Europe. However, its use as a vegetable crop is not globally widespread.

Culinary and Agricultural Uses

- **Forage and oilseed**: Primarily used as forage and for oil production.
- **Vegetable crop**: Young leaves are harvested and cooked as a pot-herb vegetable dish or made into sauces to accompany starchy staples like rice, ugali, and cassava. In rural areas of Zimbabwe, rape leaves are dried for long-term preservation to provide vegetables during the dry season.

Economic and Nutritional Importance

In Tanzania, rape is a valuable relish for urban dwellers and also grown for income generation. The plant is rich in vitamin C, iron, and several other minerals (Holland et al. 1991; Table 6.20). Rape kale contains high levels of glucosinolates, which form compounds with antioxidant and anticancer activities during preparation. Rape seeds are primarily known for their oil, which is used for cooking and as an ingredient in various processed foods. The production of canola oil presents opportunities for revenue generation through both local consumption and export. Beyond oil extraction, rape can be processed into animal feed, enhancing its economic viability. The versatility of this crop supports diverse agricultural and commercial activities.

Table 6.20 Nutritional values of rape per 100 g edible portion

Composition	Content[a]
Energy	37.0 kcal
Water	88.2 g
Protein	3.8 g
Carotene	1990.0 μg
Thiamin	0.07 mg
Riboflavin	0.06 mg
Vitamin C	55.0 mg
Fat	0.3 g
Carbohydrates	4.8 g
Fiber	1.8
Ca	250 mg
Mg	85 mg
P	81 mg
Fe	1.7 mg

Reproduced from Holland, B., Unwin, I.D., Buss, D.H., 1991. Vegetables, Herbs and Spices. Supplement to the Composition of Foods, fourth ed. Royal Society of Chemistry, Cambridge, UK, p. 163

[a] g, gram; IU, international unit; kcal, kilocalorie; mg, milligram; μg, microgram

6.5.41 Production Systems: Land Selection and Preparation

Rape properly blooms in well-drained, fertile soils with a high content of organic matter and good moisture-retaining ability. Sandy loam soils with a pH range of 5.5–7.7 are particularly suitable for rape cultivation.

Land Selection

When selecting land for rape cultivation, farmers consider several factors:

- **Soil fertility**: Fertile soils rich in organic matter are essential.
- **Drainage**: Well-drained soils prevent waterlogging, which can damage the crop.
- **Proximity to markets**: As rape is a perishable crop, choosing land close to markets minimizes losses and damage during transportation.
- **Climate and soil characteristics**: The land should have favorable climate and soil conditions to support healthy crop growth.

Land Preparation

Rape is generally grown in pure stands on well-prepared field beds (Fig. 6.49). Land preparation varies depending on the season, soil type, and water availability:

- **Tillage systems**: Raised or flat beds, furrows, and basin-making techniques are used. Proper ploughing is crucial to firm the growing bed, as rape seeds and seedlings are very small.
- **Clearing and ploughing**: The land is first cleared of weeds and debris. Ploughing ensures the soil is fine and firm, providing an optimal environment for seedling growth.

A well cleared land of residues before tillage, properly hoed and the rows adequately prepared at least a week before planting, ensures good plant establishment. A firm, moist warm, and well-aerated and well-structured seedbed is necessary for a rapid seed germination and seedling growth.

Planting Material

Rape is propagated by seeds, which are initially sown in a nursery. The seedlings are then raised and later transplanted to the main field bed (Fig. 6.50).

Nursery Bed Preparation

Nursery beds are prepared with a width of 90–120 cm and an appropriate length depending on land availability. The beds are watered and kept wet to allow weed seeds to germinate. Once germinated, the beds are well-drained and tilled to create

Fig. 6.49 Ridges prepared in a farmer's field for planting rape seedlings

Fig. 6.50 A farmer's field grown with rape in the Eastern Province of Zambia

a fine tilth. It is essential that the land used for the nursery has not been planted with any *Brassica* crop for at least 3 years to prevent disease buildup.

Sowing Seeds

After 2 weeks, weeding is done before final bed preparation. Seeds are sown thinly at a depth of 0.3–0.5 cm below the ground and spaced 5 cm between rows. Direct sowing at a rate of 3–5 kg per hectare is also practiced, with thinning after establishment. About 500 g of seed is sufficient for 1 ha. The beds are watered again and covered with grass to prevent evaporation and enhance infiltration. After seed distribution, the seedbed is harrowed using a hoe to ensure good seed-soil contact.

Germination and Transplanting

Rape seeds germinate within 5–8 days. Once germinated, the cover is removed to allow seedlings to access light for photosynthesis. If not during the rainy season, watering is done daily, depending on weather conditions. In the last 10 days before transplanting, watering is reduced to harden the seedlings. Transplanting typically occurs 3 weeks after germination. Before pulling out the seedlings, beds are watered thoroughly to prevent excessive root damage.

Methods of Seed Sowing in the Nursery

There are two methods of sowing rape seeds in the nursery, namely:

- Line sowing
  This is where seeds are sown in line with specified distances mainly 50 cm apart and at a depth of 0.3–0.5 cm (Fig. 6.51).
- Broadcasting
  Broadcasting of rape seeds is done by spreading them all over the field, and thereafter, a handful of soil is spread to cover the seeds. Rape seed sowing in line is commonly used although broadcasting is also used, but due to its high seed rate requirement, several farmers usually avoid it.
- Transplanting
  At about 1 1/2 months after sowing when the seedlings have attained the height of about 10–15 cm high and develop almost five to six leaves (depending on the variety), seedlings are transplanted to the main field. The nursery seedbed is irrigated before transplanting in order to allow easy pulling of seedlings off the soil without breaking the roots. Transplanting is carried out when the environmental temperature is not too hot to prevent seedlings from wilting.

Time of Planting

Rape can be sown year-round, but the best sowing period in regions with cool and warm areas is between December and April. Seedlings can also be sown from March to October. The timing of sowing is crucial as early sowing (before mid-November) can significantly reduce aphid attacks and damage. Late planting typically results in lower yields due to lower temperatures near the harvest period.

Spacing and Planting Depth

Proper seedling spacing is essential for good yields:

- **Spacing**: Plant seedlings 40–60 cm apart between rows and 15–40 cm apart within rows.
- **Planting depth**: Transplant seedlings at a depth of 5 cm in fine, firm beds.

After planting, irrigate the field at least twice daily for 5–7 days until the plants remain turgid. Continued irrigation is necessary for healthy plant growth and development. Thinning may be done to remove unhealthy plants and reduce competition for light, which can cause etiolation (excessive elongation).

Fig. 6.51 Newly transplanted rape seedlings in ridges in a farmer's field

Weeding

Weeding is vital to avoid competition for space, nutrients, water, and sunlight and to control pests during early crop establishment. Weeding in rape production is performed at least twice during the growing season using hand tools like hoes and garden forks (Fig. 6.52). In addition, hand weeding is often done when the land is wet. Once the crop canopy expands, weeding is no longer necessary as the crop can suppress weeds on its own.

Fertilization

Rape profusely develops in a variety of soils, with its nutrient needs determining the yield and quality of its leaves. Continuous leaf harvesting leads to high mineral uptake, necessitating the addition of organic matter, especially on sandy soils. For farmers using inorganic fertilizers, applying 700–800 kg/ha of D compound as basal dressing is recommended. One to 2 months after plant establishment, top-dressing with 40–45 kg/ha of nitrogen or 200–300 kg/ha of ammonium nitrate helps maintain vegetative growth and promotes leaf development. Many farmers, however, rely on green manure crops and compost to improve soil fertility.

Fig. 6.52 Weeding in this rape field is no longer necessary because the crop canopy controls the weeds

Irrigation

In regions with insufficient rainfall, regular irrigation is crucial for rape cultivation. Water should be supplied to the land before transplanting, and the nursery bed should be irrigated several hours before pulling out the transplants. After transplanting, the field should be irrigated immediately and daily until the seedlings regain their turgidity. Subsequent irrigation should be applied at regular intervals, depending on soil type, weather, and seedling growth stage.

6.5.42 Tomato Cultivation

Tomato (*Lycopersicon esculentum* Mill.) is an annual plant from the Solanaceae family and one of the most significant vegetables globally (Fig. 6.53). Its short growing duration under field conditions and high yield have contributed to its widespread cultivation, including in Southern Africa. The crop is preferred by subsistence farmers, resource-poor households, and home gardeners due to its high nutritional value and economic benefits. The fruits come in various shapes and colors, ranging from yellow to red, depending on the cultivar. The plant's profuse foliage provides natural protection for the fruits against sunburn, allowing for slower ripening and improved taste.

Fig. 6.53 Matured growing staked tomato plants in field (left) and harvested tomato fruits (right) in the Copperbelt Province of Zambia

Tomato Varieties and Characteristics

- **Fruit shapes**: Tomatoes can be round, oval, or pear-shaped.
- **Color range**: Depending on the cultivar, tomatoes can range from yellow to deep red.

Three different types of tomato plants are usually cultivated by small-scale farmers in the region, and these include the following:

- The tall or indeterminate type
- The semi-bush or semi-indeterminate type
- The bush or determinate type

Importance of Tomato

Tomato cultivation is widespread among smallholder farmers for both home consumption and as a cash crop. Tomatoes are incredibly versatile, consumed in numerous ways including raw fruits, as key ingredients in relishes, and in drinks like tomato juice. The fruits are also popular fresh in salads, cooked in various dishes, and widely available in marketplaces such as purées, ketchup, and canned whole or diced tomatoes. Additionally, tomatoes can be transformed into sweetened candies, dried fruits, and even wine. Their utility extends to soup flavoring, making them a staple for a healthy and balanced diet.

Nutritional Value

Tomatoes are rich in minerals, vitamins, essential amino acids, sugars, and dietary fibers (Table 6.21). They have high levels of vitamins A, B, and C, as well as iron and phosphorus. These nutrients make tomatoes a valuable crop with a long production period and compatibility with different cropping systems.

Table 6.21 Tomato nutritional values per 100 g for edible portion of raw material

Nutrient	Unit[a]	Value/100 g	Large whole, 182 g	Medium whole, 123 g	Slice, medium thick, 20 g	Small whole, 91 g	Slice, thick large, 27 g	Slice, thin/small, 15 g
Water	g	94.52	172.03	116.26	18.90	86.01	25.52	14.18
Energy	kcal	18	33	22	4	16	5	3
Protein	g	0.88	1.60	1.08	0.18	0.80	0.24	0.13
Total lipid (fat)	g	0.20	0.36	0.25	0.04	0.18	0.05	0.03
Carbohydrate, by difference	g	3.89	7.08	4.78	0.78	3.54	1.05	0.58
Fiber, total dietary	g	1.2	2.2	1.5	0.2	1.1	0.3	0.2
Sugars, total	g	2.63	4.79	3.23	0.53	2.39	0.71	0.39
Calcium, Ca	mg	10	18	12	2	9	3	2
Iron, Fe	mg	0.27	0.49	0.33	0.05	0.25	0.07	0.04
Magnesium, Mg	mg	11	20	14	2	10	3	2
Phosphorus, P	mg	24	44	30	5	22	6	4
Potassium, K	mg	237	431	292	47	216	64	36
Sodium, Na	mg	5	9	6	1	5	1	1
Zinc, Zn	mg	0.17	0.31	0.21	0.03	0.15	0.05	0.03
Vitamin C, total ascorbic acid	mg	13.7	24.9	16.9	2.7	12.5	3.7	2.1
Vitamin B-6	mg	0.080	0.146	0.098	0.016	0.073	0.022	0.012
Vitamin A, IU	IU	833	1516	1025	167	758	225	125
Vitamin E (alpha-tocopherol)	mg	0.54	0.98	0.66	0.11	0.49	0.15	0.08
Vitamin K (phylloquinone)	µg	7.9	14.4	9.7	1.6	7.2	2.1	1.2
Fatty acids, total saturated	g	0.028	0.051	0.034	0.006	0.025	0.008	0.004

Reproduced from USDA (2016). Food Composition Databases of Nutrient Database for Standard Reference. Available at: https://ndb.nal.usda.gov/ndb/search/list

[a] g, gram; IU, international unit; kcal, kilocalorie; mg, milligram; µg, microgram

Nutrient Content

According to Holland et al. (1991), fresh ripe tomatoes provide the following nutrients per 100 g: **water**, 93.1 g; **energy**, 17 kcal; **protein**, 0.7 g; **fat,** 0.3 g; **carbohydrate**, 3.1 g; **calcium (Ca)**, 7 mg; **magnesium (Mg)**, 7 mg; **phosphorus (P)**, 24 mg; **iron (Fe)**, 0.5 mg; **zinc (Zn)**, 0.1 mg; **carotene**, 0.64 mg; **thiamin**, 0.09 mg; **riboflavin**, 0.01 mg; **niacin**, 1.0 mg; **folate**, 17 µg; and **ascorbic acid**, 17 mg.

Most of the carbohydrate content is in the form of sugars. Although tomatoes may rank low in comparative nutritional value, their extensive consumption in diverse ways makes them a leading contributor to human nutrition. Tomato-based products are predominant sources of antioxidants, such as phytochemicals with health benefits including anticancer properties. Notably, carotenoids like lycopene found in tomatoes may help protect against carcinogens.

6.5.43 Production Systems

Tomatoes are cultivated by many small-scale farmers for home consumption and as a cash crop. Most farmers use seeds from local landrace cultivars, which often produce small, sour, or bitter-tasting fruits. These cultivars, however, offer advantages such as disease and pest resistance and higher yields under drought and low rainfall conditions. Some farmers purchase seeds of improved cultivars from certified shops, which, despite being costly, provide higher yields.

Land Selection and Preparation

Tomatoes thrive in well-drained sandy loam soils with good water-holding capacity and aeration. They prefer deep soils rich in organic matter, with a depth of 20–50 cm. The crop is sensitive to waterlogging and flooding and grows best in soils with a pH range of 5.5–7.0. Additional organic matter may be needed to support good plant growth.

Land preparation involves deep ploughing to improve soil structure, tilth, and water-holding capacity. Farmers use hoes to break clods and remove weeds and residues from previous crops. In water-limited areas, deep ploughing enhances water conservation and reduces soil-borne pests by exposing the soil to the hot sun. Smallholder farmers often create ridges or mounds to ensure good drainage for rainwater or irrigation.

Raising Seedlings

Tomato seedlings are initially sown in nursery beds before being transplanted to the field. Seedbeds are prepared wide enough to accommodate the required number of seedlings. About 200–300 g of seeds are needed per hectare. The nursery is prepared with a fine tilth by removing clods and stubble and adding well-decomposed farmyard manure. Seeds are sown thinly in lines, pressed gently to firm the soil, and covered with straw or grass. The seedbeds are watered twice daily to ensure sufficient moisture for germination. Once germinated, the covering is gradually removed to expose seedlings to sunlight.

Transplanting

Seedlings are transplanted to the field 3–6 weeks after sowing. The plants are hardened by reducing water supply and thoroughly watering the seedbeds 12–14 h before transplanting to minimize root damage. A clump of soil is left attached to the roots when removing seedlings from the soil. Transplanting is done late in the afternoon or on cloudy days to reduce transplanting shock.

Planting

After transplanting, the plants are immediately watered. Planting holes are deep enough to allow the lowest leaves to be close to the ground, and the soil is pressed firmly around the roots before watering around the plant base to settle the soil. Plant spacing is usually 30–50 cm in the line and 20–40 cm between rows, depending on the cultivar and local cropping systems. Mulch is applied to protect plants from heat during the first week after transplanting.

Fertilization

Tomato cultivation involves various practices to ensure optimal growth and yield. The amount and timing of fertilizer application vary with soil type and tomato cultivars. Small-scale farmers often use farmyard manure and compost, which consist of animal droppings (goats, sheep, pigs, and chickens) and kitchen remains. Compost is mainly made of decomposed field residues and grasses. Farmers with sufficient income may also use chemical fertilizers like phosphorus (P), nitrogen (N), and potassium (K):

- **N**: 40–180 kg/ha
- **P**: 60–200 kg/ha
- **K**: 30–180 kg/ha
- **Farmyard manure**: 20–25 tons/ha

Farmyard manure is applied before final ploughing, while P is mixed with the topsoil. N and K are applied in three doses: before transplanting (basal dressing), 2–3 weeks after transplanting, and about 2 weeks later (top-dressing).

Irrigation

Tomatoes do not tolerate drought, requiring regular watering, especially during early growth, flowering, and fruit setting stages to achieve acceptable yields. Adequate water supply also prevents blossom-end rot, a disorder linked to water stress and calcium deficiency. Smallholder farmers typically use watering cans with small droplet openings.

Weed Control

Weed management involves hand weeding at least three times during the tomato growth cycle. Deep ploughing before transplanting exposes weeds to the sun, while soil mulching used to cover transplants also helps suppress weeds.

Pruning

Pruning helps balance plant and fruit growth. Removing yellowish or diseased leaves and lower side-shoots improves light penetration and aeration, enhancing fruit size and quality. Pruning is done based on plant type and size to achieve uniform fruits. Care is taken to maintain sufficient leaves to prevent sun scalding and cracking of fruits.

Staking

Staking improves fruit quality by keeping plants and fruits off the ground. It involves supporting plants with wooden stakes and twine to train them to grow vertically. In the region, staking often uses bush poles, twigs, or woven split bamboo. Staking prevents rotting fungi by keeping fruits and foliage off the ground and stops the growth of fruit-bearing branches by pinching the growing tips.

6.6 Kitchen Gardening

In Zambia and the DR Congo, many households turn to kitchen gardening to grow additional vegetables like amaranth, cabbage, tomato, onion, pepper, rape, hibiscus, sorrels, sweetpotato, cassava, cocoyams, pumpkins, squash, and basil, among others (Peša 2020a). This practice is especially prevalent due to increased poverty levels and serves as a reliable source of fresh produce.

6.6.1 The Role of Kitchen Gardens

Dilrukshi Hashini Galhena et al. (2013) highlight the importance of increasing food production as the global population is projected to exceed 9 billion by 2050. To combat hunger and food scarcity, developing countries are focusing on local food production to mitigate the effects of global food shocks and price volatilities. Home gardens are increasingly seen as a strategy to enhance food security and nutrition.

6.6.2 Importance of Home Gardens

Home gardens play a crucial role in local food systems, particularly in developing countries. They are integrated into the agricultural landscape and have stood the test of time. Mugisa et al. (2016) noted that home gardening, also known as backyard or kitchen gardening, is a vital form of urban and peri-urban agriculture (UAP). It involves growing crops in backyards, balconies, vacant spaces, or small gardens, particularly in areas where land is limited and expensive.

6.7 Home Gardens in Zambia

In Zambia, Greenspace Zambia promotes sustainable gardening practices to inspire citizens to create greener cities and protect the environment (https://www.greenspacezambia.com/about-greenspace-zambia). Sustainable gardening focuses on minimizing environmental impact, conserving resources, and promoting biodiversity. Key principles include the following:

1. **Soil health and fertility**: Enhance soil quality with organic amendments like compost and mulch. Avoid synthetic fertilizers and pesticides.
2. **Water conservation**: Use water-wise techniques such as drip irrigation and rainwater harvesting. Choose drought-tolerant plants.
3. **Indigenous and adapted plants**: Select plants native or well-adapted to the local climate. Indigenous plants require less water and maintenance.
4. **Biodiversity promotion**: Plant a variety of species, including those that attract pollinators and beneficial insects. Avoid invasive species.
5. **Integrated pest management (IPM)**: Control pests using environmentally friendly methods like biological controls and crop rotation.
6. **Composting and recycling**: Compost organic waste and recycle materials to reduce waste.
7. **Energy efficiency**: Use energy-efficient tools and incorporate solar-powered lighting.
8. **Wildlife habitat conservation**: Provide habitats for wildlife with features like birdhouses and water sources. Avoid harmful chemicals.
9. **Responsible garden practices**: Manage gardens responsibly with proper pruning and waste disposal.
10. **Education and community involvement**: Share knowledge and promote sustainable practices within the community.

6.8 Home Gardens in Lubumbashi

Mushagalusa Balasha et al. (2019) conducted a survey in Lubumbashi, the DR Congo, and found that 72% of households had home gardens averaging $11.5 \pm 6.4\ m^2$. Crops included amaranth, okra, sorrel, sweetpotato, corn, and medicinal plants. Women played a significant role in deciding which vegetables to grow. Home gardening contributes to food security by providing direct access to vegetables, strengthens community relationships, conserves agricultural traditions, and introduces children to manual work. It also generates supplemental income.

Chapter 7
Atmospheric Air Quality in the Region

Keywords Air quality · Pollution · Emissions · PM2.5 and PM10 · Toxic gases · Health threat

7.1 Context of Air Quality Importance

Examining the impacts of mining pollution on air quality in Zambia and the DR Congo is not only relevant but essential. The air quality in the Central African Copperbelt region affected by mining activities serves as a critical indicator of the broader environmental and public health consequences of such industrial operations. In both Zambia and the DR Congo, mining plays a pivotal role in their economies, particularly through the extraction of numerous resources like copper, cobalt, zinc, lithium, and gold. However, these activities often lead to significant air pollution due to dust, emissions from processing plants, and the release of harmful pollutants such as sulfur dioxide and particulate matter. This chapter underscores the contrast between these adverse impacts and the importance of maintaining or restoring clean, breathable air for local communities.

Moreover, highlighting good air quality emphasizes the interconnectedness of environmental health and human well-being. Air pollution from mining operations in Zambia and the DR Congo has been linked to respiratory issues, cardiovascular diseases, and other health problems among nearby populations. A chapter on good ambient air advocates for stringent environmental regulations, better emission controls, and sustainable mining practices that prioritize air quality. This not only offers a framework for reducing harm but also inspires actionable steps for policymakers, companies, and communities.

The discussion on healthy air also situates the mining impacts within the global context of climate change and environmental sustainability. Air quality is closely tied to carbon emissions and the global fight against climate change. Zambia and the DR Congo, as nations rich in natural resources but vulnerable to environmental degradation, stand at a crossroads where their policies and practices can influence

A. Muimba-Kankolongo et al., *Mining Impacts on the Environment in the Central African Copperbelt of Zambia and the Democratic Republic of Congo*, https://doi.org/10.1007/978-3-031-96786-3_7

global sustainability efforts. This section provides a platform to discuss these broader implications, making the book a comprehensive resource for readers interested in mining's local and global environmental impacts. By emphasizing the significance of clean air, the book will shine a spotlight on the often-overlooked aspects of mining's footprint, fostering awareness and advocacy for more balanced and responsible development.

7.2 Background

Air pollution is a critical environmental issue that poses significant health risks and contributes to premature deaths globally (Earth Organization, 2024). The main sources of air pollution include industrial emissions from factories and power plants which emit pollutants like sulfur dioxide, nitrogen oxides, and particulate matter into the atmosphere, the agricultural activities which generate large amounts of methane and nitrous oxide emissions, significantly contributing to air pollution, and the incineration of solid waste that releases various pollutants into the air. Moreover, it also emphasized that high levels of air pollution are associated with breathing sicknesses, heart complications, and premature deaths. Guidelines for ambient air quality (Table 7.1) have been provided by the WHO (2021).

The Earth Organization (2024) has identified outdoor air pollution as one of the most pressing environmental issues globally (Fig. 7.1). According to data from the World Health Organization (WHO), air pollution is responsible for the deaths of an estimated 4.2–7 million people each year worldwide. Moreover, it disturbs nine out of ten people, who respire air full of hazardous pollutants.

In 2017, outdoor air pollution resulted in the deaths of 258,000 people in Africa, a significant increase from 164,000 deaths in 1990, as reported by UNICEF (2017). The primary sources of air pollution in Africa include the following: emissions from factories and power plants release pollutants such as sulfur dioxide and nitrogen

Table 7.1 World Health Organization Air Quality Guidelines

Pollutant	Averaging time	μg/m^3	Pollutant	Averaging time	mg/m^3
$PM_{2.5}$	1 year	5	CO	24 h	4
	24 h (99th percentile)	15	O_3	8 h, daily maximum	100
PM_{10}	1 year	15		Peak season	60
	24 h (99th percentile)	45	SO_2	24 h	40
NO_2	1 year	10			
	24 h (99th percentile)	25			

Source: references License: CCBY-NC-SA 3.0IGO

$PM_{2.5}$ particulate matter with aerodynamic diameter of less than 2.5 μg, *PM_{10}* particulate matter with aerodynamic diameter of less than 10 μg, *NO_2* nitrogen dioxide, *O_3* ozone, *SO_2* sulfur dioxide, *CO* carbon monoxide

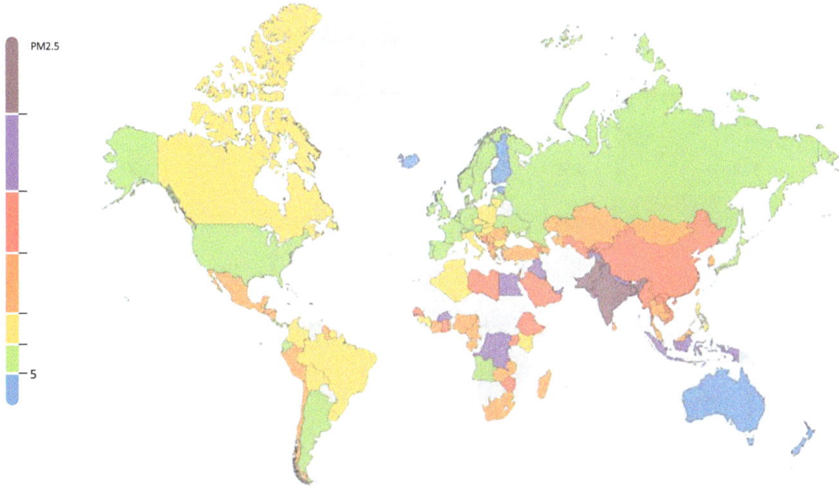

Fig. 7.1 2023 global map color coded by annual average PM2.5 concentration. (Source: Reproduced with permission from the IQAir AG)

Table 7.2 Names and colors for the six AQI categories

For this AQI…	Use this descriptor…	And this color
0 to 50	Good	Green
51 to 100	Moderate	Yellow
101 to 150	Unhealthy for sensitive groups	Orange
151 to 200	Unhealthy	Red
201 to 300	Very unhealthy index	Purple
301 to 500	Hazardous	Maroon

Note: Values above 500 are considered "beyond the AQI." Follow recommendations for the hazardous category

oxides into the atmosphere; exhaust emissions from cars, trucks, and buses contribute significantly to air pollution, releasing particulate matter and other harmful pollutants; the practice of burning biomass, such as agricultural waste and firewood, generates significant amounts of smoke and particulate matter; and natural events like dust storms can drastically reduce air quality, adding to the overall burden of pollutants in the air.

The U.S. Environmental Protection Agency (EPA) stressed the importance of reporting the air quality index (AQI) to keep the public informed about air quality (EPA, 2018). According to its guidelines, areas with a population of more than 350,000 are required to report the AQI daily, so as to help residents understand the air quality in their region and take necessary precautions to protect their health (Tables 7.2 and 7.3).

Table 7.3 Air quality index (AQI) color formulas

Color	R	G	B	C	M	Y	K
Green	0	228	0	40	0	100	0
Yellow	255	255	0	0	0	100	0
Orange	255	126	0	0	52	100	0
Red	255	0	0	0	100	100	0
Purple	143	63	151	51	89	0	0
Maroon	126	0	35	30	100	100	30

Reproduced with permission from U.S. Environmental Protection Agency (EPA)
Notes: RGB is traditionally used for screen colors, while CMYK is traditionally used for printing. The color models are based on a 0–255 scale (e.g., 50% is 126)

7.2.1 What Goes in the AQI Report?

It takes a full 24 h to obtain an AQI value (that's 24 hourly values for PM or the max 1-h or 8-h value in a 24-h period for other pollutants), so you are in effect required to report yesterday's AQI, including:

- The reporting area
- The reporting period
- The critical pollutant (the pollutant with the highest AQI value)
- The AQI value
- The category descriptor and color (if your report uses color), as shown in Table …
- The sensitive groups for all pollutants with an AQI over 100, as shown in Table …

7.2.2 How Are the AQI Colors Defined?

The colors are defined by the formulas RGB (red, green, and blue) and CMYK (cyan, magenta, yellow, and black) as shown in Table 7.3.

7.3 Air Quality in Africa

Africa is grappling with severe air pollution issues, resulting in some of the most significant health impacts globally. In 2019, air pollution was identified as the second leading risk factor for death across the continent, trailing only malnutrition.

Unsafe water, sanitation, and hygiene ranked as the fourth largest risk factor for deaths.

7.3.1 Air Pollution Sources and Health Impacts

Africa is home to five of the world's ten most polluted countries in terms of ambient fine particulate matter. In sub-Saharan Africa, approximately 75% of the population relies on solid fuels like coal, wood, charcoal, and dung for cooking. This practice exposes over 800 million people to high levels of harmful pollutants in their homes daily. The combined effects of household use of solid fuels and fossil fuel sources exacerbate air pollution, impacting the population's overall health (WHO, 2021). This results in uneven health impacts across countries, with chronic noncommunicable conditions such as heart disease and diabetes rising alongside communicable diseases like diarrheal disease, lower respiratory tract infections, and malaria.

7.3.2 Life Expectancy and Mortality

The negative impact of air pollution on life expectancy is most pronounced in less-developed areas, where many people face a dual burden of high ambient particulate matter and household air pollution. In countries like Niger and Somalia, air pollution has been estimated to reduce life expectancy by over 3 years. In 2019, air pollution was responsible for 1.1 million deaths in Africa, with more than 63% linked to household air pollution (UNICEF 2019).

7.3.3 Economic Costs

The economic toll of air pollution is substantial, with health-related damages costing an average of 6.5% of GDP across Africa. In countries like Egypt, Ghana, the Democratic Republic of the Congo, Kenya, and South Africa, the annual cost of health damages from PM2.5 exposure exceeds 5.4 billion USD.

7.3.4 Future Trajectory

Africa's rapid urbanization and industrialization will influence air pollution trends and their impacts. By 2100, the continent is expected to have 13 megacities (cities with over ten million residents). While economic development and growth can improve the quality of life, unchecked pollution from household, industrial, and transportation sources could worsen air quality. Addressing these issues requires proper environmental regulations and equitable access to clean energy.

7.3.5 *Air Quality Guidelines*

Most people in Africa live in areas where PM2.5 levels exceed the WHO air quality guideline of 5 μg/m^3. In 2019, five African countries—Niger, Nigeria, Egypt, Mauritania, and Cameroon—were among the top ten globally for highest PM2.5 exposures.

7.3.6 *Africa Regional Air Pollution*

Western Africa suffers from the highest levels of PM2.5 pollution, averaging 64.1 μg/m^3, whereas Southern Africa has the lowest levels at 26.5 μg/m^3. In Southern Africa, countries like Namibia, Zimbabwe, and Tanzania generally meet the WHO's least stringent interim target of 35 μg/m^3 for PM2.5. In contrast, over 90% of Western Africa's population lives in areas that do not meet this target.

Urban areas in Africa are particularly affected by fine particle pollution. All of the ten most populous cities in Africa exceed the WHO air quality guideline of 5 μg/m^3 for PM2.5, and four of these cities even exceed the least stringent interim target of 35 μg/m^3.

7.4 Air Pollution in Zambia

7.4.1 *Overview and Sources*

A report by Chipungu and Kunda (1996) highlighted that the air quality in Zambia is not well documented, necessitating studies to identify and quantify pollutants in the ambient air. The National Council for Scientific Research has conducted limited studies on sulfur dioxide (SO2) in Kitwe, precipitation acidity, and levels of lead (Pb) and cadmium (Cd) in the soil in Kabwe. Major sources of air pollution in Zambia include industries, motor vehicles, mining, quarrying activities, construction, and rampant burning.

Notable industries contributing to air pollution in Zambia include Zambia Consolidated Copper Mines (ZCCM), Nitrogen Chemicals of Zambia (NCZ), Ndola Lime Company Limited, Chilanga Cement Works, tanneries, and food processing industries. Key pollutants from these sources include sulfur dioxide (SO2), nitrogen oxides (NOx), particulate matter (PM), carbon monoxide (CO), carbon dioxide (CO2), dust, vibrations, noise, offensive odors, and black smoke.

7.4.2 Health Impacts

Mwanga et al. (2019) noted that mine air pollution in Zambia, primarily due to SO2 and fine and ultrafine particulate matter (PM10, PM2.5, and PM0.1) emissions, has been a significant issue since the 1920s in the Copperbelt region. Ncube et al. (2012) expressed concerns over the health impacts on people living near mines and smelting industries. These pollutants, particularly SO2 released during copper smelting, are associated with respiratory problems. Children, the elderly, and those with pre-existing respiratory conditions are especially vulnerable.

7.4.3 Environmental and Health Effects

Air pollution in Zambia contributes to climate change and the loss of biodiversity through the extinction of various species. Ningabo et al. (2022) reported other negative impacts, including soil contamination, surface water pollution, siltation in the Kafue River, and metal deposition in vegetables, fruit, and fish.

7.4.4 Historical Context and Industrial Practices

A review by Iva Peša (2021) examined changing environmental values among various stakeholders in the Copperbelt, including mine engineers, government officials, mineworkers, doctors, and farmers. She referenced a 1936 study on fumes from Lubumbashi's smelter, where engineers considered recovering sulfur dioxide to produce sulfuric acid for copper refining. Despite emitting 2500 tons of sulfur dioxide monthly in Lubumbashi, colonial engineers often dismissed complaints about the fumes. Instead of implementing chemical methods or building higher chimneys to minimize the nuisance, they proposed transferring operations to Kipushi, a smaller town with fewer critics (UMHK, 1936). This indicates that mine officials were aware of the environmental impacts of copper extraction and processing.

The focus on production often overshadowed environmental concerns for communities around the Northern Rhodesian Copperbelt. The 1934 Smoke Damage (Prohibition) Act demonstrated this by prohibiting claims for damage from smelter operations within a 5-mile radius of a copper smelter, regardless of pollution levels. Representatives from the Rhokana Corporation in Luanshya claimed that the public accepted land near smelters, knowing the environmental risks, and saw benefits in being close to industrial centers. In 1937, Rhokana argued that landowners in Luanshya township had no moral right to protection against smelter fumes, fearing legal claims would disrupt mining operations to the detriment of the entire territory.

This legislation indicated that maximizing production and profit was paramount, with government agents protecting mines against environmental claims. Complaints about pollution were often dismissed as attempts to exploit the legal system. This attitude persisted into the 1970s and 1980s, where Zambia Consolidated Copper Mines (ZCCM) representatives frequently downplayed the severity of air pollution, proposed technological fixes, or analyzed the economic costs and benefits of smoke abatement. Despite significant sulfur dioxide emissions from smelters in Mufulira and Luanshya, engineers claimed these did not cause ecological harm and that pollution levels were not alarming. They even praised natural conditions for dispersing fumes.

Occasional air pollution monitoring began in the 1970s, but enforceable air quality standards were lacking. In 1976, pollution levels in Luanshya exceeded adopted maximums five times, but officials concluded that overall averages were not excessive. To address sulfur dioxide emissions, engineers suggested building tall smokestacks and converting sulfur dioxide into sulfuric acid, which could be used to recover copper from low-grade ores, making the process economically attractive. This economic logic framed pollution control as a means to increase revenue, not just reduce emissions.

Throughout the 1990s, air pollution remained a concern, with discussions often reflecting the economic and technological perspectives of mining companies. By 1997, the Mufulira smelter lacked a sulfur dioxide abatement facility, emitting 5000 tons of sulfur dioxide monthly. Media reports highlighted the dangers to human and animal health, noting respiratory problems and hindered vegetation growth near smelters.

Journalists, despite recognizing the environmental impacts of air pollution, often prioritized economic profit over strict regulations, arguing that the nation's economic stability needed to continue. They expressed concern that enforcing stringent air pollution standards could lead to the shutdown of industries like mining in Zambia. Facing economic challenges, international pressures, and impending privatization, Zambia Consolidated Copper Mines (ZCCM) pledged to change operational practices and reduce emissions after 1990. However, stricter environmental standards and new abatement technologies were adopted inconsistently. Although laws to curb air pollution existed, controlling the problem proved difficult. Newspapers criticized government agencies for their "lukewarm approach" to the "potentially explosive issue" of unchecked toxic emissions. One opinion piece highlighted a general lethargy toward Zambia's environmental problems, which persisted over time.

7.4.5 Community Impact and Response

Residents of Mufulira and Likasi have long been aware of air pollution problems, commonly discussing sulfur dioxide emissions (known locally as senta in Bemba and kachoma in Swahili). In neighborhoods like Mufulira's Kankoyo and Likasi's Usines Chimiques de Shituru, located downwind from smelters or chemical plants, these emissions have caused significant health hazards and frequent complaints.

Sulfurous smoke causes choking sensations, high rates of pulmonary diseases and skin rashes, wilted vegetation, and corroded roofing materials. Rather than outright protest, the response has often been what Lora-Wainwright (2017) calls "resigned activism," where pollution is seen as a normal, unavoidable part of the environment, and people take routine measures to counter or avoid it. For example, women move agricultural fields away from smelters, mine management transfers workers showing bronchitis symptoms to "dust-free" environments, and urban residents place trust in new smoke abatement technologies.

7.4.6 Cultural Perception

Mining communities have tied air pollution to concepts of development, where economic welfare—indicated by operational smelters and smoking chimneys—was expected to ensure communal welfare. Smoke was accepted as an inevitable side effect of employment and prosperity. Congolese mineworkers referred to sulfur dioxide emissions as "the cough that comes from the mine" (kikozi ya ku mungoti in Swahili), indicating their awareness of pulmonary diseases originating from mine work.

7.4.7 Changing Attitudes

This resigned attitude has been shifting due to recent high-profile pollution cases, such as the death of Mufulira's District Commissioner from inhaling sulfur dioxide fumes, which sparked protests (Lusaka Times, October 18, 2013). Many respondents see present air pollution levels as a new phenomenon, worse than in the 1960s, despite less effective historical technologies. A Zambian newspaper article lamented the lack of solutions, stating that the corporate world prioritizes profit over human health (Lusaka Times, 2 May 2012). The privatization of Copperbelt mining companies in the 1990s, leading to job losses and reduced welfare services, may have influenced this more critical stance, allowing people to express long-standing environmental grievances more freely.

7.4.8 Past Context and Technological Optimism

Throughout the twentieth century, the mining industry often downplayed the severity of air pollution, proposed technological fixes, or balanced economic costs and benefits of smoke abatement. The 1934 Smoke Damage (Prohibition) Act exemplifies how business interests and technological optimism overshadowed environmental concerns. Discussions about air pollution, abatement technology, and the value of clean air have continued, with various actors contesting environmental values. Disputes over water quality further illustrate the region's environmental consciousness.

7.5 Air Quality in the DR Congo

Air pollution in the DR Congo poses a serious environmental threat, particularly affecting communities near mining operations. Mining and metallurgical industries release particulates, sulfur dioxide, and other harmful substances into the air, resulting in poor air quality. This contributes to respiratory diseases such as pneumonia, bronchitis, and asthma, especially among children and the elderly. Air pollution is also linked to an increased risk of heart disease, stroke, and lung cancer.

7.5.1 Impact in Haut-Katanga Province

In Haut-Katanga Province, air pollution levels around mining hubs such as Lubumbashi, Likasi, and Kipushi are five to ten times higher than national air quality standards. This pollution primarily results from mining and processing cobalt ore, which releases high levels of particulate matter (PM), heavy metals, and other pollutants. Sources of PM pollution include dust from mining activities, ore transportation, waste piles, and smelting and refining processes that generate sulfur dioxide, nitrogen oxides, and other toxic emissions.

7.5.2 Environmental and Health Impacts of Cobalt

Cobalt, once considered a miracle metal, is now becoming a significant environmental and health hazard due to toxic dumping. According to Earth Organization (2022), high cobalt concentrations have devastated landscapes, polluted water, and contaminated crops. High levels of cobalt have been linked to the death of crops and worms, essential for soil fertility. Fish from Tshangalele Lake near mining towns have been found to contain high levels of cobalt, posing health risks to humans who consume the fish or drink the lake's water. Cobalt is classified as a possible carcinogen and a radioactive element, posing immense health hazards.

The air around cobalt mines is often hazy with dust and grit, making it toxic to breathe. Studies indicate a higher risk of birth defects, such as limb abnormalities and spina bifida, among children of parents working in cobalt mines due to high levels of toxic pollution.

7.5.3 Social and Economic Impacts

Cobalt mining in the DRC, often referred to as "blood cobalt," has severe social implications. Workers, including children, often labor in subhuman conditions using rudimentary tools to extract cobalt. This situation has led to modern-day slavery, human trafficking, and child labor. For more insights, Siddharth Kara's book *Congo Red: How the Blood of the Congo Powers Our Lives* provides a detailed examination.

7.5.4 Broader Environmental Concerns

Banza Lubaba et al. (2009) highlighted that air pollution in Haut-Katanga Province affects a growing number of urban dwellers in developing countries. Industrial pollution is a particular concern, as many environmental risks are disproportionately experienced in these regions. Research on this issue often focuses on developed countries, where legislative and technical measures have reduced risks.

High levels of toxic metals were found in urine samples, especially in children, in Katanga Province. For some metals, particularly cobalt (Co), levels often exceeded the maximum tolerable limits for occupational exposure, even among those not exposed through their work. Demographic factors and residential area were found to be the most significant predictors of metal concentrations in urine. Smoking did not significantly impact metal levels, likely due to low tobacco consumption among the population studied. Occupational exposure was also not a major factor, possibly because there were few current workers in the sample.

- ***Environmental Exposure***
  The elevated levels of cobalt and other toxic metals in urine indicate significant environmental exposure. The exact routes of exposure—whether through air, diet (including staple foods, specific vegetables, meat, or fish), soil, or suspended dust—need further investigation. Higher metal concentrations in children's urine suggest that ingestion of contaminated dust is a key route of exposure. Ongoing epidemiological studies aim to explore the health effects of these exposures in different segments of Katanga's population.
- ***Impact on Likasi Residents***
  According to Friends of the Earth Netherlands (2015), the Groupe Bazano mining company has operated a smelting plant in Likasi's Kamatanda residential quarter since 2007. Residents complain about dust from the company's traffic and smoke from the plant furnaces. They sought a resolution in 2009 but have not received any solution. During the rainy season, roads become muddy from traffic, while the dry season covers houses in dust. Groupe Bazano claims the factory existed before the residential area, but the Congolese NGO Premicongo states that Kamatanda was established long before the plant. Premicongo has observed

an increase in respiratory diseases and a decline in fruit tree productivity but lacks the means to confirm if these are directly related to the plant's emissions.

- ***Other Affected Areas***
  Similar severe air pollution cases exist elsewhere in Katanga. In Lubumbashi, the Chemaf mining company's operations have caused significant air pollution, impacting local communities. Residents report that smoke from the operations dries out plants, causes leaves to fall off trees, and makes people, especially children and the elderly, sick.
- ***Dust and Air Quality***
  Mwitwa et al. (2012) found that mining operations generate dust from various sources, including opencast activities, waste rock stockpiles, tailings disposal facilities, crusher and processing plants, and ore transportation. This dust, often containing heavy metals and pollutants, degrades air quality. Emissions from these activities, such as sulfur dioxide, carbon dioxide, and hydrocarbons, have significantly declined air quality since the mine's establishment in 1926. These pollutants are attributed to copper processing fumes and dust from mining operations.

The high levels of toxic metals in urine samples from Katanga's general population indicate significant environmental exposure, primarily through dust and pollutants from mining activities. The resulting air pollution poses serious health risks, including respiratory diseases, to the local communities. The case of Groupe Bazano's smelting plant in Likasi exemplifies the ongoing struggle between industrial operations and community health, highlighting the need for stringent environmental regulations and effective pollution control measures.

7.6 Threat of Poor-Quality Air to Human Life

The Health Effects Institute (2022) highlighted that pollution from fine particulate matter, household burning of solid fuels, and ozone is responsible for millions of premature deaths each year, accounting for approximately one in nine deaths globally. Decades of scientific research have shown that long-term exposure to air pollution significantly increases the risk of chronic diseases such as ischemic heart disease, lung cancer, chronic obstructive pulmonary disease (COPD), stroke, and type 2 diabetes.

7.6.1 Health Impacts of Air Pollution

- **Prenatal exposure**: Exposure to air pollution in the womb is associated with being born too small or too early.

- **Early childhood**: Breathing polluted air during early years can contribute to deadly lower respiratory infections in young children.
- **Cognitive development**: Emerging evidence suggests that air pollution may impair cognitive development in children and contribute to brain disorders, including dementia.

These findings underscore air pollution as a crucial and potentially modifiable risk factor for major health issues, disabilities, and deaths worldwide.

7.6.2 Air Pollution in the DR Congo

The Mineworkers' Union in the DR Congo reported that approximately 40% of miners are regularly diagnosed with chest pains monthly. Air pollution has increased significantly with the rise of mines in and around Lubumbashi, leading to more cases of respiratory problems among residents, particularly during the dry season due to the continuous release of smoke by the Gécamines smelter.

7.6.3 Sources of Pollution

Mining operations in the DR Congo generate dust from various activities, including opencast mining, waste rock stockpiles, tailings disposal facilities, crushers, and mineral processing plants. Additionally, road transportation of the ore contributes to dust pollution. Other sources include:

- **Gaseous emissions**: Sulfur dioxide, carbon dioxide, vehicle exhaust fumes, and hydrocarbons.
- **Smelting operations**: Without adequate safeguards, smelting releases heavy metals, sulfur dioxide, and other pollutants into the air.

7.6.4 Health Hazards

In Lubumbashi, there is a notable increase in respiratory diseases, especially among children. The continuous release of smoke and dust from mining operations, particularly during the dry season, exacerbates coughing and lung problems. Children are especially vulnerable due to their habit of playing close to the ground (Fig. 7.2) and putting dirty objects in their mouths, increasing their exposure to contaminated dust (Fig. 7.3). Several residents have symptoms of pulmonary sicknesses including asthma, lung cancer, pneumonia, bronchitis, acute chest syndrome, etc. (Table 7.4, Fig. 7.4). Smoking from the mines releases various pollutants into the air, such as

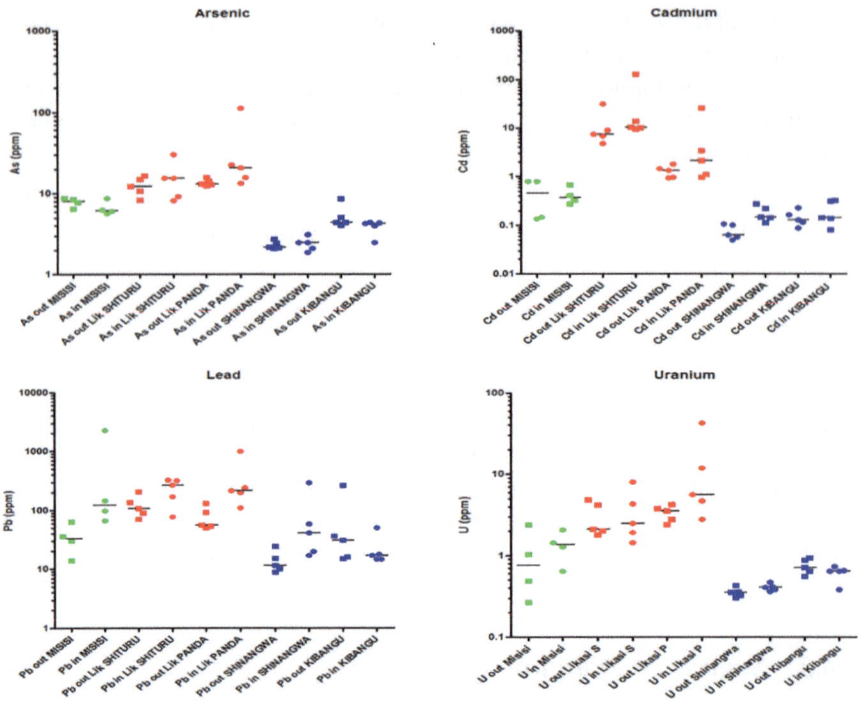

Fig. 7.2 Collected and analyzed dust around mining and metallurgic industries operations is among the major sources of commodities including human contamination with toxic pollutants. (Source: Courtesy Professor C. Banza Lubaba Nkulu, Lubumbashi University DR Congo)

particulate matter, heavy metals, sulfur dioxide, and other compounds, and these are damaging when inhaled, especially for prolonged periods of time. Under these circumstances, residents near the mining industries often close their noses to curtail breathing in the polluted air (Fig. 7.5) and to protect against health worries (Table 7.5).

The rise in mining activities and inadequate pollution control measures in the DR Congo has led to severe air quality issues, posing significant health risks to local communities. Effective strategies to mitigate pollution and protect public health are urgently needed.

Residents in mining communities often close their noses for several reasons:

1. **Dust and particulate matter**: Mining activities generate significant dust and particulates, which can harm respiratory health. By covering their noses, residents try to avoid inhaling these harmful particles.
2. **Unpleasant odors**: The extraction process in mining releases various chemicals and substances, causing unpleasant odors. Covering their noses helps residents block out these smells and reduce discomfort.
3. **Air pollution**: Mining operations contribute to air pollution, which negatively impacts human health. Closing their noses helps residents reduce their exposure to airborne pollutants.

Fig. 7.3 Children in residential communities in Lubumbashi City, Haut-Katanga Province, in the DR Congo playing on the ground with dust always tend to put their fingers in the mouth, thus ingesting contaminated soil dust. (Source: Courtesy Professor C. Banza Lubaba Nkulu, Lubumbashi University DR Congo)

Table 7.4 Population groups most affected by each air and water pollution from the mines in Katanga Province, Democratic Republic of Congo

Negative impact	Groups most affected
Pollution	Residents near Gécamines mines
Toxic waste in Likasi and Kipushi	Mostly the population in rural areas and wildlife such as fish.
Pollution and waste damping by Forrests mining	A river which was used by local people has disappeared

4. **Allergies and sensitivities**: Some residents have allergies or sensitivities to substances in the air due to mining. Covering their noses helps protect them from potential allergic reactions.
5. **Awareness of health risks**: Many residents are aware of the health risks associated with living near mining operations. Closing their noses is a precautionary measure to minimize the risk of health issues related to poor air quality.

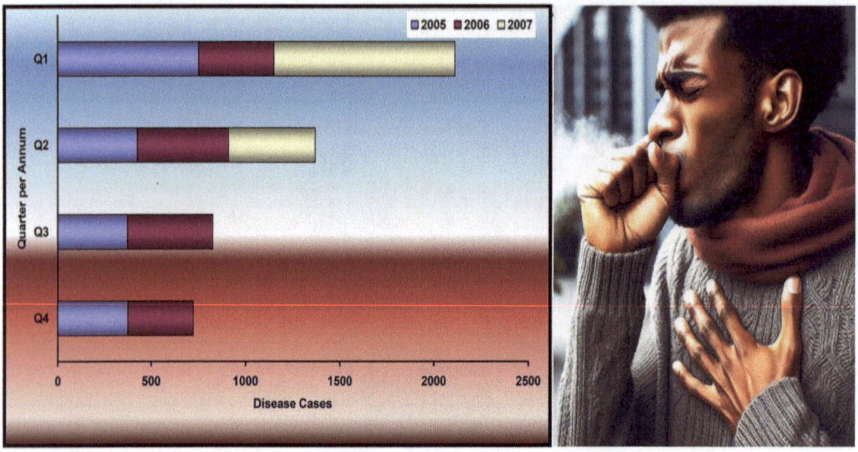

Fig. 7.4 Trends of coughing/pneumonia in children under 5 years of age in 2005 to 2007 in Chingola District, Copperbelt Province of Zambia. (Data were provided by the Chingola District Health Board)

Fig. 7.5 A woman in Lubumbashi City in Haut Katanga Province of the DR Congo with her nose covered with hands to protect against poor-quality air due to smoke pollution from mines

Table 7.5 Most common human diseases in Chingola District, Copperbelt Province of Zambia

Common disease	Affected population group	Residential areas most affected	Proportion of population affected	Relationship with the mine
Diarrhea	Babies	Low income, high density	30–40%	Not known
		High income, low density	10–20%	Not known
Aspiration pneumonia common form of coughing	Babies/adults	Unclassified	Common with town	Mining dust
HIV/AIDS	All			
Malaria	All		15,545 cases in 2006 (9% of population) 10,163 cases in 2007 (6% of population)	

NB. Information supplied by the District Health Board of Chingola in the Zambian copper belt

Table 7.5 Most common human diseases in [illegible]

[illegible]

Chapter 8
Impacts of Mining on Biodiversity in the Region

Keywords Ecosystems · Flora and fauna · Habitat · Species endangerment · Conservation · Biodiversity loss

8.1 Contextual Perspective

Addressing biodiversity loss induced by the environmental impacts of mining in Zambia and the DR Congo is crucial because of the immense ecological significance of the Copperbelt region. Both countries boast rich and diverse ecosystems, ranging from the miombo woodlands in Zambia to the Congo Basin forests in the DR Congo, which is the second largest rainforest in the world. These ecosystems are not only home to countless species, many of which are endemic and endangered, but they also provide essential services such as carbon sequestration, water purification, and climate regulation. Mining activities—often occurring in areas of high ecological value—pose a direct threat to these critical ecosystems through deforestation, habitat destruction, and pollution. Dealing with biodiversity loss would highlight the urgent need to balance economic development with the preservation of the natural heritage that sustains life and livelihoods in the region.

Moreover, emphasizing biodiversity loss allows the readers to explore the broader implications of mining beyond the immediate environmental damage. The decline of biodiversity has cascading effects on local communities that depend on these ecosystems for food, medicine, and cultural practices. For example, the loss of pollinators due to habitat destruction can impact agricultural productivity, while contaminated water sources can disrupt aquatic ecosystems that support fishing communities. This topic brings attention to these often-overlooked socioeconomic dimensions, advocating for more inclusive and sustainable mining policies that consider the well-being of both nature and people.

A. Muimba-Kankolongo et al., *Mining Impacts on the Environment in the Central African Copperbelt of Zambia and the Democratic Republic of Congo*, https://doi.org/10.1007/978-3-031-96786-3_8

In addition, the global significance of biodiversity loss ties this issue to international conservation efforts and climate change mitigation. Zambia and the DR Congo play a pivotal role in global biodiversity conservation due to the unique species and ecosystems they harbor. Mining activities that accelerate biodiversity decline not only threaten local ecological balance but also undermine global commitments such as the Convention on Biological Diversity and the Sustainable Development Goals (SDGs). Further, the discussion on the issue extends to strategies for mitigating these impacts, such as implementing biodiversity offsets, strengthening protected areas, and promoting responsible sourcing of minerals. This serves as a compelling call to action, urging stakeholders at all levels to recognize the value of biodiversity and act to protect it in the face of mounting pressures from mining operations.

8.2 Background

8.2.1 Understanding Biodiversity and Its Importance

Definitions of Biodiversity

Biodiversity, as defined by Swift et al. (2004), encompasses all species, including their genetic variations, across terrestrial and aquatic organisms. This includes invertebrates, protists, bacteria, fungi, vertebrates, and plants, both above and below ground. The FAO (2020) further defines biodiversity for food and agriculture as the variety and variability of animals, plants, and microorganisms at the genetic, species, and ecosystem levels. These components sustain ecosystem structures, functions, and processes around production systems like crops, livestock, forests, fisheries, and aquaculture. This includes genetic resources for food and agriculture, forest genetic resources, associated biodiversity, and wild foods.

Biodiversity Loss

Biodiversity loss refers to the reduction of life on Earth at various levels, from genetic diversity to the collapse of entire ecosystems. Biodiversity supports ecosystem services, forming the backbone of the global economy (Lai 2022). Scientists believe we are presently experiencing a sixth mass extinction, similar to past events that led to the disappearance of dinosaurs and over 95% of all species.

Types of Biodiversity

Earth Organization (2022) labels three foremost categories of biodiversity:

1. **Genetic diversity**: This refers to the variety of genes within a species, affecting physical traits and allowing for variations within populations. For example, humans have small genetic differences that contribute to unique physical features and high levels of variation among different regions.
2. **Species diversity**: This pertains to the variety and distribution of species within a habitat. Habitats like coral reefs and rainforests have high species diversity, while deserts have low biodiversity due to harsh environmental conditions. Species diversity also involves the roles of species, such as predators and prey, within the ecosystem.
3. **Ecosystem diversity**: This is the largest scale of biodiversity, reflecting the variety of ecosystems or habitats in a geographic area. It considers both biological and non-biological factors. High ecosystem diversity helps communities better withstand drastic changes.

Benefits of Biodiversity

Biodiversity is crucial for the survival of living organisms on Earth. Its loss affects essential mechanisms needed for food production, health maintenance, and climate regulation. The main benefits of biodiversity include:

- Supporting food production and providing raw materials.
- Maintaining health through medicinal resources.
- Regulating climate and contributing to resilience against environmental changes.

Understanding the importance of biodiversity and its benefits helps underscore the need for conservation efforts to maintain the balance of ecosystems and support human well-being. Biodiversity plays several crucial roles in maintaining the balance of ecosystems and supporting life on Earth including:

Disease Resistance

Populations with a wide range of genetic diversity are more likely to withstand disasters like pandemics. This diversity means some individuals have genetic traits that protect them from certain threats. When these resilient individuals reproduce, they pass on their advantageous traits, enhancing the entire population's ability to survive and thrive despite the challenges.

Carbon Sequestration

Carbon sequestration involves capturing and storing carbon dioxide from the atmosphere. This process helps lower atmospheric carbon dioxide levels and aims to mitigate climate change. Ecosystems such as forests, peatlands, grasslands, seabeds, wetlands, and kelp beds serve as natural carbon sinks, absorbing carbon dioxide and storing it in vegetation and soil.

Storm, Flooding, and Coastal Erosion Regulation

Coastal sea levels are rising, and the World Economic Forum warns that up to 410 million people could be affected by the end of the century. While the majority of this rise (59%) is expected in tropical Asia, countries like China, France, Senegal, Nigeria, and the United States are also at risk. Protecting and restoring coastal ecosystems such as salt marshes and mudflats will play a crucial role in flood prevention for low-lying coastal areas. Salt marsh plants and microbes help stabilize and bind the soil, and their extensive root systems provide stronger resistance to soil erosion.

Food Security

Biodiversity is crucial for our food systems and agriculture. A diverse range of species works together to provide us with grains, vegetables, fruits, and animal products. Food production relies on several services that biodiversity offers, such as pollination, soil fertility maintenance, pest and disease resistance, climate regulation, and water filtration. Losing key pollinators like bees could have severe financial impacts, potentially costing the United Kingdom's agricultural industry up to £700 million (US$812 million) annually and threatening the country's food supply.

Deena (2024) at Earth Organization reported that in the past 50 years, rapid growth in human consumption, population, global trade, and urbanization has led to the overuse of Earth's natural resources. A WWF report (2020) highlighted a significant decline in the population sizes of mammals, fish, birds, reptiles, and amphibians, averaging a 68% reduction between 1970 and 2016. The primary driver of this biodiversity loss is land use change, such as converting forests, grasslands, and mangroves into agricultural land.

Illegal wildlife trade heavily impacts animals like pangolins, sharks, and seahorses, with pangolins becoming critically endangered. More broadly, scientists have found that the sixth mass extinction of wildlife on Earth is accelerating. Over 500 species of land animals are on the brink of extinction and are likely to vanish within 20 years—a rate of loss that would have naturally taken thousands of years without human interference. Additionally, climate change is drastically affecting ecosystems, such as the melting sea ice in Antarctica, which threatens emperor penguins with potential population wipeout by 2100 (2023 research).

Lei Nguyen (2024) at Earth Organization emphasized the global importance of Africa's biodiversity. The continent is home to 22% of the world's mammalian species, nearly 20% of avian species, and one-sixth of all plant species. Of the world's

36 biodiversity hotspots, eight are in Africa, featuring numerous endemic species and being among the richest and most biologically endangered regions. Factors like population growth, extensive farming, rapid urbanization, infrastructural development, and illegal trafficking are driving an unprecedented decline in Africa's biodiversity.

These findings highlight the critical relationship between biodiversity and food systems, as well as the urgent need to address human impacts on biodiversity to ensure a sustainable and healthy future for our planet.

8.3 Biodiversity in Zambia

The GRZ Ministry of Lands, Natural Resources, and Environmental Protection (2015) provided a complete full account of biodiversity standing in the country. The report is outlined in the following sections:

8.3.1 Status of Biodiversity

National Protected Area System

Zambia has an extensive network of statutory protected areas (PAs) that cover significant portions of the country:

- **National parks (NPs)**: 20 parks covering over 63,580 km^2
- **Game management areas (GMAs)**: 36 areas covering approximately 167,557 km^2
- **Forest reserves (FRs)**: 490 reserves covering around 74,361 km^2

Together, more than 18% of Zambia's land area is made up of national parks, and nearly 10% is composed of forest reserves.

Despite this extensive network, there is a need for continuous monitoring and follow-up actions to ensure the effectiveness and sustainability of these protected areas. Recent projects such as the reclassification of protected areas have primarily focused on national parks and game management areas (Fig. 8.1). Additionally, the integrated land use assessment has concentrated on developing methodologies for the national inventory of forests. However, it has not adequately included species diversity within the sampled areas for the forest inventory.

Since the Fourth National Report, several important changes have occurred in Zambia's protected area system. These include the creation of Lusaka National Park in 2011, the degazetting of some forest reserves, and the identification of gaps in plant and animal species representation within existing national parks and game management areas, based on the reclassification conservation plan.

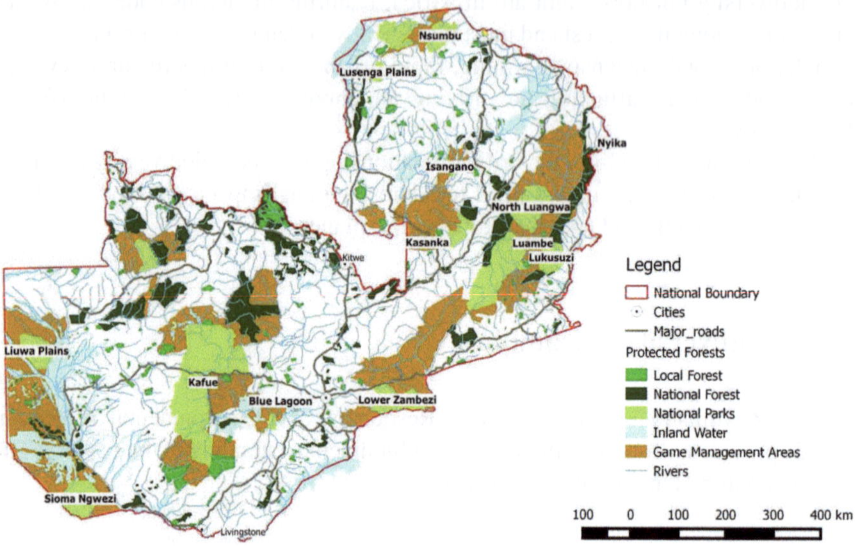

Fig. 8.1 National parks, game management areas, and forest reserves in Zambia. (Source: Government of the Republic of Zamia MLNR, 2015, reproduced with permission from Ministry of Lands and Natural Resources)

8.3.2 Key Changes

1. **Creation of Lusaka National Park**: Established as a national park in 2011.
2. **Degazetting forest reserves**: Some forest reserves have been removed from the protected area system.
3. **Identification of gaps**: The reclassification conservation plan highlighted the need to address gaps in species representation within national parks and game management areas.

8.3.3 Recommendations

- **Comprehensive gap analysis**: Conducting a thorough analysis of species to identify and fill gaps in biodiversity representation.
- **Improved management**: Enhancing the effectiveness of management within the national protected area system.

Wetlands

Zambia is home to eight internationally recognized wetlands, designated as Wetlands of International Importance under the Ramsar Convention. These include:

- **Kafue Flats**: Comprising Lochinvar and Blue Lagoon National Parks
- **Bangweulu Swamps**: Including Chikuni
- **Lukanga Swamps**
- **Busanga**
- **Lake Tanganyika**
- **Luangwa Floodplains**
- **Lukanga Swamps** (listed twice, possibly a typo)
- **Barotse Floodplain**

These wetlands provide critical habitats for numerous important flora and fauna species, some of which are endemic and endangered (ZAWA 2015).

8.3.4 Important Bird Areas (IBAs)

At present, about 82% of the area covered by Important Bird Areas receives some form of protection (national park, 60%; game management area, 19%; national forest, 2%; privately owned, 1%). The remaining 18% are in open areas with no legal status; a few of these have developed some local by-laws with the help of chiefs and traditional leaders. Several biological species including animals, birds, fish, and trees are within these protected habitats (Fig. 8.2).

Presently, approximately 82% of Zambia's Important Bird Areas (IBAs) are under some form of protection. This protection is distributed as follows:

- **National parks**: 60%.
- **Game management areas**: 19%.
- **National forests**: 2%.
- **Privately owned lands**: 1%.

The remaining 18% of IBAs are in open areas without legal protection, although some of these areas have local by-laws developed with the help of community leaders. These protected habitats support various biological species, including animals, birds, fish, and trees (Fig. 8.2).

Fig. 8.2 Diverse biodiversity life species comprising wild animals, insects, vertebrates, birds, fish, flowers, vegetation, and forest trees in Zambia

8.3.5 Flowering Plants in Zambia

A comprehensive assessment of flowering plant diversity in Zambia was conducted using field inventory data. The estimated total number of wild flowering plant species in Zambia is 3543, comprising:

- **Sedges**: 273 species.
- **Herbaceous plants**: 2660 species.
- **Woody plants**: 1610 species.

Most of the data available are on tree species, which means the overall diversity of flowering plants is likely underestimated.

8.3.6 Areas of High Woody Plant Species Richness

The regions with the highest richness in woody plant species (36–48 species per hectare) are located in northern and northwestern Zambia, with the largest continuous forest area in the Kafue Headwaters. Three smaller regions with high woody plant species richness include Mansa, Mkushi, and Siavonga.

8.3.7 Mammal Species

A stocktaking assessment focusing on critical mammal species (threatened, endangered, and vulnerable) faced challenges such as limited time and difficulties accessing data (Table 8.1). The diversity of mammals in Zambia is estimated at around 224 species. The REMNPAS report highlights that about 43 species of large mammals are particularly important for several reasons:

Table 8.1 Critical mammal species in Zambia and their status

		Conservation status	
Common name	Scientific name	Zambian criteria	IUCN criteria
Pangolin	*Manis temminckii*	Rare	Threatened
Aardwolf	*Proteles cristatus*	Rare	Threatened
Aardvark	*Orycteropus afer*	Rare	Rare
African elephant	*Loxodonta africana*	Endangered	Endangered
Black rhinoceros	*Diceros bicornis*	Endangered	Critically Endangered
White rhinoceros	*Ceratotherium simum*	Endangered	Endangered
Giraffe	*Giraffa camelopardalis*	Threatened	Threatened
Springhare	*Pedetes capensis*	Vulnerable	Vulnerable
Blue monkey	*Cercopithecus mitis*	Rare	Vulnerable
Wild dog	*Lycaon pictus*	Endangered	Endangered
Hyena	*Crocuta crocuta*	Vulnerable	Vulnerable
Serval cat	*Leptailurus serval*	Vulnerable	Vulnerable
Caracal	*Caracal caracal*	Vulnerable	Vulnerable
Cheetah	*Acinonyx jubatus*	Endangered	Endangered
Leopard	*Panthera pardus*	Vulnerable	Vulnerable
Lion	*Panthera leo*	Vulnerable	Vulnerable
Sitatunga	*Tragelaphus strepsiceros*	Vulnerable	Vulnerable
Eland	*Tragelaphus oryx*	Vulnerable	Vulnerable
Blue duiker	*Cephalophus monticola*	Rare	Vulnerable
Yellow-backed duiker	*Cephalophus silvicultor*	Threatened	Threatened
Puku	*Kobus vardoni*	Threatened	Threatened
Red lechwe	*Kobus leche leche*	Threatened	Threatened
Roan antelope	*Hippotragus equinus*	Threatened	Threatened
Sable antelope	*Hippotragus niger*	Vulnerable	Vulnerable

Source: Government of the Republic of Zambia MLNR 2015. Reproduced with permission from Ministry of Lands and Natural Resources

1. **Economic potential**: They generate income through photographic and consumptive tourism.
2. **Local household economies**: They provide protein and income through legal and illegal market structures.
3. **Aesthetic value**: Their existence is appreciated globally.

According to the 2014 Red List of Threatened Species, over 28 animal species and subspecies in Zambia are classified as threatened, endangered, or vulnerable.

8.3.8 Fish Species

Zambia's water bodies are home to 490 fish species, spanning 24 families. Most fish biodiversity studies have been conducted to support fisheries, particularly in commercial fishing areas on large water bodies. However, only a few sections of these areas have been thoroughly sampled. Key families in artisanal, commercial fisheries, and aquaculture are well studied, while smaller fish which are not important to these sectors have not been extensively investigated.

8.3.9 Crop and Livestock Genetic Resources

Approximately 15% of cultivated crops in Zambia are indigenous, originating from Africa. These include sorghum, millets, cowpea, Bambara groundnuts, sesame, and various vegetables. The gene bank conserves both interspecific and intraspecific diversities of crop species but needs better support with equipment, manpower, and financial resources.

Some crops, particularly their wild relatives (CWR), have received more attention for collection, characterization, and conservation:

- Cowpea (*Vigna unguiculata*).
- Rice (*Oryza sativa*).
- Finger millet (*Eleusine coracana*).
- Pearl millet (*Pennisetum glaucum*).
- Sorghum (*Sorghum bicolor*).

The traditional subsector in Zambia owns the majority of livestock, with 83% of cattle, 97% of goats, 64% of sheep, and 90% of pigs being indigenous breeds or crosses with exotic breeds. Due to challenges in accessing feeds and drugs, many rural small pig producers prefer indigenous breeds, which require less management and have natural disease resistance compared to exotic breeds.

8.3.10 Invertebrates

Zambia is home to an estimated 6135 species of invertebrates. Out of these, 69 species are endemic, and 14 species, mainly freshwater mollusks, are threatened. The country has 57 identified species of mollusks, with significant centers of molluskan endemism located in Lake Tanganyika, Mweru Wantipa, and the Barotse Plains. Notable endemic mollusks include:

- ***Gabbiella stanleyi***: Found in Chitipa, Eastern Zambia.
- ***Gabbiella zambica***: Found in Mankoyo, Northwestern Zambia.
- ***Mutela zambesiensis*****:** Primarily found in the Zambezi River basin and its tributaries.

In addition, more than 200 species of crustaceans are known to inhabit various ecosystems in Zambia, with over half of these species being endemic to the country.

8.3.11 Vertebrate Diversity in Zambia

The diversity of amphibians, including frogs and toads, in Zambia is estimated at 74 species, with no notable geographical gradient in species richness. Thirteen species are considered rare, having been recorded in only one locality. The country also boasts a diversity of reptiles, including lizards, snakes, and tortoises, estimated at 156 species. Forty-five reptile species are deemed rare as they have been recorded in only one location. Overall, the total number of biological species recorded is shown in Table 8.2.

Campbell et al. (2011) and Mudenda Siachoono (2018) also provided detailed observations about Zambia's biodiversity. Zambia has eight Ramsar-designated wetlands covering over four million hectares and 39 Important Bird Areas (IBAs), 15 of which overlap with national parks. To conserve floral biodiversity, 59 botanical reserves have been established within the national forest system. Unfortunately, 29 of these reserves face issues such as encroachment or depletion due to reasons like unmaintained boundaries and inadequate capacity within the Forestry Department.

In 2018, Mudenda Siachoono referenced Harkonen et al. (2015), who classified Zambian fungi into three major groups:

- **Glomeromycota**: About 200 species, important in ecological processes, living as mycelium in the soil without visible fruit bodies.
- **Ascomycota**: The largest group with approximately 64,000 species.
- **Basidiomycota**: Comprising about 30,000 species, named for their spore-producing cells called basidia.

Table 8.2 Total number of recorded species in Zambia by 2015

Group		Number of species	
Main	Subgroup	1998	2015
Microorganisms		563	563
Plants	Algae	147	147
	Mosses	129	129
	Ferns	142	147
	Flowering plants	3410	3543
Animals	Invertebrates	1808	6135
	Amphibians	67	74
	Reptiles	150	156
	Fish	409	490
	Birds	733	757
	Mammals	224	242
Domesticates	Crops	86	106
	Animals	16	16
Total		**7884**	**12,505**

Source: NIRAS and Government of the Republic of Zamia MLNR 2015. Reproduced with permission from Ministry of Lands and Natural Resources

8.3.12 *Threats to Biodiversity*

Nkula Mwanza (2006) identified several human and natural factors threatening Zambia's biodiversity:

- Land use conflicts.
- Human settlement expansion.
- Climate change.
- Pollution.
- Overexploitation of resources.
- General deforestation.
- Introduction of non-native species.
- Biotechnology impacts.
- Lack of biodiversity knowledge.

8.3.13 *Environmental Impacts of Mining*

Mabeta et al. (2018) noted that mining effluents have negatively affected terrestrial and aquatic ecosystems, including the Kafue River and the Mushishima stream. Monitoring these impacts is challenging due to the limited reach of the Zambia Environmental Management Agency (ZEMA). Rising metal demand in Asian markets could worsen biodiversity loss in Zambia and other mineral-rich regions. In northwestern province, about 350,000 hectares of protected forest areas (PFAs) has been converted to mining concessions since 2005, with further deterioration as new mines open (Matakala et al. 2015).

8.4 Biodiversity in the DR Congo

8.4.1 Overview

Located in the southeast of the Democratic Republic of Congo (DR Congo), the former Katanga province, with Lubumbashi as its capital, covered 21.3% of the nation's land area, totaling 496,877 km^2 (UNDP 2009, Fig. 8.3). This region bordered four other provinces: South Kivu in the northeast, Maniema in the north, and the two Kasai provinces in the northwest. Additionally, Katanga shared international borders with Angola to the southwest, Zambia to the south and southeast, and Tanzania to the east.

8.4.2 Climate

Katanga experienced two main seasons: the rainy season and the dry season. The climate varied across the province, with a humid tropical climate prevailing in some areas and a temperate climate in others. Temperatures ranged from −3 °C in colder regions to an average annual temperature of 20 °C in the south and 24 °C in the north. Overall, temperatures across the province ranged between 19 °C and 25 °C (Omasobo Tshonda et al. 2018).

8.4.3 Geographical Features

Until 2015, the former Katanga province featured a vast plateau formed by slow erosion from an old lake, now reduced to a marshy plain with the Upemba lagoons, crossed by the Lualaba (Congo River, from its source to Bukama) and the Lufira River.

8.4.4 Administrative Divisions

The capital of Katanga was Lubumbashi, complemented by the cities of Kolwezi and Likasi. Additionally, the province had 27 urban centers. Rural Katanga was divided into five districts, further subdivided into 22 territories, 92 sectors or chiefdoms, and 798 groups.

Fig. 8.3 Map of Katanga region with its administrative division that includes Haut-Katanga, Lualaba, Haut-Lomami, and Tanganyika provinces in the southern part of the DR Congo. (Source: Courtesy of Professor Banza Lubaba Nkulu, Lubumbashi University, DR Congo)

8.4.5 *Mining Activities*

In 1906, Union Minière du Haut Katanga (UMHK), now Gécamines, started detailed exploration of the Kasonta and Lupoto deposits, focusing on regional soil geochemistry and mapping. By 1921, UMHK's exploration included surface and underground efforts at Lupoto and Karavia. In 1930, the Kasonta C1 Polygon (now Kasonta-Lupoto Polygon) was established for detailed geological investigations.

8.4.6 *Timeline of Key Mining Activities*

- **1923–1967**: Intermittent exploration by UMHK, including drilling at Lupoto and Karavia
- **1959–1963**: Mining of the high-grade Karavia prospect by UMHK
- **1967**: Nationalization of UMHK, leading to the end of its exclusive rights to the Kalumines
- **1977**: Final definition of the Lupoto resource
- Copper-Cobalt Project.

Today, various public, parastatal, and private companies and informal artisanal miners operate in Katanga's mining sector (Global Witness 2006; KPMG Global Mining Institute 2014; Banza Lubaba Nkulua et al. 2018).

The typical life cycle of a copper mining operation includes several stages:

- **Prospecting**: Searching for ore deposits.
- **Development**: Preparing the site for mining.
- **Construction**: Building the necessary infrastructure.
- **Operation**: Extracting and processing the ore.
- **Staff housing and support**: Providing accommodations and services for workers.
- **Product stockpiling**: Storing the mined products.
- **Mineral processing**: Refining the ore to extract the desired minerals.
- **Waste management**: Handling the waste generated from mining and processing.
- **Rehabilitation**: Restoring the site after mining activities have ceased.
- **Closure**: Officially ending the mining operations and ensuring the site is safe and stable.

8.4.7 *Environmental and Social Impacts*

Each stage of a copper mine's life cycle can drive environmental changes on local, national, and even international scales, such as air pollution plumes affecting the southern African region (Ashton et al. 2001). Mining activities, while essential for resource extraction, often lead to significant environmental changes during various stages of mining cycles. They include:

Exploration

During the exploration phase, scientists and geologists evaluate areas for potential mining which often leads to environmental impacts such as:

- **Habitat disruption**: Initial surveys can lead to the disturbance of ecosystems, disrupting wildlife habitats and leading to potential species displacement.
- **Soil erosion**: Clearing land for exploration can increase soil erosion, negatively affecting local water quality and biodiversity.
- **Water quality impact**: The introduction of chemicals or heavy machinery can alter local water sources, leading to contamination.
- **Noise and air pollution**: Exploration activities often produce noise and emissions, which can affect nearby communities and wildlife.

Extraction

The extraction phase involves the actual removal of minerals or resources from the Earth. The impacts here are often the most profound:

- **Deforestation**: Substantial land clearing for mining operations results in loss of forests, contributing to biodiversity loss and climate change.
- **Water resource depletion**: Mining often requires significant amounts of water, depleting local water bodies and impacting surrounding communities.
- **Soil contamination**: The use of toxic substances (like cyanide in gold mining) can lead to soil contamination, resulting in long-term ecological damage.
- **Air pollution**: Dust from mining operations, emissions from machinery, and processes can contribute to air quality degradation, affecting human health and ecosystems.
- **Tailings**: The disposal of waste materials (tailings) can lead to leaching of harmful substances into the environment, posing risks to water sources.

Processing

After extraction, ore undergoes processing to separate valuable minerals. This phase can further exacerbate environmental issues:

- **Chemical pollution**: The use of hazardous chemicals in processing (e.g., arsenic, mercury) can lead to soil and water contamination.
- **Energy consumption**: Processing requires substantial energy, often derived from fossil fuels, contributing to greenhouse gas emissions and climate change.

- **Waste generation**: The production of waste and by-products during processing can pose significant disposal and pollution challenges.

Transportation

Transporting mined resources to processing facilities or markets has its own set of environmental implications:

- **Infrastructure development**: Building roads and transport networks can lead to habitat destruction and fragmentation of ecosystems.
- **Carbon emissions**: Transportation of materials contributes to carbon emissions, exacerbating global warming.
- **Spills and leaks**: Transporting hazardous materials can result in spills, leading to immediate local environmental crises (e.g., contamination of waterways).

Closure/Reclamation

After mining ceases, the closure and reclamation phases involve restoring the mined areas:

- **Land degradation:** If not managed properly, former mining sites can remain degraded and uninhabitable, leading to long-term ecological damage.
- **Acid mine drainage**: Abandoned mines can lead to acid mine drainage, where water flows over acidic rock, leading to the release of toxic metals into the environment.
- **Biodiversity loss:** Reclaimed areas may not return to their original ecological state, meaning long-term impacts on local biodiversity.
- **Community impact:** Neglecting local communities during closure can lead to socioeconomic issues, as jobs and resources are lost.

The overview highlighted above underlines the complex interactions between mining operations and potential environmental impacts on ecological alterations, emphasizing the strong need for careful management and mitigation strategies at each stage of the mining life cycle.

Mining activities can have profound and cumulative environmental impacts, leading to potentially irreversible damage to both aquatic and terrestrial ecosystems. To address these impacts, a project in the Katanga region is leveraging local expertise to reverse biodiversity loss and improve ecosystem health. This initiative aligns with the DR Congo's environmental and biodiversity strategies (Ministère de l'Environnement, Conservation de la Nature et Développement Durable 2022) and supports bilateral missions' objectives to promote biodiversity for food security and better livelihoods at national and household levels. The project aims to raise awareness about the environmental risks of trace metal pollution, directly benefiting the region and the nation.

8.5 Biodiversity Analysis in the DR Congo

The DR Congo is one of the world's ten mega-biodiverse countries, boasting high rates of endemism and valuable freshwater, forest, and soil resources, including mineral-rich subsoils. However, demographic and economic pressures, along with unsustainable resource use, have negatively impacted the environment, particularly forest resources. Forest cover decreased from 55% in 1990 to 52% by 2000, with the highest deforestation rates in densely populated areas (Tchatchou et al. 2015).

8.5.1 Threats to Habitat and Species

Habitat destruction is a major threat to biodiversity in the DR Congo. As of 1994, six of the country's 19 habitat types were continually degraded due to human activities, climate change, and aging plants. This degradation threatens various species, including:

- **White rhinoceroses**: Garamba National Park.
- **Mountain gorillas and hippopotamuses**: Virunga National Park.
- **Bonobos**: Equator province.

The country holds 50% of Africa's freshwater resources and diverse aquatic species, with strong potential for hydroelectric energy development. However, pollution and human pressures are undermining this potential, leading to increased susceptibility to natural disasters (Mwitwa et al. 2012a).

8.5.2 Resource Management and Biodiversity Loss

Effective measures are needed to manage activities related to fishing and agrobiodiversity, which have contributed to biodiversity loss. The benefits from biological resources are inequitably distributed, with a small segment of the population enjoying them while the majority lives in extreme poverty and deteriorating conditions (https://www.cbd.int/countries/profile/?country=cd#facts).

For more than 70 years, the former Katanga province in the country has been one of the central producers of copper, cobalt, and several other minerals for global markets (http://www.katangamining.com/operations/history.html). Mining activities affect biodiversity at multiple levels—local, landscape, regional, and global—through direct processes like mineral extraction and indirect processes involving supporting industries and external stakeholders accessing biodiversity-rich areas. In the former Katanga province, for example, Gécamines (the state mining company) and its subsidiaries, along with other private and artisanal miners, have significant mining concessions covering around 30,000 km^2 near Kolwezi, Likasi, and

Lubumbashi. This extensive mining activity contributes to considerable biodiversity loss.

8.5.3 Environmental Pollution from Mining

Mining operations release harmful substances into the environment:

- **Heavy metals**: Released into aquatic environments, impacting aquatic life.
- **Particulates**: Emitted into the air and water, causing pollution.
- **Mine effluents**: Discharged onto land, polluting soil and water sources.

Plumes of smoke and dust carry dangerous particles into the air, while runoff from cleaning chemicals and mineral alloys pollutes streams.

8.5.4 Freshwater Ecosystem Threats

According to Brummett et al. (2008), a major threat to freshwater ecosystems is the loss of riparian habitats due to pollution from mining activities.

8.5.5 Research on Mining Impacts

There is limited research on the effects of mining and metallurgical industry pollution on biodiversity. However, several studies (De Putter et al. 2011; Mwitwa et al. 2012a; Cheyns et al. 2014) have shown that trace metals from mining continuously pollute agricultural land, ecosystems, vegetation, trees, water sources, and food crops. This pollution compromises land suitability for agriculture and the safety and nutritional quality of food, posing serious health risks, such as cancer, lung and kidney dysfunction, and heart damage (Saline and Griswold 2009; Nemery 2011).

8.5.6 Community Concerns

Households near mines in Katanga have raised concerns about land degradation and food spoilage (Mwitwa et al. 2012a). The extensive environmental damage from mining activities necessitates studies to address these impacts.

8.6 Biodiversity Fundamental Values

Shift et al. (1994) highlighted several fundamental values of biodiversity, including:

- **Utilitarian value**: This direct use value benefits land managers (farmers, local communities, government) through agricultural crops, forest trees, fish, and animals.
- **Bequest value**: The belief in future, yet unknown, benefits of biodiversity for future generations, such as undiscovered microorganisms with potential industrial applications.
- **Functional value**: Biodiversity supports ecosystem life, preserving ecological structure and integrity.

Moreover, Thomas Lewinsohn (2019) noted that our geographical understanding of biodiversity is uneven, being more comprehensive in Europe and North America but less so in other continents (http://fellows.rfiea.fr/dossier/climat-plus-que-deux-ans-pour-reagir/article/preserver-la-biodiversite-necessite-plus-que-la?language=en).

8.7 Ecosystems in the DR Congo

The DR Congo, the second largest country in sub-Saharan Africa, treasures diverse ecosystems, including:

- Nearly half of Africa's rainforests (IUCN 1992).
- Forest-savanna ecotones, savannas, and afro-mountainous forests.
- Large and small lakes, rivers, and swampy forests (Inogwabini et al. 2005).

Since the colonial era, biodiversity preservation efforts have focused on creating protected areas. However, these efforts have often excluded local communities and been enforcement-oriented, making the protected areas unpopular and politically fragile (Inogwabini 2014).

8.8 DR Congo's Natural Heritage

The DR Congo features the world's second largest tropical forest and ranks fifth globally for animal and plant diversity. It is home to:

- 450 species of mammals
- 1150 species of birds
- 400 species of fish
- Over 15,000 species of plants (Lydall and Auchterlonie 2011; Wolfire et al. 1998).

Notable species include the northern white rhino, mountain gorilla, endemic okapi, and great apes such as the eastern lowland gorilla and bonobo. About 6% of mammals and 10% of plants are endemic to the DR Congo. The country has five natural World Heritage sites, including Virunga National Park, which houses the famous mountain gorilla.

8.9 Threats to Biodiversity

Widespread poverty, conflict, and reliance on mineral extraction pose significant threats to biodiversity. Issues such as poaching, pollution, deforestation, and soil erosion endanger wildlife and habitats (Ashton et al. 2001). For instance, Grauer's gorillas have seen a 77% decline since 1995 due to illegal hunting, civil unrest, and habitat loss from mining (MFNP 2002). With over 50% of Congo's basin forests, the DR Congo is a top ten country globally in terms of biodiversity.

Recent findings by Rhett A. Butler (2020) identified the following biodiversities (https://rainforests.mongabay.com/congo/conservation.html#):

- More than 11,000 species of plants.
- 450 mammals
- 1500 fish
- 1150 birds
- 300 reptiles
- 200 amphibians.

The Congolese coastal zone is defined as the area stretching from the Congolese marine waters in the Atlantic to the international port of Matadi upstream of the Congo River estuary. This coastal region encompasses the entire Bas-rivière district. It is bordered to the northwest by the Angolan province of Cabinda, to the northeast by the Cataractes district, and to the southwest by Angola. The Congolese Atlantic coastline spans 40 km and features significant mangrove areas, including the Mangrove Marine Park, extending to its northern border with Cabinda. This coastal zone covers approximately 110,000 hectares.

8.9.1 Mangrove Marine Park

Established on May 2, 1992, by ministerial decree No. 044/cm/ECN/92, the Mangrove Marine Park is located at the Congo River estuary, within the Muanda territory of the Bas River district, Bas-Congo province. The park covers about 66,000 hectares, with geographical coordinates between 5°45′–6°55′ southern latitude and 12°45′–13° east longitude, at an altitude of less than 500 meters.

The Mangrove Marine Park is divided into two zones:

1. Zone A (full protection):
   - **North**: Bounded by Banana Cove at the Tshikayoto river mouth.
   - **East**: From the Tshikayoto river mouth to Mompanga Island.
   - **South**: From Mompanga Island to Pointe de Bulambemba.
   - **West**: Between Bulambemba and Rosa islands.
2. **Zone B (partial protection)**:
   - **North**: From the Tshikayoto river mouth to its intersection with the Lukunga river.
   - **East**: From the Lukunga intersection to the national road mouth.
   - **South**: From the Lukunga river mouth along the coastal strip from Cabinda enclave to Banana point.

   Additionally, the marine park includes the Luki Biosphere Reserve.

8.10 Haut-Katanga Province Biodiversity

8.10.1 Edible Insects

The former Katanga province in the DR Congo is recognized as one of Africa's significant biodiversity hotspots. The region is notable for its wide diversity of edible insects, with approximately 11 species from four families. Additionally, the area boasts 694 taxa, including 664 species, 15 subspecies, and 15 varieties of trees and shrubs.

8.10.2 Key Vegetation Types

Upper Katanga has three primary woody vegetation types, each with distinct phytogeographic and taxonomic compositions:

1. **Miombo woodland**: Contains 254 taxa, 62% of which are Zambezian, with a notable presence of Caesalpinioideae.
2. **Riverine and swamp forest**: Comprises 196 taxa, with 36% being Guineo-Congolian and 17% Afromontane species.
3. **Zambezian dry evergreen forest**: Includes 117 taxa, 37% of which are Guineo-Congolian, with an overrepresentation of Rubiaceae.

8.10.3 Endemic and Unique Species

- **Endemics**: Twenty taxa (3%) are strictly endemic to Upper Katanga, mostly found in the Katanga-Zambian sector of the Zambezian region. However, there are taxonomic uncertainties, with 13 endemics known from only one or two collections.
- **Termite mound associations**: Forty-nine taxa are almost exclusively associated with termite mounds.
- **Indicator species**: Eight ecological groups of indicator species are proposed for the miombo woodlands.

Plant Communities and Reptiles

Efforts to conserve and restore copper-cobalt (Cu-Co) ecosystems in Katanga began in the early twentieth century, with significant discoveries and descriptions made over 50 years ago (Sharp 1956; Duvigneaud and Denaeyer-De Smet 1963; Malaisse et al. 1979, 1983; Malaisse et al. 1994). These ecosystems are found on highly mineralized Cu-Co outcrops scattered throughout the miombo forest landscape. They support unique plant communities that thrive along copper-cobalt gradients from the top to the bottom of the outcrops.

8.10.4 Biodiversity in cu-co Outcrops

Approximately 600 plant species, known as metallophytes, inhabit the metalliferous soils of these outcrops. Many of these species are endemic to the region. However, the ongoing mining activities pose a significant threat to these unique plant communities, emphasizing the need for urgent conservation and restoration efforts (Faucon et al. 2010).

Reptile Diversity

The area is also home to at least 119 species of reptiles, including:

- **Chelonians**: Seven species.
- **Lizards**: 37 species.
- **Amphisbaenians**: Four species.
- **Snakes**: 69 species.
- **Crocodilians**: Two species.

Of these, 15 species (12.6%) are endemic to the region.

Mushrooms

The people in the former Katanga province highly value edible mushrooms, especially ectomycorrhizal species such as amanita, chanterelle, and lactaria, which grow in association with trees in open forests. Another preferred type is *Termitomyces*, which is found on termite mounds (De Kesel et al. 2017). During the dry season, when these mushrooms are scarce, people turn to less abundant and less appreciated saprotrophic species found in forest galleries along rivers. The province also harbors other wild mushroom species in various ecosystems.

Mammals

Thirty-eight large wild mammals are hunted for meat in Katanga and Zambia (Malaisse and Parent 1986). Historically, large ungulates like buffaloes, cobes, hippotragus, and zebras were abundant but have significantly declined due to human population growth, poaching, and overhunting (Hasson 2015). The scarcity of large fauna has led locals to hunt smaller game, particularly monkeys (mainly *Cercopithecus*), despite their low energy value. Presently, bushmeat consumption focuses on small antelopes and around 30 species of rodents, such as the killer whale, Gambian rat, and porcupine (Malaisse and Parent 1982), along with birds like green pigeon, turtledove, francolin, and guinea fowl.

Birdlife

DR Congo, including the former Katanga province, is home to a diverse bird population, with several threatened or endangered species such as the shoebill, wattled crane, and spotted ground thrush (Table 8.3). The province supports 1189 bird species, with 21 endemics and 44 globally threatened species. Katanga's varied habitats—from dense forests to open savannas and wetlands—offer ideal environments for numerous bird species, making it a significant area for ornithological studies and bird-watching.

Habitat Types

1. **Forests**: Provide nesting and foraging habitats for numerous birds, including endemics and migrants.
2. **Savannas and grasslands**: Support ground-nesting birds and raptors.
3. **Wetlands and rivers**: Vital for aquatic and semiaquatic birds like herons, storks, and ducks.

Table 8.3 Bird species recorded in three provinces of the Democratic Republic of Congo—order by taxonomic family

Order	Family (common name)	Species (common name, where available)	Bandundu	Bas-Congo	Equateur
Ciconiiformes	Ardeidae (herons)	1.*Bubulcus ibis* (cattle egret)	+		+
		2.*Egretta alba* (great white egret)	+		+
		3.*Egretta ardesiaca* (black heron)	+	+	+
		4.*Egretta garzetta* (little egret)	+		
		5.*Egretta gularis* (western reef heron)		+	
		6.*Ardea purpurea* (purple heron)	+		
	Ciconiidae (storks)	7.*Ciconia episcopus* (woolly-necked stork)	+ M		
Coraciiformes	Meropidae (bee-eaters)	8.*Merops pusillus* (little bee-eater)	+	+	+
	Bucerotidae (hornbills)	9. *Tockus* sp.	+		
		10. *Tockus fasciatus* (African pied hornbill)			+
	Alcedinidae (kingfishers)	11. *Alcedo cristata* (malachite kingfisher)	+		
		12.*Ceryle rudis* (pied kingfisher)	+	+	+
		13.*Halcyon senegalensis* (woodland kingfisher)	+		+
		14.*Halcyon leucocephala* (gray-headed kingfisher)			+
		15.*Halcyon* sp.		+	
Gruiformes	Rallidae (rails)	16. *Porphyrio alleni* (Allen's gallinule)			+
Piciformes	Capitonidae (barbets)	17. *Pogoniulus bilineatus* (yellow-rumped tinkerbird)	+	+	
		18. *Pogoniulus* sp.		+	

(continued)

Table 8.3 (continued)

Order	Family (common name)	Species (common name, where available)	Bandundu	Bas-Congo	Equateur
Pelecaniformes	Anhingidae (anhinga)	19. *Anhinga rufa* (African darter)	+		+
	Phalacrocoracidae (cormorants)	20.*Phalacrocorax africanus* (long-tailed cormorant)	+	+	
Psittaciformes	Psittacidae (parrots)	21. *Psittacus erithacus* (gray parrot)			+
Falconiformes	Accipitridae (hawks/eagles)	22. *Gypohierax angolensis* (palm-nut vulture)	+	+	+
		23.*Milvus migrans* (black kite)	+		+
Galliformes	Phasianidae (pheasants and partridges)	24. *Francolinus* sp.		+	
Columbiformes	Columbidae (pigeons and doves)	25. *Treron calvus* (African green pigeon)	+ M		
		26. *Turtur afer* (blue-spotted wood dove)	+		+
		27. *Streptopelia semitorquata*		+	+
Charadriiformes	Jacanidae (jacanas)	28. *Actophilornis africanus* (African jacana)	+		+
	Charadriidae (plovers)	29. *Vanellus albiceps*	+		
	Scolopacidae (sandpipers and snipes)	30.*Tringa* sp.	+		
Musophagiformes	Musophagidae (turacos and allies)	31. *Corythaeola cristata*	+		
Cuculiformes	Cuculidae (cuckoos, coucals, and anis)	32. *Centropus senegalensis* (Senegal coucal)	+	+	+
Caprimulgiformes	Caprimulgidae (nightjars and allies)	33. *Macrodipteryx vexillarius* (pennant-winged nightjar)			+
Apodiformes	Apodidae (swifts)	34. *Rhaphidura sabini (Sabine's spinetail)*			+

(continued)

Table 8.3 (continued)

Order	Family (common name)	Species (common name, where available)	Bandundu	Bas-Congo	Equateur
Passeriformes	Passeridae (Old World sparrows)	35. *Passer griseus* (gray-headed sparrow)	+	+	+
	Ploceidae (weavers)	36. *Ploceus melanocephalus* (black-headed weaver)			+
		37. *Ploceus nigerrimus* (Vieillot's black weaver)	+		
		38. *Ploceus pelzelni* (slender-billed weaver)			+
		39. *Ploceus aurantius* (orange weaver)	+		
		40. *Ploceus cucullatus* (village weaver)	+		+
		41.*Quelea quelea* (red-billed quelea)			+
		42. *Brachycope anomala* (bob-tailed weaver)			+
	Estrildidae (waxbills)	43. *Estrilda* sp.			+
		44. *Lonchura cucullata* (bronze mannikin)	+	+	
		45. *Uraeginthus angolensis* (blue-breasted cordon-bleu)		+	
	Pycnonotidae (bulbuls)	46. *Pycnonotus barbatus* (common bulbul)		+	+
		47. *Pycnonotus* spp.	+		+
	Corvidae (crows, jays, and allies)	48. *Corvus albus* (pied crow)	+	+	+
	Hirundinidae (swallows and martins)	49. *Hirundo abyssinica* (lesser-striped swallow)		+	
		50. *Hirundo senegalensis* (mosque swallow)		+	

(continued)

Table 8.3 (continued)

Order	Family (common name)	Species (common name, where available)	Bandundu	Bas-Congo	Equateur
		51. *Hirundo rustica* (barn swallow)		+	
		52. *Delichon urbica* (house martin)		+	
	Nectariniidae (sunbirds and spiderhunters)	53. *Nectarinia* sp.	+		
		54. *Nectarinia chnloropygia*	+		
		55. *Anthrepes collaris* (collared sunbird)			+
	Viduidae (indigo birds)	56. *Vidua macroura* (pin-tailed whydah)	+		+
	Motacillidae (pipits and wagtails)	57. *Motacilla aguimp* (African pied wagtail)		+	
		58. *Motacilla flava* (yellow wagtail)	+		
	Sylviidae (Old World warblers)	59. *Cisticola natalensis* (croaking cisticola)		+	
	Turdidae (thrushes)	60. *Saxicola torquatus* (south African stonechat)			+
	Total species by axis		**34**	**23**	**32**

Source: Shumway et al. 2003, reproduced with permission from Drs. C. Shumway et al.

Notable Bird Species

- **African fish eagle (*Haliaeetus vocifer*)**: Common near wetlands and large water bodies.
- **Great blue heron (*Ardea herodias*)**: Found in coastal and freshwater habitats.
- **African pygmy kingfisher (*Ispidina picta*)**: Small, colorful bird near water bodies.
- **Shoebill (*Balaeniceps rex*)**: Rare sightings in specialized wetland habitats.

Fish Diversity

The region encompasses a variety of water bodies, including rivers, lakes, and wetlands, which support numerous fish species. They include:

(a) **Nile tilapia** (*Oreochromis niloticus*) is a widely distributed and economically important freshwater fish. Nile tilapia is recognized for its adaptability to various habitats, including lakes, rivers, and swamps.
(b) **Catfish** (various species, notably *Clarias gariepinus*) is often referred to as the African catfish; *Clarias gariepinus* is a robust species commonly found in African waters. This fish is notable for its barbels and adaptability to different aquatic conditions.
(c) ***Bagrus* spp.** (e.g., *Bagrus bajad*) is a genus which includes several species of catfish known for their elongated bodies and strong predatory behavior. *Bagrus bajad* is one of the more common species found in the region.
(d) **African lungfish** (*Protopterus* spp.) is a unique freshwater fish known for their ability to breathe air. *Protopterus annectens* is often found in the region's shallow waters and ponds.
(e) ***Barbus* spp.** (e.g., *Barbus filamentosus*) is a genus which includes several species of cyprinids, which are generally characterized by their slender bodies and are common in the rivers of Katanga.
(f) ***Mormyrus* spp.** is known as elephant fishes; members of this genus possess elongated bodies and can produce electric signals. They are adaptable and thrive in various environments.
(g) **Cichlids** are a diverse group of freshwater fish, many of which are endemic to African lakes and rivers. They exhibit a wide range of colors and behaviors.

According to Shumway et al. (2003), 54% of Congo's fish species belong to the Otophysi lineage. Fig. 8.4 shows the percentage of species in each of the 11 orders that we recorded. The most abundant orders include Osteoglossiformes, Siluriformes, Characiformes, Perciformes, and Cypriniformes. The graph indicates that the most prevalent family of fish is the Mormyridae, also known as the electric fish family (order Osteoglossiformes), making up 23% of the species identified and 27% of those known for the Congo River. Following them are the Cichlidae, accounting for 11% of the sample, the Characidae at 10%, and the Distichodontidae at 9%. Two catfish families, Mochokidae and Bagridae, represent 9% and 7% of the sample, respectively. Cyprinidae make up 7% of the sample. Additionally, the Mormyridae family is dominant in all provinces, making up over 20% of the sample. Other significant families across the provinces, each constituting more than 5% of the sample, include Characidae, Mochokidae, Bagridae, Distichodontidae, and Cichlidae. In Bas-Congo, Cyprinidae are notably prominent at 11%, while in Equateur, Clariidae stands out at 6%.

In addition to the fish species described earlier, Shumway et al. (2003) indicated an overview of some fish families, listed by common name from the DR Congo including the Katanga region as:

1. Lungfishes: *Protopterus* is an air-breathing fish. It can live in anoxic or hypoxic water.
2. Bichirs: *Polypterus* is characteristic of ancient fishes, with lobed fins and a hard coating on its scales.

Percentages indicate the proportional representation of species in each family.

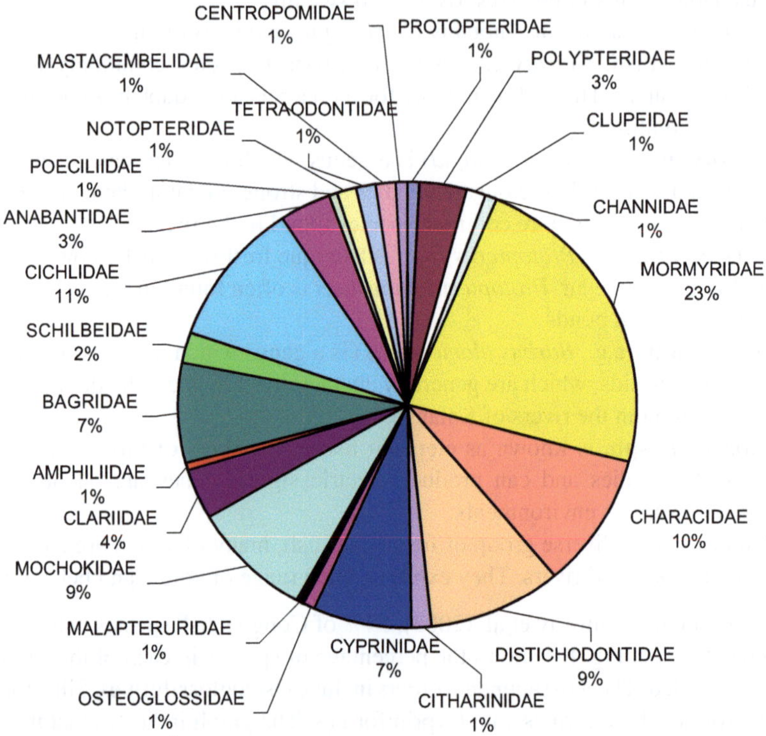

Fig. 8.4 Percentage of different families of fish in Katanga, DRC. (Source: Shumway et al. 2003, reproduced with permission from Drs. C. Shumway et al.)

3. Herrings: *Odaxothrissa* is a common freshwater species of a largely marine herring family (Clupeidae).
4. Elephant fishes: Mormyrids are weakly electric fish, with an electric organ in their tail, used for location and communication. They are primarily active at night.
5. Characins: Characids can be identified by their small dorsal adipose fin.
6. Killifishes (in the order Cyprinodontiformes): These top minnows are small insectivorous species that are an important prey item for other fish. They feed on insect larvae, such as mosquitoes, and may be important for disease control.
7. Cichlids are recognized by a single pair of nostrils.

They further reported that the DR Congo is known to have 929 bird species, of which 24 are endemic and 26 threatened. With our rapid investigation, using only one mist net, we recorded 60 species of birds, belonging to 15 families. The most abundant order was the Passeriformes, followed by Coraciiformes and Ciconiiformes. They observed the following five threatened bird species (on the CITES list): *Treron calvus* (African green pigeon; Bandundu), *Turtur afer* (blue-spotted wood dove;

Equateur), *Streptopelia semitorquata* (Bas-Congo and Equateur), *Corythaeola cristata* (great blue turaco; Bandundu), and *Pycnonotus* spp. (bulbuls; Bandundu and Equateur).

Mining in the DR Congo is presently centered in the former Katanga province where mineral products such as copper, cobalt, zinc, manganese, uranium, cassiterite (tin ore), coal, gold, and silver are produced. Although it is generally perceived as a catalyst for economic growth and development, mining sector activities often undermine economic development and environmental diversity. A variety of pollutants from smelters are continuously released into the atmosphere and in rivers resulting in environmental negative impacts of major concern. A range of environmental consequences from mining and metallurgical industries include:

- Direct loss of vegetation cover in the mined area and other areas associated with deposition of tailings or smoke discharge on air.
- Pollution especially of rivers aggravated by accidents (e.g., tailing dams).
- Impacts on other protected area values from noise and visual intrusion, arising from both mining and secondary activities, including transportation.
- Land degradation and loss of vegetation cover in the mined area and other parts directly affected by associated activities such as deposition of tailings or consequences such as
- Secondary effects of human immigration in association with real or perceived livelihood opportunities (e.g., on water supplies, illegal hunting, harvesting of vegetation, alien invasions, illegal land settlements).
- Impacts on other protected area values from noise and visual intrusion, arising from both mining and secondary activities, including transportation.

Several of the above biological species have reduced considerably or are threatened by mining and metallurgical industry pollution in various terrestrial and aquatic ecosystems in the Katanga region. The biodiversity—which is made up of ecosystems, species, and genetic diversity in Katanga region—is steadily disappearing due to human activity such as mineral exploitation and operations of metallurgical industries. All kinds of pollution, air, water, noise, plastic, and even light pollution, affect biodiversity, and pollutants—such as from mining and metallurgical industries—alter the environmental conditions directly or indirectly and negatively affect the flow of nutrients and constitution of the ecosystem. This leads to changes in abundance of species and hence biodiversity (https://4circularity.com/what-is-causing-biodiversity-loss/). Declining biodiversity lowers the ecosystems' productivity—the amount of food energy that is converted into the biomass—and lowers the quality of the ecosystems' services comprising of maintaining the soil, purifying water that runs through it, and supplying food and shade. In addition, biodiversity loss greatly threatens the structure and proper functioning of ecosystems and affects economic systems and human society. Mankind needs biodiversity for various services including provision of food, clean water and air, fertile soil, medicines, raw materials, protection from storms and floods, a stable climate, and recreation.

The above situation analysis is in line with statements by the WWF director general (WWF, 2020) that “This highlights a deep cultural and systemic shift is urgently needed, one that so far our civilisation has failed to embrace: a transition to a society and economic system that values nature, stops taking it for granted and recognises that we depend on nature more than nature depends on us.”

8.11 Environmental Impacts of Cobalt Mining in the DR Congo

Cobalt mining in the DR Congo has garnered significant attention due to its critical role in the global supply chain for electronic devices, electric vehicles, and renewable energy technologies. While cobalt is an essential mineral, the environmental impacts of its extraction and processing are profound and multifaceted. Cobalt, once hailed as a miracle metal, is now causing severe environmental damage due to toxic dumping, which devastates landscapes, pollutes water, and contaminates crops (Davey 2023). High cobalt concentrations are linked to the death of essential soil organisms like crops and worms, crucial for soil fertility.

A Congolese resident Heritier Maloba recounts that fish have long vanished from his childhood fishing spot, killed by acids and waste from nearby mines. This is a widespread issue across Congo’s cobalt-rich regions. A study on fish from Tshangalele Lake, near mining towns, revealed significant cobalt contamination, posing health risks through fish consumption or drinking the lake’s water. Cobalt is classified as a possible carcinogen and a radioactive element, making it a significant health hazard.

8.11.1 Health Hazards and Air Pollution

The toxic materials released during mining activities and the exposure to contaminated water and soil often induce severe health risks for local populations. Communities may experience increased incidences of lead poisoning, respiratory diseases, and other health issues linked to exposure to pollutants. The air around cobalt mines is filled with toxic dust and grit, making it hazardous to breathe. Studies have found that working in cobalt mines greatly increases the risk of birth defects, such as limb abnormalities and spina bifida, due to high levels of toxic pollution. Dust and emissions from mining activities, including equipment and transportation, contribute to air pollution. The burning of fossil fuels in mining operations further exacerbates this issue. Airborne pollutants can affect not only the immediate mining environment but also neighboring communities, leading to respiratory issues and other health problems for local populations.

8.11.2 Human Rights Violations

The term "blood cobalt" reflects the severe human rights abuses linked to Congo's cobalt mining industry. Workers, including children, toil in degrading conditions, using primitive tools to extract cobalt. These conditions have led to modern-day slavery, human trafficking, and child labor. In addition, mining operations lead to the displacement of local communities and land acquisition for mining forcing people to abandon their homes and livelihoods, often without fair compensation or resettlement options. This displacement creates social tensions and exacerbates poverty, as displaced individuals may struggle to find new means of subsistence.

8.11.3 Environmental Degradation and Exploitation

Congo's cobalt boom illustrates how the clean energy revolution, intended to combat climate change, is entangled in environmental degradation and exploitation. Large-scale mining operations often require significant land clearing, leading to deforestation. In the DR Congo, where dense rainforests play a crucial role in carbon sequestration and biodiversity, this can result in loss of habitat for various species, including endangered flora and fauna. The destruction of forests contributes to increases in carbon emissions and decreases in biodiversity, disrupting local ecosystems and altering water cycles. Moreover, cobalt mining activities disturb soil structures resulting in considerable soil erosion. The removal of vegetation exposes soils to wind and rain, which can wash away topsoil and lead to sediment displacement into nearby waterways. Soil erosion can impact agricultural productivity in surrounding areas, while sedimentation can clog rivers and streams, negatively affecting aquatic life and degrading water quality.

Mining processes often involve the use of hazardous chemicals, such as sulfuric acid and other reagents, which can leach into nearby water bodies. Additionally, waste rock and tailings can release heavy metals and other pollutants causing contamination of water sources which also poses serious threats to local communities and ecosystems. Polluted waterways can affect fish populations, harm animals that depend on these water sources, and compromise drinking water quality.

Chapter 9
Findings

Keywords Bioaccumulation · Pollution · Exposure · Health · Safety · Children · Community health · Public health risks

Some of the findings reported below are portions of the results of more than 10-year-long DelPhE project investigations in the Central Africa Copperbelt region of Zambia and the DR Congo. They are reproduced from the article published in the journal *Environmental Advances* by Ambayeba Muimba-Kankolongo, Célestin Banza Lubaba Nkulu, Jacob Mwitwa, Florence M. Kampemba, Misery Mulele Nabuyanda, Vincent Haufroid, Erik Smolders, and Benoit Nemery, Vol. 6, 100103, Copyright Elsevier Inc., December 2021. They also include results reproduced from Ambayeba Muimba-Kankolongo, Celestin Banza Lubaba Nkulu, Jacob Mwitwa, Florence M. Kampemba, and Mulele M. Nabuyanda (2022), Impacts of Trace Metals Pollution of Water, Food Crops, and Ambient Air on Population Health in Zambia and the DR Congo, *Journal of Environmental and Public Health*, Volume 2022, Article ID 4515115, Copyright © 2022 A. Muimba-Kankolongo et al., 14 pages.

9.1 Contamination of Food Crops

Recent toxicological analyses of food crop tissues from the region reveal substantial contamination with heavy metals and metalloids. A significant disparity in pollutant concentrations was observed ($p < 0.001$) between the DR Congo and Zambia, as well as among the six surveyed locations within both countries (refer to tables and figures below).

The study examined the distribution of nine trace elements in food crops from Zambia ($n = 128$) and the DR Congo ($n = 145$), revealing a wide variation between the minimum and maximum concentrations (Table 9.1). Notably, concentrations of all elements, except for manganese (Mn), were significantly higher in plant samples

A. Muimba-Kankolongo et al., *Mining Impacts on the Environment in the Central African Copperbelt of Zambia and the Democratic Republic of Congo*, https://doi.org/10.1007/978-3-031-96786-3_9

Table 9.1 Concentration of trace elements (μg/g dry weight) in samples of food crops from Zambia and the DR Congo

	Zambia (n = 128)	DRC (n = 145)		
	P50 (P25-P75)	P50 (P25-P75)	P50 DRC /	P*
	μg/g dry weight	μg/g dry weight	P50 Zambia	
Mn	**78.1** (32.9–170)	**66.1** (39.4–128)	**0.8**	0.34
Co	**1.49** (0.80–2.96)	**6.44** (1.63–16.4)	**4.3**	<0.001
Ni	**7.06** (2.50–12.0)	**10.4** (4.90–17.9)	**1.5**	<0.001
Cu	**33.4** (19.9–59.0)	**62.1** (24.8–166)	**1.9**	<0.001
Zn	**72.8** (40.5–137)	**140** (59.5–324)	**1.9**	<0.001
As	***0.800*** *(0.800–0.800)*	***0.800*** (*0.800*–2.60)	**1.0**	<0.001
Cd	**0.156** (0.056–0.373)	**0.867** (0.359–2.25)	**5.4**	<0.001
Pb	**0.464** (*0.150*–0.997)	**3.87** (1.11–10.4)	**8.3**	<0.001
U	**1.96** (0.931–3.69)	**3.09** (1.47–5.79)	**1.6**	<0.001

P50, P25, and P75 are 50th (median), 25th, and 75th percentiles, respectively
*p value for difference between Zambia and DRC (Mann–Whitney test)
values in italics are imputed values (half the value of limit of quantification)
See Supplemental Table D for more details

from the DR Congo compared to those from Zambia. Although arsenic (As) levels remained below the limit of quantification (LOQ), they were still higher in crop samples from the DR Congo than those from Zambia.

The analysis performed using the Kruskal–Wallis test followed by Dunn's multiple comparisons procedure for the geometric means of trace elements and their ratios can be found in Table 9.2. Significant differences were noted for six elements—cobalt (Co), copper (Cu), arsenic (As), cadmium (Cd), lead (Pb), and uranium (U)—when comparing crops grown near mining areas to those from distant sites, with geometric means ranging from approximately 2 for As to over 5 for Co. An exception was arsenic (As), which exhibited significantly higher levels in Kipushi Territory, in the DR Congo.

Food crops from Lubumbashi City were found to be even more contaminated than those from Likasi City, Kipushi Territory, and any in Zambia. Generally, concentrations of nickel (Ni), lead (Pb), and cadmium (Cd) were higher in nearly all analyzed food crops. Copper (Cu) levels were particularly higher in samples of *Cucurbita maxima* (pumpkin) and *Amaranthus hybridus* (amaranth). Despite the presence of both Cd and Pb in crops from both countries, their levels significantly exceeded the allowable standards for food plants in the DR Congo.

Overall, the trace metal content (per dry weight) was higher in leafy vegetables (amaranth, pumpkin, sweet potato, and cassava leaves), as well as in tomatoes and beans, compared to crops such as maize and tuberous roots, for which fewer samples were available for analysis.

Trace element concentrations in all food crops are shown by location in the figures below and by village or urban neighborhood as well. Spearman correlations (Table 9.3) between the concentrations of various trace metal elements in food crops varied significantly, with the exception of manganese (Mn), which only correlated

Table 9.2 Geometric means (GM) and their 95% confidence intervals of trace metal concentrations (mg/kg dry weight) in food crops in Zambia (Z) and the DR Congo (C) and according to distance from mining (FAR or NEAR)

	Z all	C all	GM ratio Z all / C all	Z FAR	Z NEAR	C FAR	C NEAR	GM ratio Z NEAR / Z FAR	GM ratio C NEAR / C FAR	GM ratio C FAR / Z FAR	GM ratio C NEAR / Z NEAR
N samples	128	145		40	88	33	112				
Mn	76.3 (61.9-94.1)	64.5 (54.0-77.0)	0.84	114 (75.0-173)	63.6 (50.3-80.4)	125 (80.2-193)	53.1 (44.5-63.4)	0.56	0.43	1.09	0.83
Co	1.54 (1.25-1.90)	5.31 (4.06-6.95)	3.45	1.19 (0.85-1.65)	1.74 (1.34-2.26)	1.42 (0.95-2.14)	7.83 (5.85-10.5)	1.46	5.50	1.20	4.51
Ni	5.29 4.38-6.40)	9.85 (8.52-11.4)	1.86	6.24 (4.52-8.62)	4.91 (3.88-6.22)	9.06 (7.01-11.7)	10.1 (8.49-12.0)	0.79	1.12	1.45	2.06
Cu	34.0 (29.0-39.7)	56.2 (42.7-73.9)	1.65	25.0 (19.6-31.8)	39.0 (32.1-47.5)	17.6 (11.7-26.6)	79.0 (58.1-108)	1.56	4.48	0.71	2.02
Zn	88.8 (72.4-109)	132 (111-158)	1.49	53.6 (40.3-71.5)	112 (86.4-144)	106 (71.6-158)	141 (115-173)	2.08	1.33	1.98	1.26
As	0.83 (0.80-0.86)	1.46 (1.28-1.67)	1.76	0.80 (0.80-0.80)	0.84 (0.81-0.88)	0.90 (0.80-1.02)	1.69 (1.43-1.98)	1.05	1.87	1.13	2.00
Cd	0.15 (0.12-0.18)	0.92 (0.73-1.16)	6.25	0.11 (0.07-0.18)	0.17 (0.13-0.21)	0.38 (0.28-0.52)	1.20 (0.92-1.56)	1.47	3.15	3.35	7.21
Pb	0.54 (0.43-0.68)	3.61 (2.81-4.65)	6.68	0.29 (0.23-0.38)	0.72 (0.53-0.96)	1.15 (0.82-1.62)	5.06 (3.81-6.72)	2.45	4.40	3.93	7.07
U	1.84	2.71	1.48	1.89	1.81	1.18	3.46	0.96	2.94	0.63	1.91

with zinc (Zn). Correlations were generally stronger for samples from the DR Congo compared to those from Zambia.

Based on the proximity to mining and metallurgic industry operations of the fields or gardens from which food crops' samples were obtained, the concentrations of trace elements across eight different produces' categories, separately for Zambia and the DR Congo, are shown in Figs. 9.1, 9.2, 9.3, 9.4, and 9.5. For manganese (Mn), significantly lower concentrations—approximately half—were found in areas near mining compared to control sites, in both Zambia and the DR Congo. In Zambia, only zinc (Zn) and lead (Pb) had significantly higher concentrations—about twofold—in vegetables collected near mining sites compared to control sites.

In the DR Congo, statistically significant differences in pollutant concentrations were observed for six elements—cobalt (Co), copper (Cu), arsenic (As), cadmium (Cd), lead (Pb), and uranium (U)—between crops near mining areas and those from remote sites, with geometric mean (GM) ratios ranging from about 2 for As to over

Table 9.3 Spearman correlations of concentrations of nine trace metals in crop samples from Zambia and the DR Congo

All ($i = 273$)	Mn	Co	Ni	Cu	Zn	As	Cd	Pb	U
Mn	1.00	0.06	0.08	−0.02	0.19	−0.06	0.06	0.00	−0.03
Co		1.00	0.27	0.83	0.46	0.54	0.62	0.72	0.64
Ni			1.00	0.30	0.11	0.19	0.36	0.37	0.31
Cu				1.00	0.41	0.54	0.61	0.71	0.73
Zn					1.00	0.40	0.50	0.45	0.34
As						1.00	0.56	0.60	0.40
Cd							1.00	0.74	0.51
Pb								1.00	0.58
U									1.00

Zambia ($n = 128$)	Mn	Co	Ni	Cu	Zn	As	Cd	Pb	U
Mn	1.00	0.10	0.02	−0.06	0.23	0.00	0.24	0.00	−0.08
Co		1.00	0.00	0.63	0.21	0.18	0.30	0.43	0.45
Ni			1.00	0.12	−0.09	−0.11	0.20	0.21	0.22
Cu				1.00	0.13	0.20	0.51	0.57	0.61
Zn					1.00	0.06	0.27	0.14	0.09
As						1.00	0.07	0.15	0.21
Cd							1.00	0.43	0.38
Pb								1.00	0.41
U									1.00

DRC ($n = 145$)	Mn	Co	Ni	Cu	Zn	As	Cd	Pb	U
Mn	1.00	0.06	0.17	0.03	0.14	−0.08	−0.07	−0.01	0.04
Co		1.00	0.33	0.93	0.50	0.53	0.67	0.84	0.77
Ni			1.00	0.35	0.18	0.16	0.29	0.28	0.29
Cu				1.00	0.49	0.62	0.67	0.83	0.81
Zn					1.00	0.47	0.60	0.61	0.47
As						1.00	0.59	0.66	0.48
Cd							1.00	0.74	0.61
Pb								1.00	0.75
U									1.00

5 for Co (Table 9.2). Even control values in the DR Congo were significantly and substantially higher than those in Zambia for Zn (twofold), Cd (threefold), and Pb (fourfold). Near mining sites, concentrations were significantly higher in the DR Congo than in Zambia for all elements except Mn and Zn, with Cd and Pb GMs more than sevenfold higher. Excluding maize and tuberous roots did not significantly alter these comparisons.

Standard limits for As, Cd, Mn, Zn, and Pb in foods for human consumption are detailed in the WHO-FAO guidelines (FAO/WHO 2011, 2017). For inorganic arsenic (As), the FAO-WHO Codex Alimentarius sets a maximum limit of 0.2 mg/kg fresh weight in rice, approximately 1 mg/kg dry weight (assuming a 5/1 wet/dry

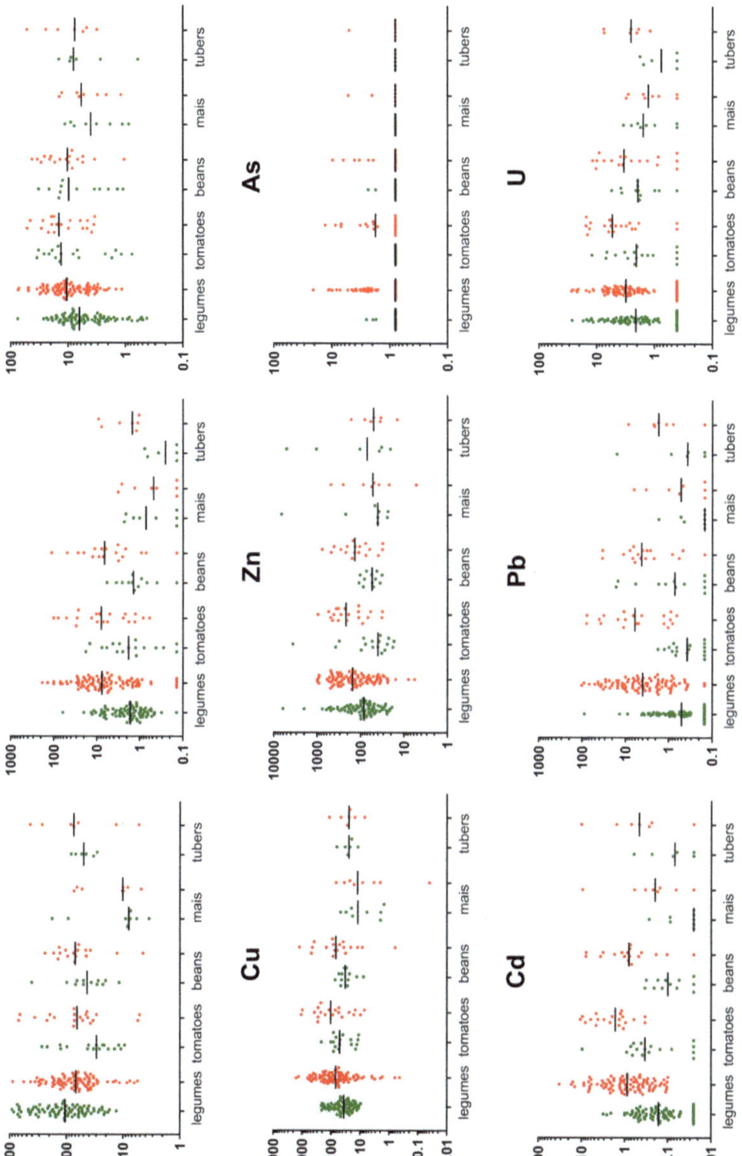

Fig. 9.1 Concentrations (µg/g dry weight) of nine trace elements in food crops by category (leafy vegetables, tomatoes, beans, maize, tuberous roots): Green, Zambia; red, DR Congo

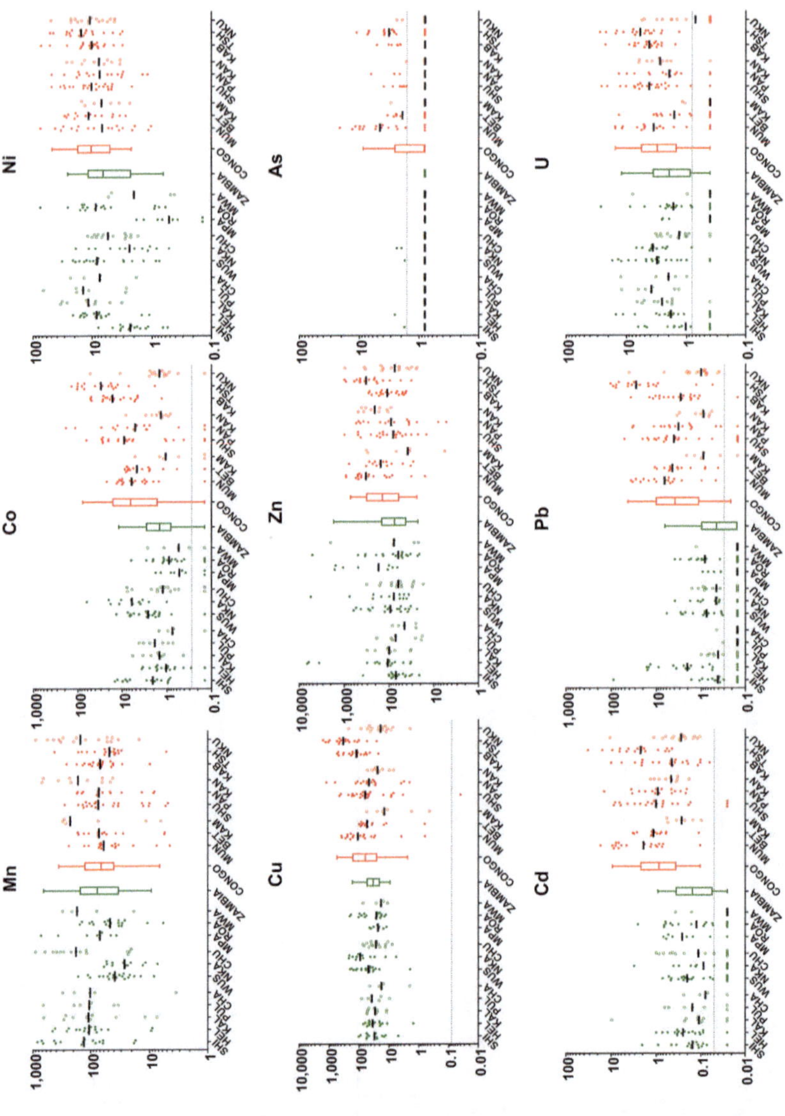

Fig. 9.2 Concentrations (µg/g dry weight) of nine trace elements in food crops by village or urban neighborhood

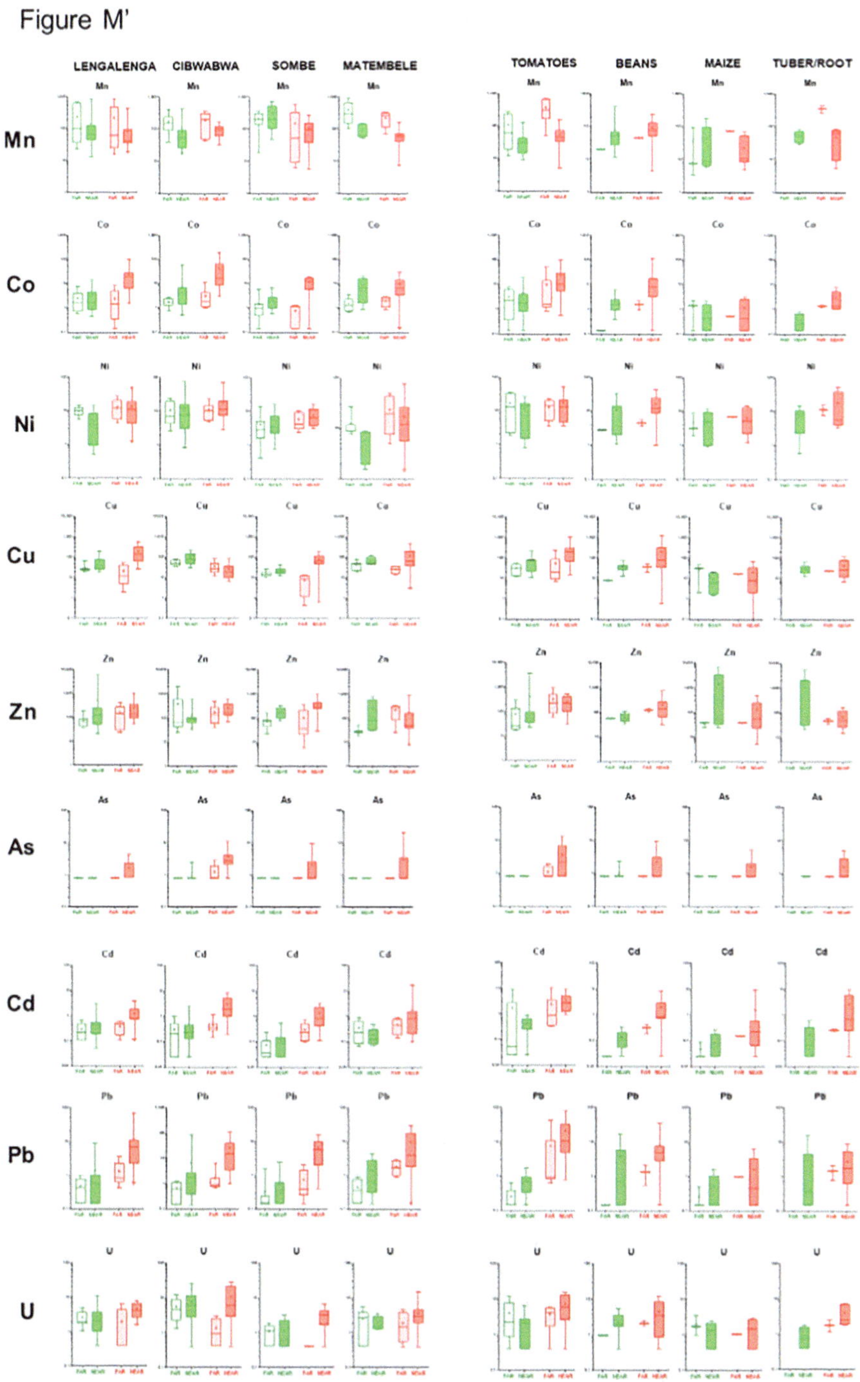

Fig. 9.3 Concentrations (μg/g dry weight) of nine trace elements in food crops by type of food crop

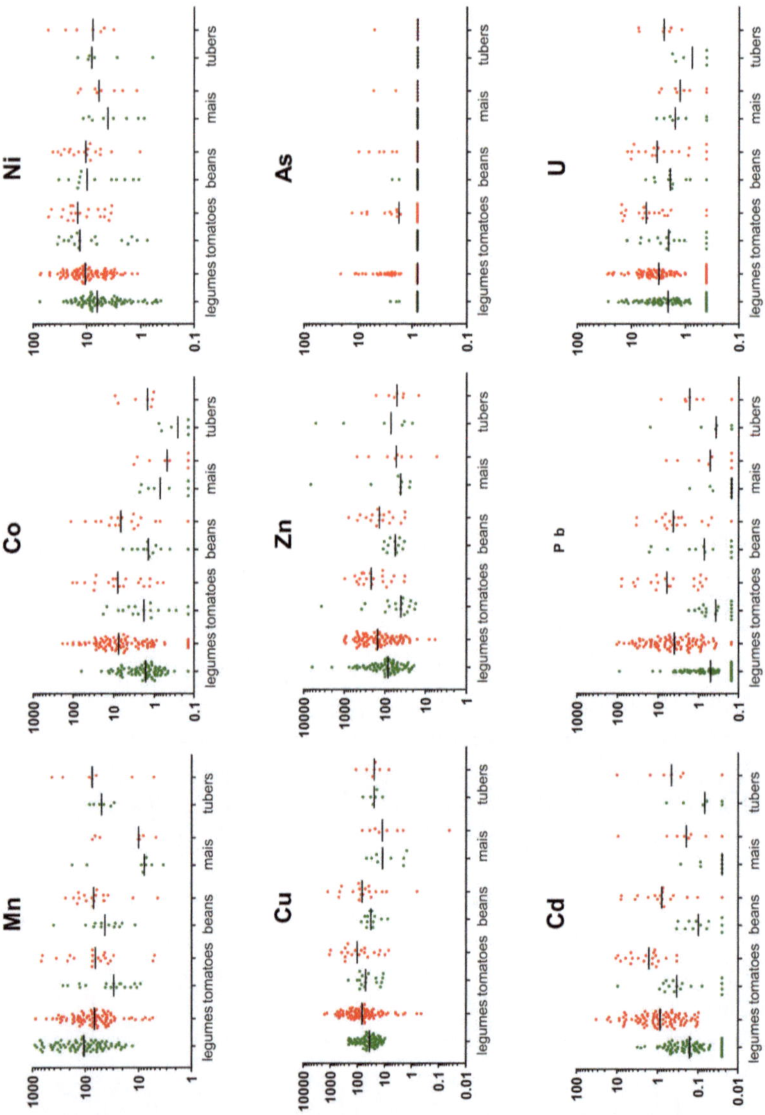

Fig. 9.4 Concentrations (μg/g dry weight) of nine trace elements in food crops by category (leafy vegetables, tomatoes, beans, maize, tuberous roots): Green, Zambia; red, the DR Congo

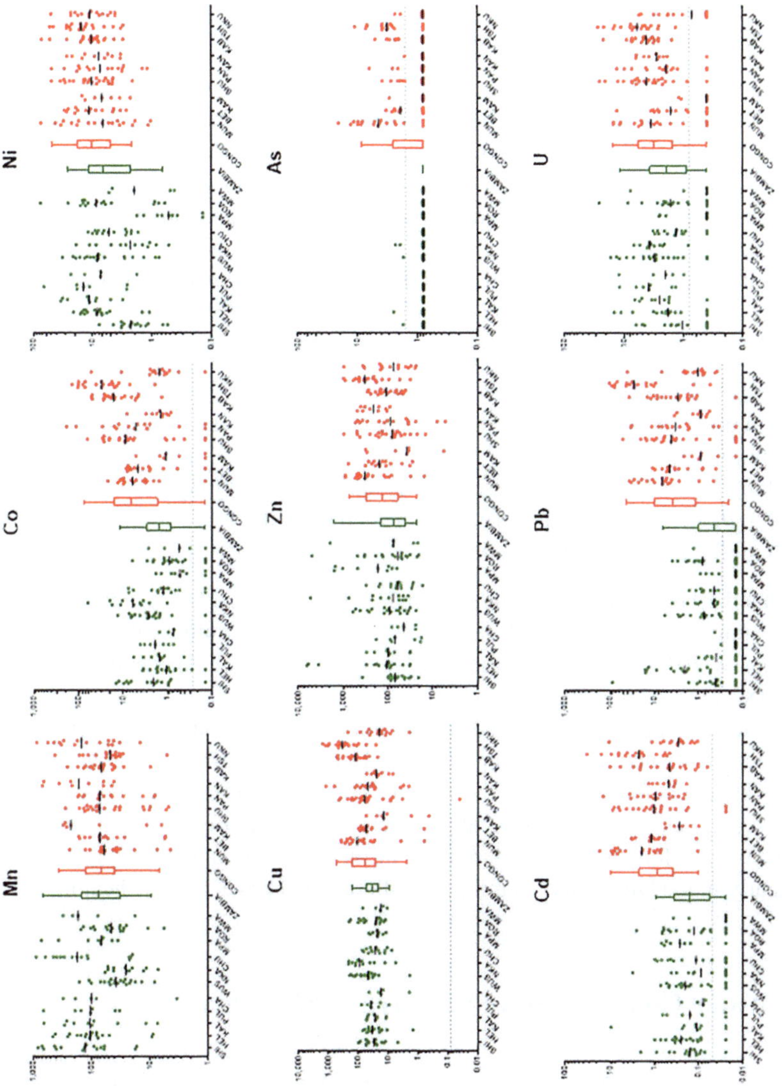

Fig. 9.5 Concentrations (µg/g dry weight) of nine trace elements in food crops by village or urban neighborhood

weight ratio). This limit was exceeded in 4/128 samples (3%) in Zambia (all near mining) and 59/145 samples (41%) in the DR Congo (all but four near mining).

For cadmium (Cd), the maximum limits vary by vegetable type:

- 0.05 mg/kg for brassica, bulb, and fruiting vegetables
- 0.1 mg/kg for legume vegetables, pulses, root and tuber vegetables, stalk and stem vegetables, cereal grains
- 0.2 mg/kg for leafy vegetables and wheat.

Using the most generous limit (0.2 mg/kg) and a 5/1 conversion factor for dry weight, 4/128 samples (3%) exceeded 1 mg/kg dry weight in Zambia, whereas nearly half (68/145, 47%) of crop samples exceeded 1 mg/kg dry weight in the DR Congo (4/33 [12%] in control sites and 64/112 [57%] in mining areas). All 20 tomato samples from the DR Congo contained over 0.25 mg Cd/kg dry weight (GM 2 mg/kg), compared to 10/17 (59%) in Zambia (GM 0.24 mg/kg). Similarly, most bean samples (13/18, 72%) from the DR Congo had over 0.5 mg Cd/kg (GM 0.82 mg/kg), while none in Zambia exceeded this (GM 0.089 mg/kg).

For lead (Pb), maximum limits also vary by vegetable type:

- 0.20 mg/kg for cereals, legumes, and pulses
- 0.10 mg/kg for vegetables (excluding brassica, leafy vegetables, fresh herbs, and fungi)
- 0.30 mg/kg for brassica, leafy vegetables, and cultivated fungi
- 0.10 mg/kg for fruit excluding berries and small fruit
- 0.20 mg/kg for berries and small fruit.

With a 0.30 mg/kg fresh weight limit and a 5/1 conversion factor, a large proportion of samples exceeded the Pb limit: 23/128 (18%) in Zambia (all but one near mining) and 101/145 (70%) in the DR Congo (9/33 [27%] in control sites and 92/112 [82%] in mining sites).

The mean concentrations of these trace metals in food samples from the DR Congo, which are notably above the WFP/FAO standards, pose potential health risks to the local population. Moreover, the stark difference in food contamination between the two countries sharing geological sediment-hosted stratiform copper highlights the urgent need for improved environmental management and stricter regulations, particularly in the DR Congo, to mitigate pollution and protect public health.

9.2 Water Contamination with Trace Elements

The analyzed data explanation clearly shows significant variation in trace metal contents between countries and the various sites where samples were collected. An ANOVA test revealed significant differences ($p \leq 0.001$) in the concentrations of all trace metals and metalloids in water samples from the DR Congo and Zambia, with higher contamination recorded in the DR Congo (Table 9.4). The concentrations of nine trace elements in water from Zambia (n = 42) and the DR Congo (n = 35) showed wide variation between the lowest and highest concentrations. Notably,

Table 9.4 Concentration of trace elements (µg/L) in unfiltered water samples from Zambia and the DR Congo

		ZAMBIA ($n = 42$)		DRC ($n = 35$)			
	WHO limit	P50 [P25-P75]	>WHO limit	P50 [P25-P75]	>WHO limit	P50 DRC /	*P**
	µg/L	µg/L	*n* (%)	µg/L	*n* (%)	P50 Zambia	
Mn	50	**12.6** [8.00–43.1]	9 (21%)	**79.8** [15.8–2606]	19 (51%)	**6.3**	0.001
Co	NA	**2.04** [0.83–8.61]		**80.0** [21.4–2624]		**39**	<0.001
Ni	70	**1.74** [1.02–6.53]	0	**7.55** [2.16–119]	11 (31%)	**4.3**	<0.001
Cu	2000	**5.09** [1.40–17.8]	1 (2%)	**139** [27.4–3416]	11 (31%)	**27**	<0.001
Zn	5000	**14.6** [4.74–27.8]	1 (2%)	**27.3** [8.31–482]	4 (11%)	**1.9**	0.021
As	10	**0.167** [0.098–0.244]	0	**1.89** [0.504–23.7]	11 (31%)	**11**	<0.001
Cd	3	***0.008*** [*0.008*–0.017]	0	**0.416** [0.049–6.12]	11 (31%)	**55**	<0.001
Pb	10	**0.080** [0.030–0.690]	1 (2%)	**14.2** [1.28–354]	21 (60%)	**177**	<0.001
U	30	**0.033** [*0.004*–0.172]	0	**0.808** [0.131–3.21]	8 (23%)	**24**	<0.001

WHO limit is upper limit for drinking water according to World Health Organization guidelines (not available for Co); P50, P25, and P75 are 50th (median), 25th, and 75th percentiles, respectively. **p* value for difference between Zambia and DRC (Mann–Whitney test); values in italics are imputed values (half the value of limit of detection). See Supplemental Table B for more details

concentrations were significantly higher in water samples from the DR Congo compared to those from Zambia, except for zinc (Zn). Consequently, the median value concentrations in the DR Congo were consistently above WHO guidelines for drinking water standards, posing serious health risks to the local population. Spearman's correlation analyses (Table 9.5) revealed a weak correlation between trace metal concentrations in water from Zambia and a strong correlation in samples from the DR Congo.

Mean contents (µg/L) of trace metals were as follows:

- **DR Congo**: Mn, 5454.6; Zn, 2552.2; Cd, 138.7; Pb, 39.7; and U, 21.4.
- **Zambia**: Mn, 108.9; Zn, 543.3; Cd, 0.3; Pb, 0.2; and U, 0.5.

9.2.1 Trace Element Concentrations in Surface Waters, Boreholes, and Taps

Trace element concentrations in water from boreholes were significantly higher in Likasi City (average = 20,791.4 µg/L) than in any other location surveyed in the DR Congo. In Zambia, concentrations of trace elements were highest in tap water from

Table 9.5 Spearman correlations of concentrations of nine trace metals in vegetable samples

ALL (n=273)	Mn	Co	Ni	Cu	Zn	As	Cd	Pb	U
Mn	1.00	0.06	0.08	−0.02	0.19	−0.06	0.06	0.00	−0.03
Co		1.00	0.27	0.83	0.46	0.54	0.62	0.72	0.64
Ni			1.00	0.30	0.11	0.19	0.36	0.37	0.31
Cu				1.00	0.41	0.54	0.61	0.71	0.73
Zn					1.00	0.40	0.50	0.45	0.34
As						1.00	0.56	0.60	0.40
Cd							1.00	0.74	0.51
Pb								1.00	0.58
U									1.00
ZAMBIA (n=128)	Mn	Co	Ni	Cu	Zn	As	Cd	Pb	U
Mn	1.00	0.10	0.02	−0.06	0.23	0.00	0.24	0.00	−0.08
Co		1.00	0.00	0.63	0.21	0.18	0.30	0.43	0.45
Ni			1.00	0.12	−0.09	−0.11	0.20	0.21	0.22
Cu				1.00	0.13	0.20	0.51	0.57	0.61
Zn					1.00	0.06	0.27	0.14	0.09
As						1.00	0.07	0.15	0.21
Cd							1.00	0.43	0.38
Pb								1.00	0.41
U									1.00
DRC (n=145)	Mn	Co	Ni	Cu	Zn	As	Cd	Pb	U
Mn	1.00	0.06	0.17	0.03	0.14	−0.08	−0.07	−0.01	0.04
Co		1.00	0.33	0.93	0.50	0.53	0.67	0.84	0.77
Ni			1.00	0.35	0.18	0.16	0.29	0.28	0.29
Cu				1.00	0.49	0.62	0.67	0.83	0.81
Zn					1.00	0.47	0.60	0.61	0.47
As						1.00	0.59	0.66	0.48
Cd							1.00	0.74	0.61
Pb								1.00	0.75
U									1.00

Chingola District (average = 223.5 μg/L) compared to those from Kitwe and Luanshya Districts, although all concentrations in Zambia were far below those in the DR Congo, despite involvement in mining activities.

Concentrations of trace metals in surface waters, and water from boreholes or taps, are depicted by location in Figs. 9.6, 9.7, and 9.8. In surface waters (17 samples), the highest concentrations of trace elements were in Likasi City (DR Congo), where water was sampled near metallurgical plant effluents, and the lowest in the Chanembo River in Chingola District, Zambia. Trace element concentrations in water from boreholes (37 samples) or taps (23 samples) were consistently higher in various DR Congo locations than in Zambia. Figures illustrate significantly higher

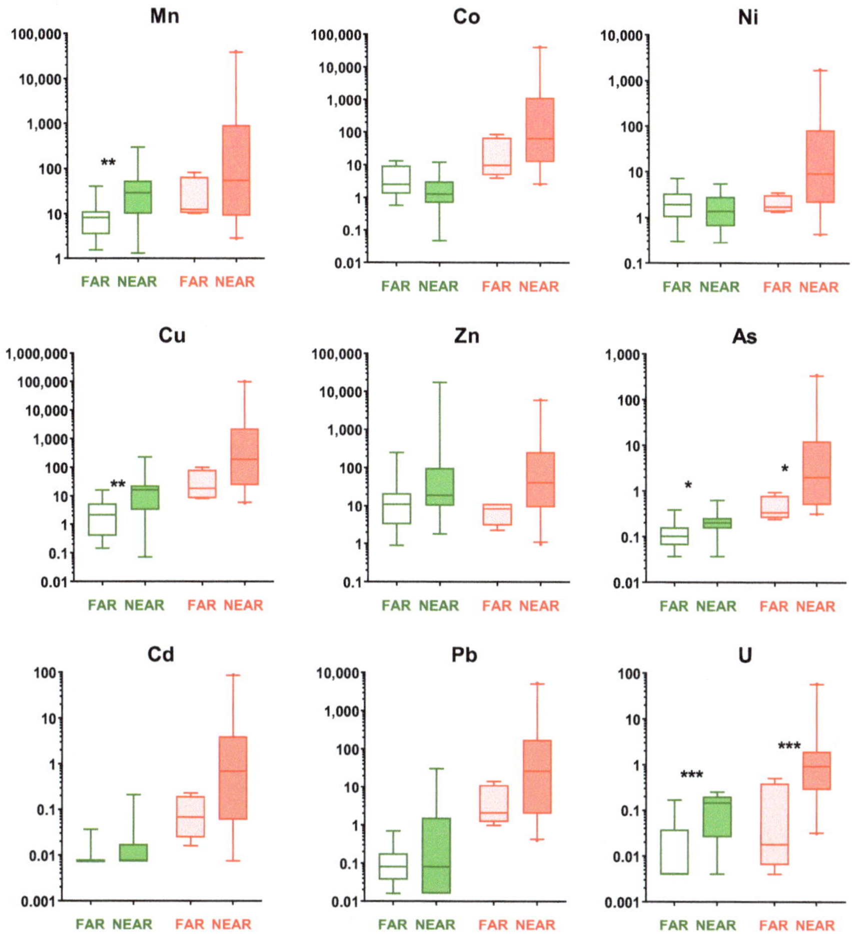

Fig. 9.6 Concentrations (μg/L) of nine trace elements in unfiltered drinking water according to proximity to mining operations, separately for Zambia and the DR Congo

concentrations of cobalt (Co), nickel (Ni), copper (Cu), arsenic (As), cadmium (Cd), lead (Pb), and uranium (U) in drinking water (boreholes and taps) near mining sites in the DR Congo compared to Zambia, with the most pronounced differences for Pb (335-fold difference for medians, 112-fold for GM), Cd (96-fold for medians, 50-fold for GM), and Co (51-fold for medians, 105-fold for GM).

In each panel, single values are shown for three locations (CHI Chingola, KIT Kitwe, and LUA Luanshya) in Zambia (left in green) and three locations (KIP Kipushi, LIK Likasi, and LUB Lubumbashi) in the DRC (right in red). Open circles are used for samples taken in sites far from mining activities; full circles are used for samples taken in sites near mining activities. The box plots (median with P25 and

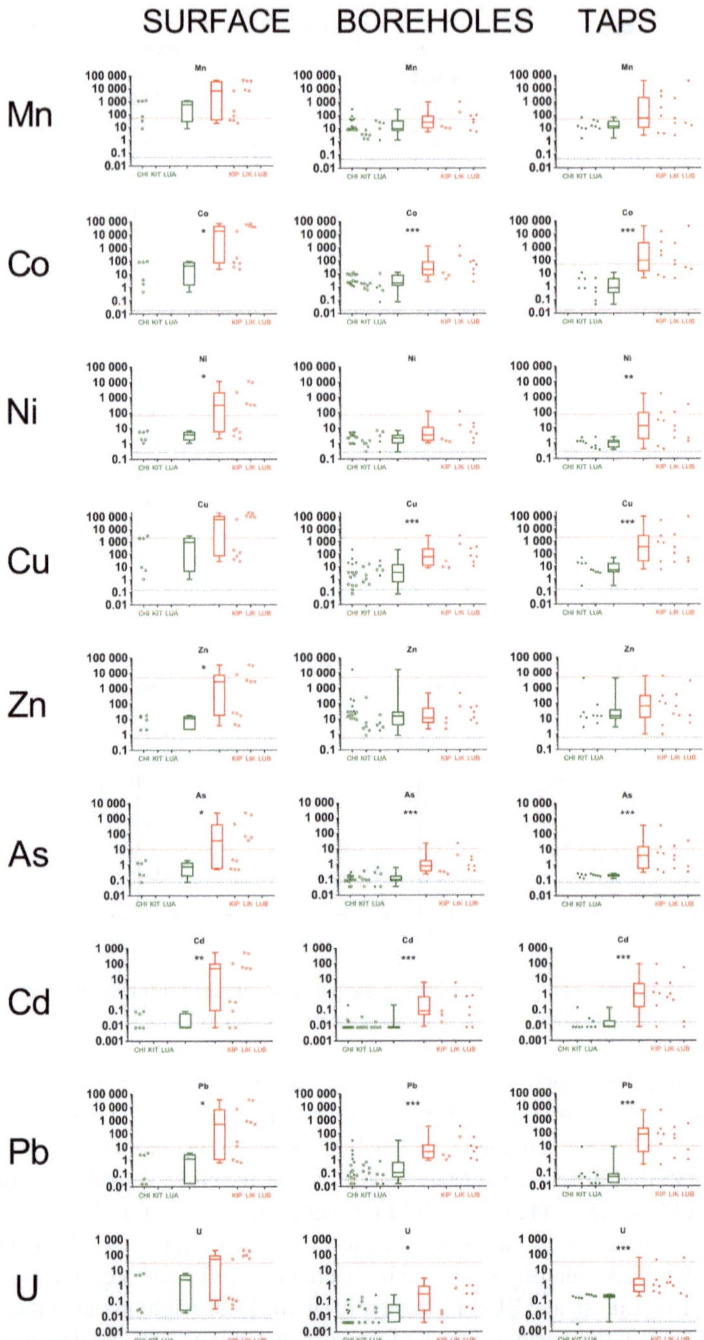

Fig. 9.7 Concentrations (µg/L) of nine trace elements in surface water, borehole water, or tap water in Zambia and the DR Congo

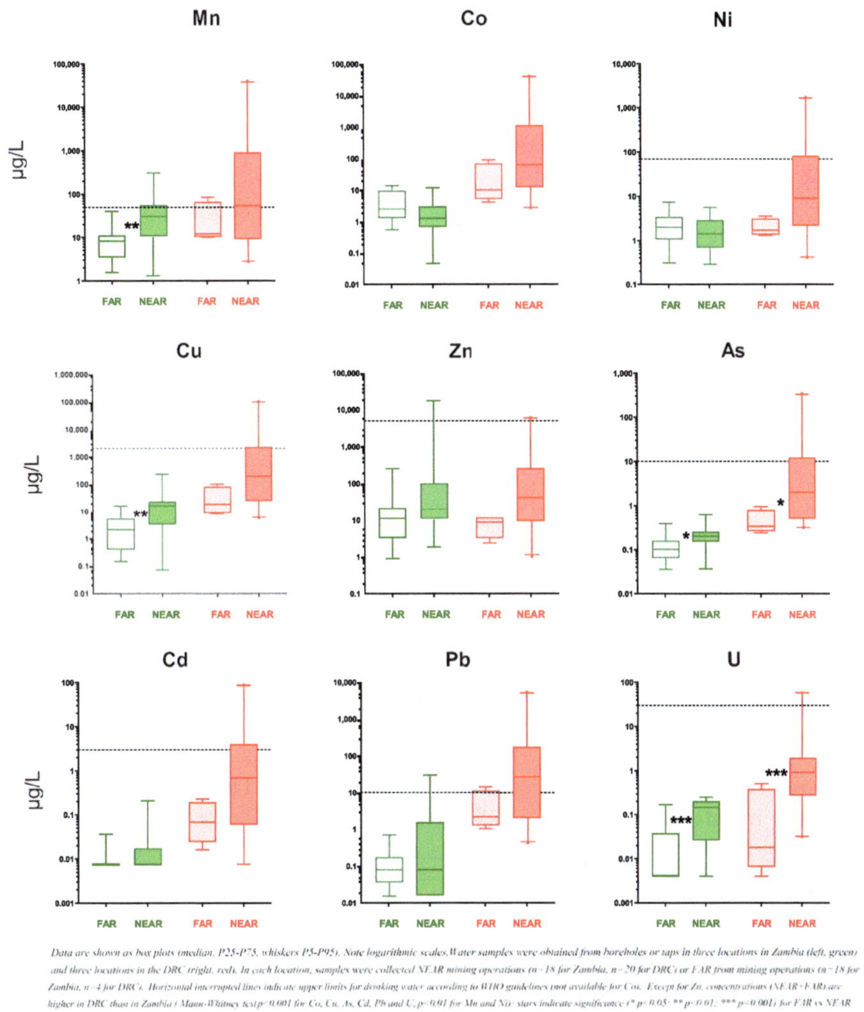

Fig. 9.8 Concentrations (µg/L) of nine trace elements in unfiltered drinking water according to proximity to mining operations, separately for Zambia and the DR Congo

P75; whiskers P5 to P95) depict the overall data for Zambia (left) and for the DRC (right). Red dotted horizontal lines represent upper limit values for drinking water according to WHO guidelines (not available for Co). Black dotted lines represent values of limit of detection (LOD); values below LOD have been assigned values of LOD/2. Stars indicate significance (* $p < 0.05$; ** $p < 0.01$, *** $p < 0.00$) for the difference between Zambia and the DR Congo (Mann–Whitney test).

9.3 Trace Metals and Metalloid in Urine

9.3.1 High Urinary Trace Metal Concentrations Near Mining Areas

Our study found that individuals living near mining and metallurgical industries in southeastern Katanga Province, the DR Congo—specifically in the Kabesha and Kawama areas of Lubumbashi City—exhibited abnormally high concentrations of trace metals in their urine. These urinary concentrations (measured in µg/g of creatinine) were especially elevated in younger children, who were the most affected group.

9.3.2 Vulnerability of Children

Children are particularly vulnerable to pollutants due to several factors, including their higher consumption of air, water, and food per unit of body weight compared to adults, often due to their outdoor activities (Etzel and Landrigan 2024).

9.3.3 Results of Laboratory Analysis of Urine Samples

Laboratory analysis of the collected urine samples revealed abnormally high levels of heavy metals and metalloids in all participants' urine (Fig. 9.9). The concentration ranges (in µg/g of creatinine) were as follows:

- Manganese (Mn): 0.03–140.94
- Cobalt (Co): 0.5–969.13
- Nickel (Ni): −0.01–31.17
- Copper (Cu): 2.63–531.35
- Zinc (Zn): 40.36–4079.08
- Arsenic (As): 2.70–219.95
- Cadmium (Cd): 0.07–4.79
- Lead (Pb): 0.64–580.87
- Uranium (U): −0.002–0.24

Comparisons Between Children and Adults

Overall, the levels of trace metals in the urine of children were higher than those in adults, with extremely high concentrations (µg/g of creatinine) in one subject (a 2-year-old boy) showing:

- Mn: 140.9

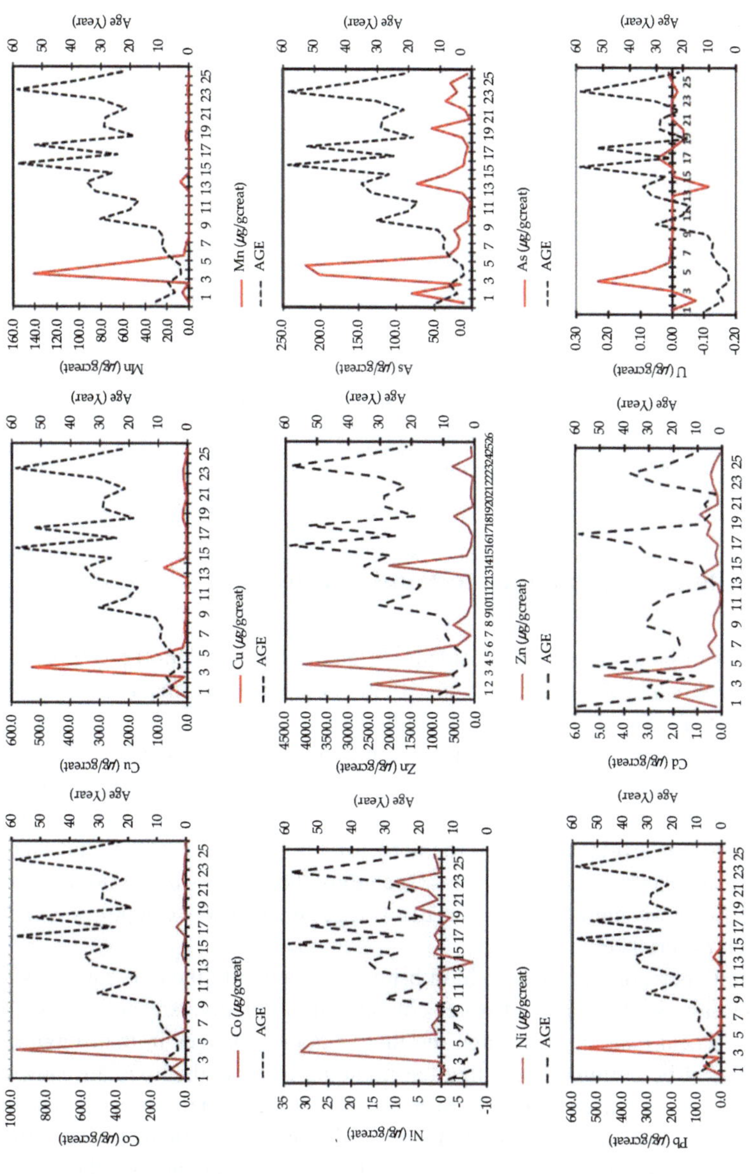

Fig. 9.9 Elements of trace metals and metalloids (µg/g creatinine) in urine samples from inhabitants in Kabesha and Kawama villages in Lubumbashi City, the DR Congo, with almost all metals (__) being greater in younger than elderly people (—)

- Co: 969.1
- Ni: 31.2
- Cu: 531.3
- Zn: 4079.1
- As: 201.9
- Mo: 1761.5
- Cd: 4.8
- Pb: 580.9
- U: 0.2

Residents reported that children's urine sometimes turned blackish and there were cases of congenital malformations.

9.3.4 Age Correlation

A strong correlation was found between urinary creatinine levels and the age of community residents ($r = 0.95$, $p \leq 0.001$), reinforcing the idea that age is a crucial factor during exposure to trace metal elements.

9.4 Population's Reactions to Safety Issues

The reactions of local community residents to mining pollution in the Copperbelt region of Zambia and the DR Congo are a vital component in understanding and addressing the environmental and social challenges posed by the mining industry in these countries. They are home to extensive mining operations that have brought economic growth while simultaneously exposing communities to significant environmental hazards. Residents' responses—ranging from strong activism to advocacy—serve as powerful testimonials to the real-life impacts of mining pollution on their health, livelihoods, and ecosystems. Including their voices in the book narrative offers a human-centered perspective that bridges the gap between policy, industry, and lived experiences.

Community perceptions are critical in bringing attention to the consequences of mining pollution, such as air and water contamination, biodiversity loss, and soil degradation. Local residents are often the first to notice changes in their environment, whether it be polluted rivers that harm fishing livelihoods, toxic air causing respiratory issues, or barren lands where crops once thrived. Their observations and lived experiences provide valuable, firsthand evidence that complements scientific data and underscores the urgency of these issues. Thus, documenting their reactions, the book readers will not only humanize the environmental narrative but will also discover the resilience and agency of these communities in demanding accountability and advocating for sustainable mining practices.

Additionally, community responses act as a catalyst for change, driving both public awareness and policy reforms. In Zambia and the DR Congo, grassroots movements and local protests have played pivotal roles in pushing governments and mining companies to acknowledge and address the environmental and social costs of their operations. Residents' collective efforts often lead to stronger environmental regulations, the adoption of mitigation measures, or even compensation for those affected. Their reactions also shed light on systemic issues such as the lack of enforcement of existing laws and the imbalance of power between mining corporations and local populations. Including these dynamics illustrates how community actions are integral to fostering accountability and achieving environmental justice.

Furthermore, community responses to mining pollution emphasize the ethical dimensions of mining and the importance of inclusivity in decision-making processes. Many residents feel excluded from discussions about mining operations that directly affect their lives, which fuels mistrust and exacerbates tensions. Their reactions underscore the need for more transparent, participatory approaches that prioritize the well-being of both people and the environment. By exploring community responses, authors intend to advocate for a more equitable balance between economic development and the protection of natural and human resources in Zambia and the DR Congo. Their voices offer hope and a path forward, reminding readers that the fight for a healthier, more sustainable future begins with empowering those on the front lines of environmental challenges.

9.4.1 *Impact of Mining and Metallurgical Industries on Central Africa Copperbelt Residents*

In the Central Africa Copperbelt region, every resident (100%) interviewed recognized the positive economic, social, and developmental benefits brought to their communities by the mining and metallurgical industries. These advantages include employment opportunities, increased household incomes, infrastructure development such as roads, improved connectivity, market access, and the construction and maintenance of medical facilities enhancing local healthcare services. Additionally, the building and support of schools contribute to better education for the community as part of corporate social responsibility (Table 9.6).

Moreover, during a previous survey in Zambia and the DR Congo (Mwitwa and Muimba-Kankolongo (2012, unpublished report to CIFOR), it was established that copper and cobalt mining had serious socioeconomic effects, both positive and negative, which are vital to analyze. On the one hand, the mining industry contributes to local and national economies by generating employment, revenue, and infrastructure development. On the other hand, it also brings challenges like social, economic, and environmental costs. Interviews with diverse stakeholders reveal a complex interplay of benefits and drawbacks impacting various groups.

Table 9.6 Population perception of the impacts of mining and metallurgic industries on the livelihood and health of different locations, artisanal miners, traders, price negotiators, and NGOs in Zambia and the DR Congo

Salient mining problems discussed with urban and rural residents	ZAM ($n = 180$) % Yes[a]	ZAM ($n = 180$) % No	DRC ($n = 318$) % Yes	DRC ($n = 318$) % No	Community reaction to mining and metallurgic industries pollution	Local health services ZAM	Local health services DRC	NGOs ZAM	NGOs DRC	Miners/traders/ mineworkersunion[d] ZAM	Miners/traders/ mineworkersunion[d] DRC
Loss of agricultural land	68.0	32.0	100	0.0	Pollution has deteriorated our land			●[c]	○		
Decrease in wild foods	100	0.0	100	0.0	Distance to get wild foods increased			●	○		
Water pollution from mine discharge	56.0	44.0	100	0.0	The water used is of poor quality	●	○	●	○	●	○
Soil is degraded	85.0	15.0	100	0.0	Crop yield has decreased	●	○	●	○	●	○
Drying of some streams	24.0	76.0	35.0	65.0	Several streams have disappeared	●	○	●	○		
Health issues among residents	100	0.0	100	0.0	We suffer from numerous illnesses[b]	●	○	●	○	●	○
Alternative to water at homestead	0.0	100	0.0	100	There is nowhere water is pure					●	○
Livelihood without quality water	0.0	100	0.0	100	This is the only water we have					●	○
Role of mining in livelihood	100	0.0	100	0.0	We except a lot from mine industries					●	○

[a]Percentage rounded to the nearest one of residents acknowledging the issue (yes) and used to deduct the no respondents
[b]Cancer, diarrhea, respiratory issues, eye and throat irritation, skin tingling, fatigue, and congenital defects
[c]Recognition of the problem being present in Zambia (ZAM) (●) and in the DR Congo (DRC) (○), except otherwise stated. dArtisanal miners and traders do not perceive any illness that could arise from their activity

Infrastructures

1. Infrastructure Development in the DR Congo

In the DR Congo, mining activities—particularly by companies like UMHK/Gécamines—played a pivotal role in establishing modern towns and infrastructure across the Copperbelt region. Gécamines developed crucial infrastructure, including administrative buildings, hospitals, schools, churches, roads, bridges, and power dams in Lubumbashi, with road networks spanning roughly 450 by 50 square kilometers. Additionally, towns such as Likasi, Kipushi, Kolwezi, and Lwena owe their creation to Gécamines' mining operations.

At its peak, Gécamines contributed around 70% of the country's foreign earnings, although its production has significantly diminished due to political instability and war. Despite this, the company remains a key player in the DR Congo's development, alongside other private firms. Mineworkers have benefited through initiatives like training programs for their dependents. Companies such as ANVIL Mining have also supported communities by constructing clinics, schools, and sinking boreholes for clean water.

2. Contributions in Zambia

In Zambia, Konkola Copper Mines (KCM) has been recognized for its socioeconomic contributions, benefiting mineworkers, urban and rural residents, and the broader society. KCM has invested in urban infrastructure, including roads, schools, health facilities, street lighting, recreational areas, and boreholes to supply clean water in peri-urban areas like Mushishima, where water sources are contaminated. Retiring employees are assisted with skill development to support small businesses, enhancing their postretirement income.

KCM also supports educational initiatives by providing computers to schools, organizing sports activities, and conducting malaria control programs with insecticidal sprays, even in remote villages. Additionally, the company fosters economic growth in Chingola by creating demand for services, contributing to national revenue, and financing universities and sports teams. As part of community development efforts, KCM funds small business startups, agricultural extension programs, and skill training initiatives.

Transportation

The exploitation of mineral resources has significantly improved transportation services, as noted by villagers in mining regions. In Zambia, public transport systems have enhanced, leading to reduced transport costs. Similarly, in the DR Congo, the time spent waiting for transport to deliver agricultural goods to markets has decreased, reflecting an improvement in efficiency.

Effects on Informal Settlements

Mining has spurred migration to mining towns in Zambia, as unskilled laborers from other areas seek employment opportunities. However, periods of low copper prices and subsequent worker retrenchment have weakened the local economy, leading to the expansion of informal settlements. Families in these settlements often live in extreme poverty and rely on informal, land-based activities to sustain themselves. This includes cultivating land for subsistence, which often involves using forested areas previously reserved for timber production for mining activities. Additionally, tree cutting for charcoal production has become a common practice to generate income for dependents, resulting in deforestation and the degradation of woodlands around copper mines.

Furthermore, increased migration is associated with rising cases of HIV/AIDS and various social challenges, emphasizing the broader societal impacts of mining-related in-migration (source: original content provided). Let me know if you'd like further refinements!

9.4.2 Negative Impacts on Livelihoods

Despite these benefits, residents reported numerous negative effects on their livelihoods. Observations in various surveyed neighborhoods revealed that grazing and pasturing lands have been severely degraded through soil erosion, loss of litter layer, acidification, and siltation. Both surface and underground water have been heavily polluted by toxic dust and granules from mining activities, becoming of poor quality as a result of deposition and leakage of toxic dust and granules causing physical disturbance.

9.4.3 Environmental and Health Impacts

Mining extractions and other industrial operations have negatively impacted water quality (through contamination with harmful chemicals and heavy metals), air quality (via dust and emissions), vegetation (removal of cover), soil (severe erosion from topsoil removal), and the broader environment (ecosystem and habitat destruction, biodiversity loss, and acid mine drainage contaminating local water bodies). These activities have significantly depleted natural resources, reducing the availability of essential wild products such as wild fruits, vegetables, mushrooms, edible insects, honey, roots and tubers, and bulbs.

Community residents testified that noise pollution from heavy mining trucks and blasts has become a major concern, causing intolerable stress, especially for those living near mining sites in both the DR Congo and Zambia (Table 9.7). Other reported issues include illegal settlements of laid-off miners due to a failing local

Table 9.7 Most salient negative impacts of the mines and metallurgic industries among residents in surveyed locations in Zambia. Negative impacts of operations of mining and metallurgic industries on communities' livelihood in Zambia and the DR Congo

Noise pollution Modern mines use heavy and loud equipment that frequently cause blasting which impose intolerable stress upon the local people	**Indigenous rights** Local people do not have recognized land rights, and mines frequently hide behind this to apply double standards in treating them	**Displacement** Displacement of people including forced resettlement in areas without adequate resources remains a common feature of mining development	**Human rights violations** Opposition to mine operations is suppressed using force and military or police deployment, violating human rights through torture, arrest, and harassment
Employment and livelihoods One of the positive incentives to allow mining operations is employment that often fails to materialize causing high poverty among the local communities	**Treatment of women and children** Often, mining jobs are restricted to men, thus alienating children's and women's livelihoods such that unaccompanied men suffer severe prostitution and high incidence of HIV and AIDS	**Health hazards** Contamination of available resources, as well as soil and air, contributes to increased levels of toxic buildup in peoples' bodies. Several chronic respiratory issues are widespread, particularly in children and old people	**Environmental degradation** Mining exploitation of natural resources has taken a toll on the quality of land, water availability and quality, forest and soil resources, and rainfall patterns, undermining local livelihoods in the process

economy, employment marginalization of community members, and human rights violations, particularly against women.

9.4.4 *Artisanal Miners' Health Risks*

Artisanal miners, although aware of their exposure to toxic metals like uranium, often disregard the potential health risks from prolonged exposure. They work without protective gear, carrying bags of ore on bicycles or their heads for washing and sifting in nearby rivers. Their main complaints include hard labor, joint pain, chronic coughs, colds, and low incomes due to low product prices.

Inhabitants further reported adverse health effects such as eye and nose irritation, lead and cadmium intoxication, diarrhea, pneumonia, HIV/AIDS, cancer, tuberculosis, coughing, asthma, sore throat, skin tingling, and persistent body fatigue.

9.4.5 *Air Quality Concerns*

Residents claimed that when smoke plumes from mining or metallurgical factories are emitted into the sky, they are forced to protect their noses and mouths to avoid breathing polluted air. Dust from tailings dams, unpaved roads, and emissions from smelter plants significantly impact air quality, posing health hazards (Table 9.7).

Additionally, the cumulative impact of multiple pollutants—often referred to as synergistic effects (Mauderly and Samet 2008; Billionnet 2012; Martin et al. 2021; Martin 2023; Tong Feng et al. 2025)—poses a significant environmental and health challenge in both Zambia and the DR Congo. When different pollutants interact, their combined effect can be far more harmful than the sum of their individual impacts, leading to complex issues for ecosystems and communities.

In Zambia, the mining sector releases various pollutants, such as heavy metals, sulfur dioxide, and particulate matter, into the air, water, and soil. These pollutants can interact chemically, amplifying their toxicity. For instance, heavy metal contamination in water sources is exacerbated when mixed with industrial acids or other chemical compounds, making the water unsafe for consumption and threatening aquatic ecosystems. Similarly, airborne pollutants from mining operations can combine, creating hazardous conditions for respiratory health, especially in communities near mining sites. Over time, these synergistic effects contribute to widespread environmental degradation and health crises among local populations.

In the DR Congo, copper and cobalt mining activities have led to the release of multiple pollutants, such as heavy metals, waste rock particles, and chemical effluents. The interaction of these pollutants has caused severe contamination of water bodies and soil, harming agricultural productivity and reducing food security for communities reliant on farming. Forests and wildlife near mining areas are also affected, as pollutants disrupt ecosystems and limit biodiversity. Additionally, the combination of air pollutants has led to increased respiratory ailments among populations living close to mining operations, compounding public health challenges.

Across both nations, these synergistic effects highlight the urgent need for sustainable mining practices and effective environmental regulations to mitigate the cumulative damage caused by multiple pollutants. Addressing these impacts is essential to protect human health, preserve ecosystems, and ensure a sustainable future for mining communities.

9.5 Urban and Rural Inhabitants' Testimonies to Mining Impacts on Livelihoods

The poor, impoverished population, delivered to unemployed and plagued by tribalism, cannot unite for the same cause. When the smoke emerges from the GCM factory, the inhabitants must block their noses and mouth so as not to breathe the contaminated air. Respiratory health problems including tuberculosis, asthma, pneumonia, dermatological (tickling the skin) and intense and persistent body fatigue are observed. The inhabitants are forced to dig the waste abandoned by the Belgians to deal with the needs of the family. Artisanal mining activities in the region often involve vulnerable groups, including women and children.

The population fought in vain to prevent the exploitation of this land. Many of them used this terrain for vegetable crops, tubers, etc. Agricultural products from

Fig. 9.10 Gécamines mine in Lubumbashi City in Haut-Katanga Province releasing smoke in the atmosphere that pollutes local ambient air quality

gardens are also contaminated and have poor growth. According to residents, health problems such as scabies have come forward and are linked to metal contamination of domestic water by the acid produced by Chemaf. The color of children's urine becomes black at certain periods; there are even cases of congenital malformation, and babies and pregnant women run a great danger. Plume of smoke is regularly released into the atmosphere, affecting the ambient air and our health (Fig. 9.10).

The local community believes that the government is more interested in money than in the health of the population, and several steps have been carried out without success.

9.5.1 A Local Lawyer's Comment

Lethargy in the processing of the dossier of the claim of the population is linked to the fact that certain authorities are shareholders in certain mining companies and do not know how to make decisions. Thus, only legal actions can lead to compensating communities and victims and to pushing businesses to the repair and restoration of the environment.

9.5.2 Commentary by the Federation of Business

The metallurgical industry Chemaf should not close the doors; he may lose capital since he has paid the costs and received the documents allowing him to operate the land. In addition, it avoids increasing the unemployment rate among the population because this company employed 2500 individuals. According to him, the company has made a lot of efforts to decrease the harmful effects of its exploitation, but it also administers eye care for both urban circles and in rural areas.

9.5.3 Comments from the Minister of Mines, Katanga

Although having tabled the environmental impact study, several companies do not fully respect these rules, despite certain warnings. The administration's weakness in the environmental sector is evident in its inability to systematically monitor industrial compliance with environmental assessment protocols, largely due to inadequate human resources. In many provincial offices, only three to four civil servants are assigned to oversee these responsibilities.

9.5.4 A Comment from the Minister of Health, Katanga

Several people (children, adults) living near mining sites suffer from cancer and throat, eye, and respiratory problems.

9.5.5 Comments by a Toxicology Professor (UNILU/Public Health Department)

The population has every reason to complain because the inhabitants living within 3 km from the mining sites are affected by metal pollution. Studies have shown a relationship between heavy metals found in the body and the pathologies from which the population suffers. A study was initiated on congenital malformations. Malformation cases are observed every 2 days.

With a significant contribution of 43% to the national budget, the mining sector remains the pillar of the DR Congo's economy. However, the pace of mining in the Katanga region worries both researchers and civil society actors. The consequences on the environment are increasingly devastating. In addition to environmental impacts on the air, soil, or vegetation, mining activity threatens the water table. Each time the mine reaches a water table, the company must empty the water to prevent the mine from being flooded. Meanwhile, water resources are running out, and when available it is unsafe for domestic consumption.

9.5.6 Comments by an Agronomist Professor (UNILU, Faculty of Agriculture)

The demographic explosion and poverty have pushed the population to anarchic urbanization and to be built on potentially contaminated sites such as Penga-Penga (tenth polluted site in the world). It is an environment that presents a serious public health problem because the population who lives there is exposed to metal contamination. An alternative to reduce the accumulation of heavy metals is to install permanent plant covers on empty spaces or plots of volunteers. Or excavate a layer of 20–30 cm high and replace it with contaminated soil. This technique requires a lot of means, and to achieve the GCM or UMHK, the main polluter must participate actively.

9.5.7 Media Comments (Nyota Chain)

Several accidents are linked to metallurgical mines and factories. All inputs and mining products are transported by road with the risk of population contamination.

9.5.8 Comments by the Minister of Unionists

Very few conventions are respected. Several companies use workers temporarily, without determined contract. The fault is the responsibility of the government which must force investors to apply the rules and respect two things: the development of mining growth and the social impact and the environment. The observation is that 10 years after exploitation, these are holes and not development and respect for the environment: investors win and the population is more detected. If the code was 100% respected, good results should be observed. On the one hand, some investors were not exploiting the mines and received titles; on the other hand, the weakness of the administration was the basis of this situation. The expectations of the people in various communities are not encountered, causing an uprising anger which often culminate in the formation of rebellious groups such as the so-called Mai-Mai. Hence the need to revise contracts with investors.

9.5.9 *Comments by the Minister of the Former Prime Minister*

Increasing the percentage of the government to 35% in the mines brings war with mining companies because, according to him, financial profitability will intervene when these companies have reached maturity. The solution would be to create processing companies to create jobs and add value to products to be competitive on the market.

9.6 Socioeconomic Impacts

Effectively, mining pollution creates detrimental and profound long-term socioeconomic consequences on local communities, particularly in the Central Africa Copperbelt region of Zambia and the DR Congo, where mining is a cornerstone of economic activity. These impacts are multifaceted, considerably affecting livelihoods and social structures.

Economically, while mining can create jobs and stimulate local economies, the benefits are often short-lived and inequitably distributed. In many cases, mining companies prioritize profits over community welfare, leaving local populations to bear the brunt of environmental degradation. For example, in Zambia, weak regulatory frameworks have allowed mining companies to operate with minimal accountability, exacerbating economic instability and environmental hazards. In the DR Congo, the focus on cobalt mining for global energy transitions has not translated into improved living conditions for local communities. Instead, many residents face poverty and displacement.

Socially, mining pollution disrupts community cohesion and traditional ways of life. Displacement due to mining activities often leads to the loss of ancestral lands and cultural heritage. In both Zambia and the DR Congo, communities have reported conflicts over land rights and resource allocation, further deepening social inequalities. The lack of sustainable practices and community engagement in decision-making processes exacerbates these challenges, leaving local populations marginalized. Several of the socioeconomic impacts of mining pollution in the region are well described by the affected community inhabitants in Table 9.7.

9.7 Ecological Degradation

9.7.1 *Mining Released Smoke Impact*

Residents living near mining and metallurgical industries have voiced significant concerns about their health and that of their livestock due to operation-related smoke and road dust. They reported that inhaling particulate matter and other pollutants from mining operations often led to various respiratory issues, such as asthma, bronchitis, lung cancer, and an increased risk of cardiovascular diseases, including heart attacks and strokes. Moreover, smoke and air pollution from mining also adversely affect livestock health, causing respiratory problems, reduced productivity, and even death. This has a profound impact on the income and food security of pastoral communities.

9.7.2 Soil Degradation

Both urban and rural communities in the DR Congo and Zambia cultivate available land to grow food crops to sustain their livelihoods. However, their lands, once fertile and abundant, have now become degraded due to mining pollution and the pressures of industrial expansion.

9.7.3 Impact on Soil Quality

Soil is a crucial natural resource that provides essential services for the functioning of terrestrial ecosystems and human societies. However, various human activities have led to the degradation of soil quality and productivity across the Central Africa Copperbelt region of Zambia and the DR Congo (Figs. 9.11 and 9.12). Pollutants from mining and metallurgical industries—such as heavy metals, organic compounds, salts, and other contaminants—enter the soil through various pathways, including industrial effluents, atmospheric deposition, and improper waste disposal. Their accumulation in the soil has resulted in decreased soil fertility, reduced plant growth, and potential harm to soil-dwelling organisms and human health.

9.7.4 Observations of Soil Salinization

One form of soil degradation observed during the survey is the increased soil salinization caused using saline irrigation water in mining operations. Soil salinization as a result of mining pollution occurs when mining activities introduce excessive salts into the surrounding environment. Mining operations, particularly those involving the extraction of minerals from underground or surface mines, often produce saline wastewater or runoff, which can seep into nearby soils. This is especially common in arid or semiarid regions, where evaporation rates are high, and the salts left behind accumulate on the soil surface. Additionally, the disruption of natural drainage patterns and the exposure of salt-laden geological layers during mining can exacerbate salinization. Over time, the increased salt content in the soil reduces its fertility, negatively affecting plant growth, agricultural productivity, and the overall health of the ecosystem.

Fig. 9.11 Type of soil degradation in Lubumbashi in Haut-Katanga Province of the DR Congo caused by mining pollution

Fig. 9.12 Type of soil degradation in Chingola District, Copperbelt Province of Zambia

9.8 Biodiversity Loss

9.8.1 *Decline in Biodiversity and Ecosystem Health*

According to the International Institute for Applied Systems Analysis (IIASA) (2020), the variety and abundance of species, along with the extent and quality of their ecosystems, have been declining at an alarming rate for many years. Our study found similar trends in the mining areas of Zambia and the DR Congo, as reported by local inhabitants. We observed the death of forest trees and vegetation cover, as well as a significant reduction in wild foods such as fruits, mushrooms, and insects due to mining pollution and changing rainfall patterns caused by atmospheric pollution. These changes have severely undermined local livelihoods.

9.8.2 *Deforestation and Forest Degradation*

The rich forest ecosystems in Zambia and the DR Congo face increasing destruction primarily due to intensified timber harvesting, agricultural expansion, and wood fuel demand from urban areas. According to the National Research Council (1991) and Njovu et al. (2004), the main threats to forest resources include pollution from industrial and transportation activities, deforestation caused by land clearance for agriculture, firewood and charcoal production, expanding urban and industrial activities, logging, and mining.

Reed (1996) observed that over 90% of deforestation results from land clearing for agriculture, though the proportion of forests cleared for fuel wood is locally higher around the copper belt. Similarly, Kanninen et al. (2007), Nsiami et al. (2006), Coalition for Rainforest Nations (2008), and Campbell et al. (2011) identified agricultural activities, wood extraction for timber, pulpwood and charcoal production, and infrastructure extension (including roads, settlements, public services, open-pit mines, and hydroelectric dams) as the main causes of deforestation and degradation in Africa. Consequently, the total forest area in Zambia and the DR Congo has been significantly affected.

9.8.3 *Vegetation Cover and Forest Quality Changes*

In a report to CIFOR in 2012, Mwitwa and Muimba-Kankolongo (unpublished) highlighted changes in vegetation cover and forest quality in the DR Congo and Zambia using Landsat Imageries. Buffers of 50 km and 200 km radii were created around each of the 12 mining towns to track land cover changes in detail. Changes were detected by comparing the percentage of land cover class within the buffer area for each decade.

It was assumed that the forest cover was relatively undisturbed a year before and in the year that copper mining began based on respondents' perceptions. Nevertheless, residents in the six surveyed mining locations estimated, on average, more than 50% decline in natural regeneration since 1926 when the mine was established. These observations are supported by satellite images from 1990, showing forest clearance and degradation in areas surrounding the mines in Chingola, Kitwe, and Luanshya in Zambia and Lubumbashi, Kipushi, and Likasi in the DR Congo (Fig. 9.13).

Following the decline in mature trees of valuable species near the copper mines in both countries, suppliers began sourcing timber from more distant forests, thus spreading the impact of mining to larger areas. KCM and Gécamines indicated that mines are supplied with timber sourced from up to 300 km away. Presently, part of the timber for Gécamines in Lubumbashi is still harvested from the forests around Likasi District, located about 120 km from the Lubumbashi copper mines.

When these changes in land cover are quantified, we find a 46% and 38% decrease in old Miombo for the 50 km and 200 km buffers, respectively. These declines are matched by increases of 93% and 139% in degraded/regenerating Miombo and by 175% and 60% increases in open/cultivated area for the 50 and 200 km buffers, respectively. These figures need to be interpreted with caution, given the multiplicity of factors shaping land use in the copper belt. While mining is clearly a predominant force, so is charcoal burning for urban towns and the opening of agricultural land for settlement. While these forces are clearly distinct, they are nevertheless related—given the influence of mining on infrastructure and urbanization and the effect that these factors in turn have on land use and deforestation (Fig. 9.14).

The changes in patterns in land cover change from the first to the third decade are primarily influenced primarily by the subsistence agriculture and infrastructure development that is connected to or influenced by copper mining. Though the map and the figures conclusively demonstrate a clear reduction of the forest vegetation cover over periods of time, it looks however impossible through spatial analysis to assess the proportion of land use change directly and indirectly attributable to mining and to identify the directionality of change and to make some inferences about the role of mining by exploring its influence within each buffer zone and the nature of land use changes at different distances from mines (Mwitwa et al. 2012a, b). Moreover, they stated that by carrying out a comparative analysis of land cover changes in the two buffer zones, a number of interesting patterns are emerging.

1. The most obvious is that forest transformation has occurred at a more rapid pace closer to mining towns.
2. The second observation is the most predominant land cover replacing mature Miombo. In the 50 km buffer, the opening of land for agriculture is the predominant pattern, whereas in the 200 km buffer mature Miombo is giving way to degraded/regenerating Miombo. This would suggest that the population pull of mining towns is contributing to a more significant and lasting transformation of Miombo woodlands, relative to outlying areas where selective logging (from the

The three mining towns in Zambia are:

CHINGOLA
50 km buffer

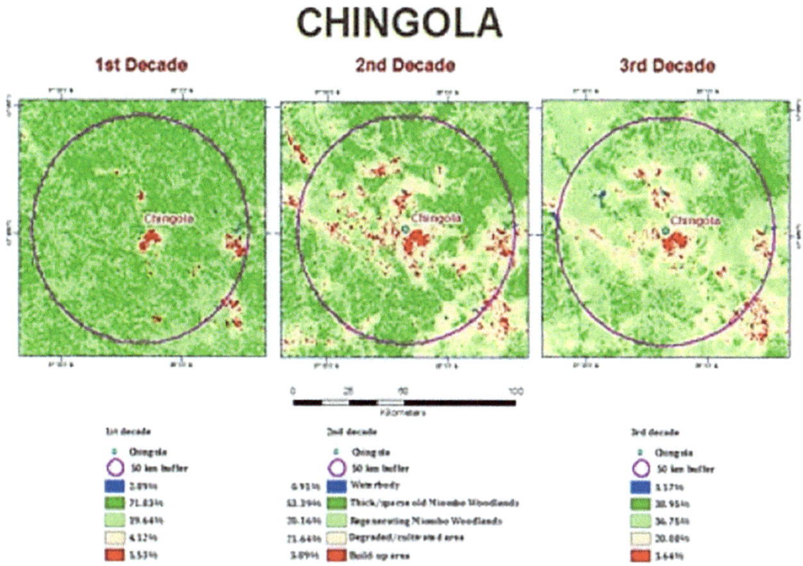

200 kms buffer

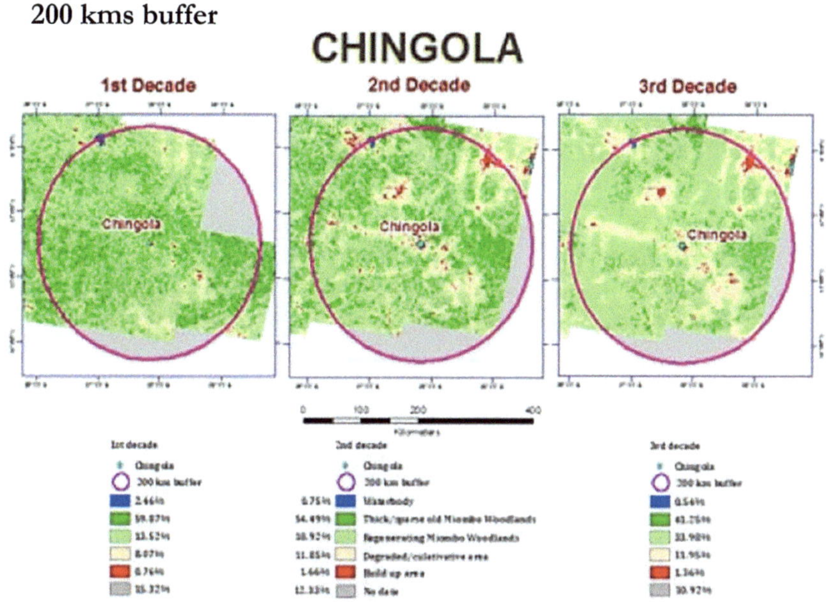

Fig. 9.13 Satellite imageries depicting vegetation depletion in the surrounding areas of the Central Africa Copperbelt region of Zambia and the DR Congo between the second decade in 1989 and 1991 onward. (Source: Mwitwa and Muimba-Kankolongo (2009) unpublished report to CIFOR)

KITWE

50 kms buffer

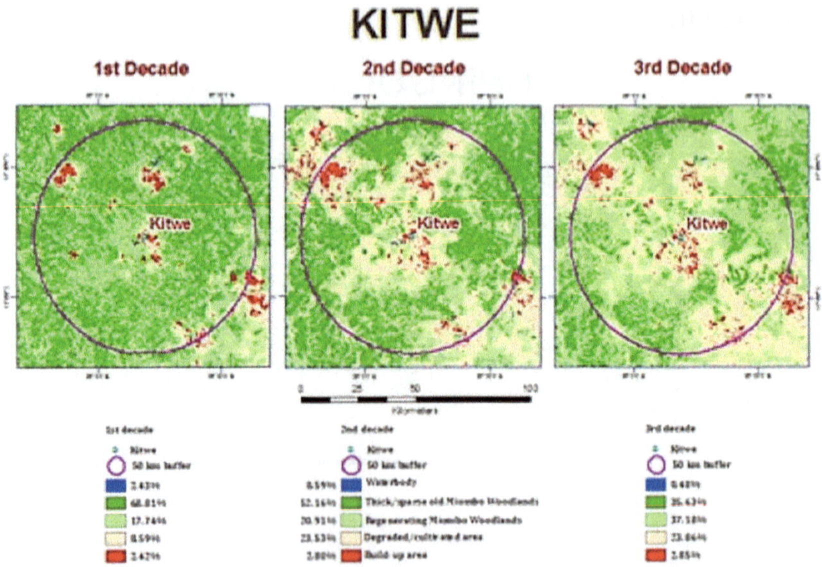

200 kms buffer

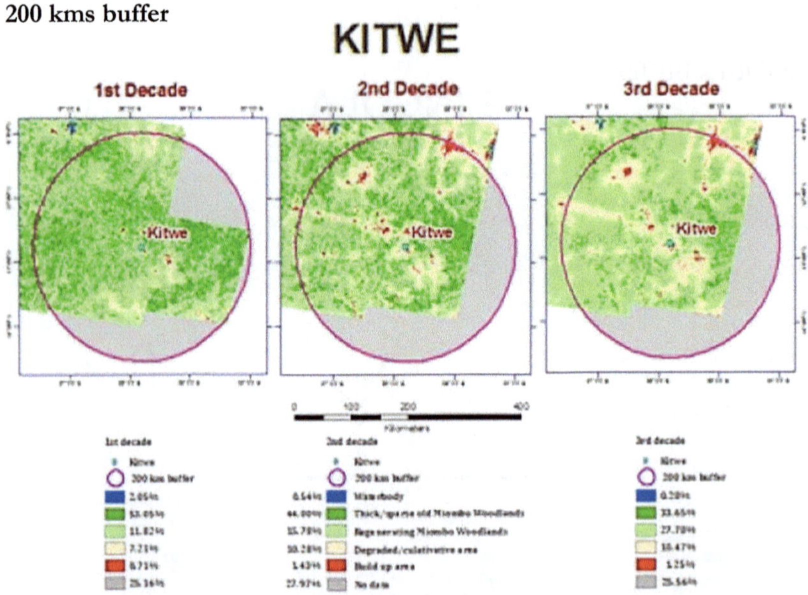

Fig. 9.13 (continued)

LUANSHYA

50 kms buffer

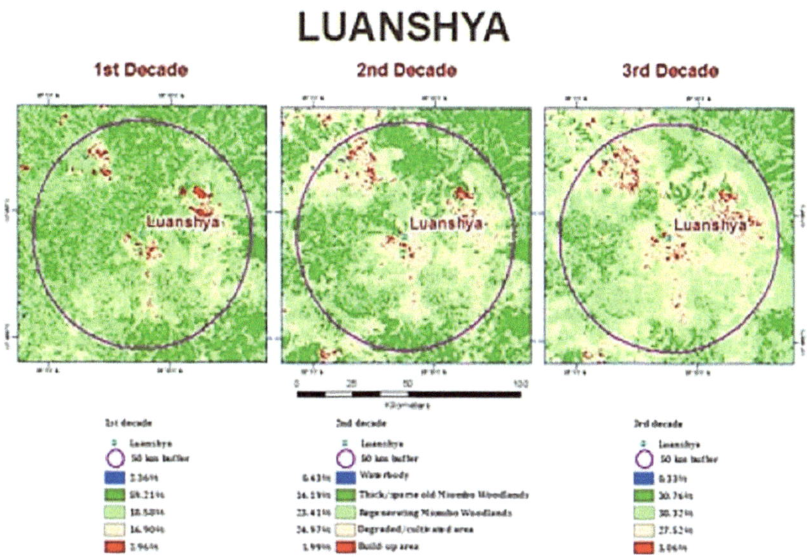

200 kms buffer

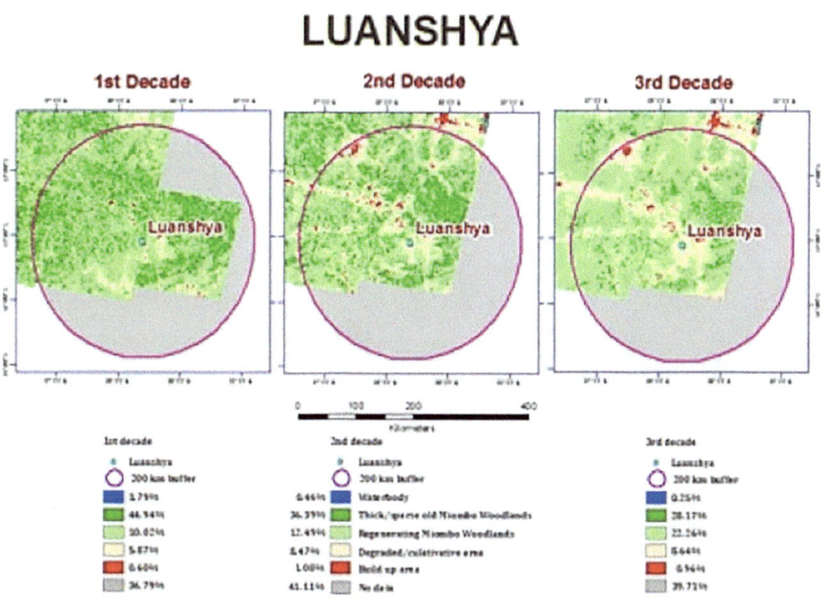

Fig. 9.13 (continued)

The three mining towns in the DR Congo are:

LUBUMBASHI
50 kms buffer

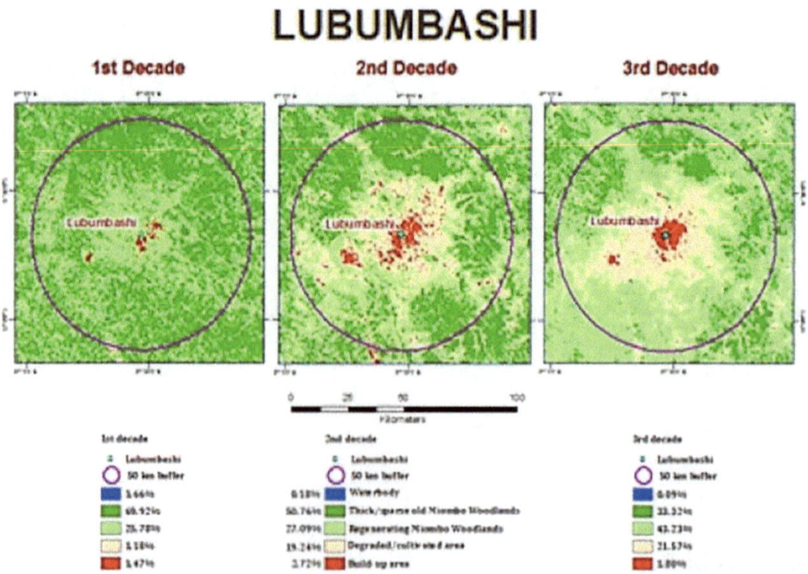

200 kms buffer

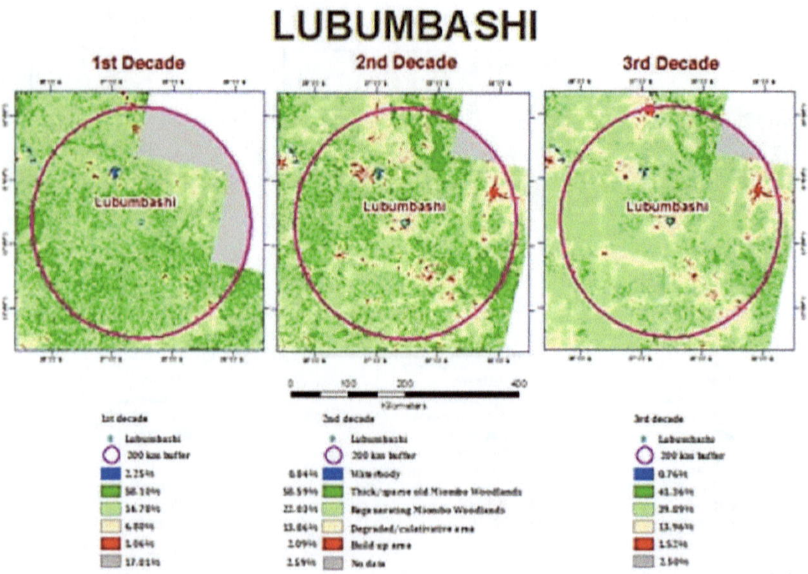

Fig. 9.13 (continued)

50 kms buffer

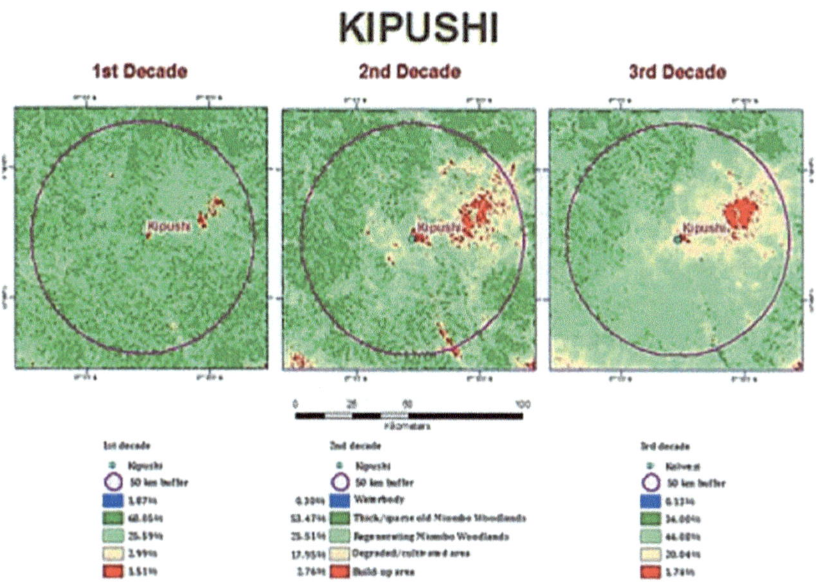

200 kms buffer

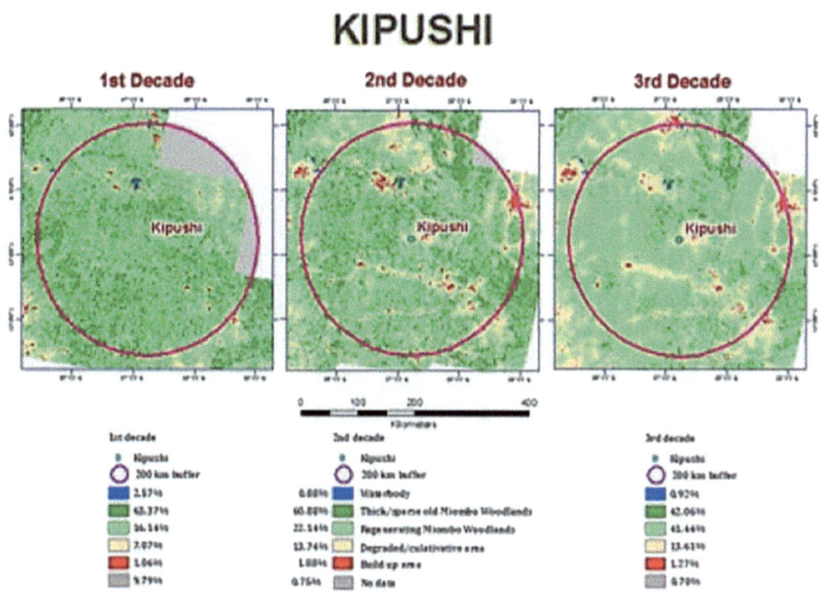

Fig. 9.13 (continued)

50 kms buffer

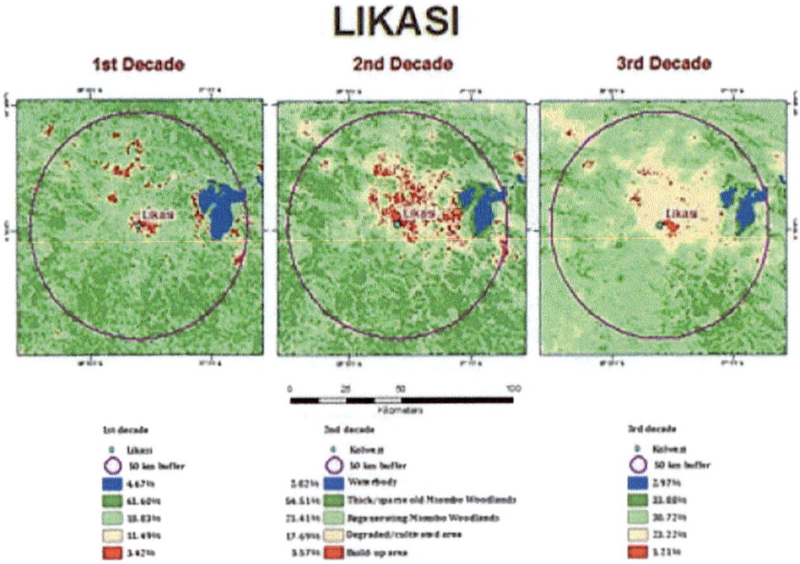

200 kms buffer

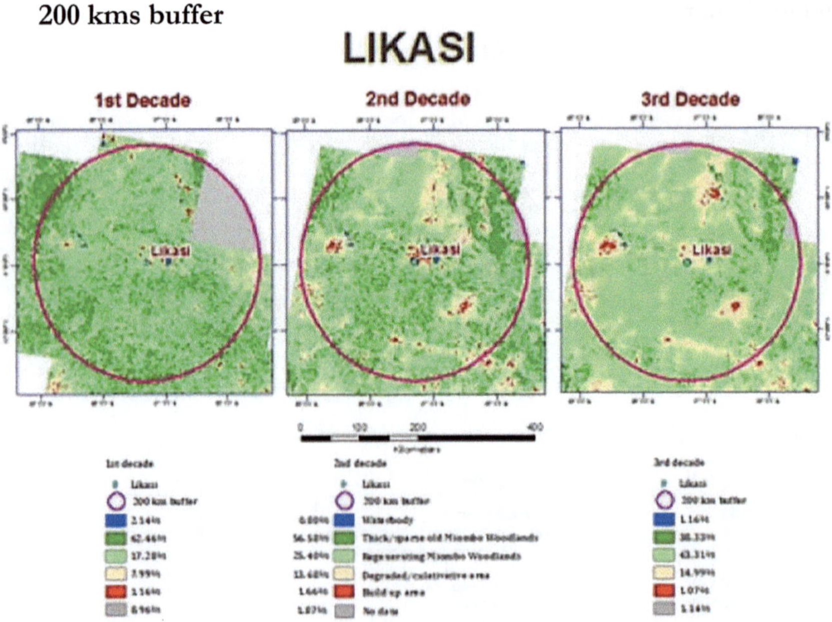

Fig. 9.13 (continued)

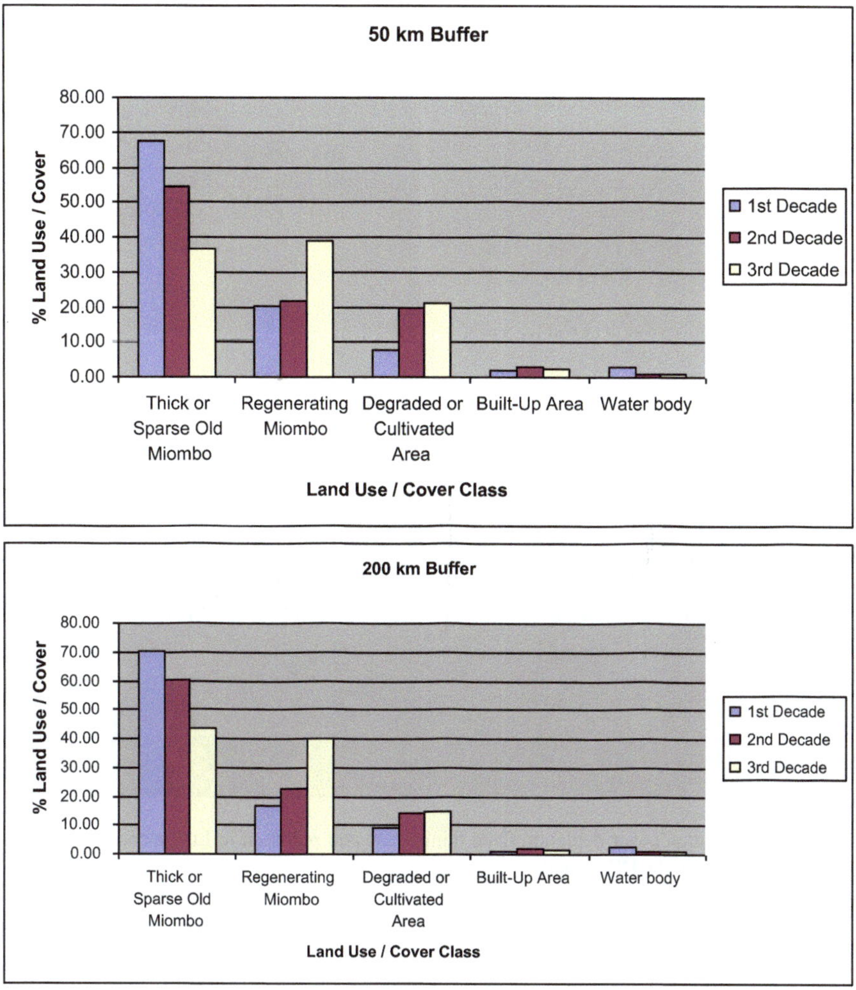

Fig. 9.14 Land use/cover over time within 50 km (**a**) and 200 km (**b**) radiuses of mining towns in the Copperbelt ($n = 12$). (Source: Mwitwa et al. (2012a) reproduced with permission from LAMBERT Academic Publisher)

mines) and charcoal burning (much of which may be unrelated to the growth of mining towns) are key factors inducing land cover change.

Changes in Vegetation Structure and Cover

According to Mwitwa et al. (2012a, b), quantitative assessments indicate significant changes in the vegetation structure and cover of the Copperbelt. In 1979, an average of 70.9% of the Copperbelt was covered in thick, old Miombo woodlands, including

Fig. 9.15 Vegetation cover depletion from mining use and pollution in Lubumbashi City, Haut-Katanga Province in the DR Congo

degraded areas and cultivated regions. In Lubumbashi, DR Congo, and Chingola, Zambia, the coverage of these woodlands declined considerably from 69.9% and 71.8% in 1979 to 50.8% and 53.4% by 1991 and further to 33.3% and 38.9% by 2002, respectively (Figs. 9.15 and 9.16).

Reasons for the Decline

The relatively slower decline in vegetation between 1991 and 2002 is attributed to declining commodity prices, poor economic performance in the DR Congo and Zambia, and the worsening war situation in the DR Congo during that period. The 60% decline in vegetation cover reported by respondents aligns well with the average estimate of about 49.1% change in thick, old Miombo woodlands observed from satellite images, although respondents noted this change in 2007, while the calculated statistics cover the period 2000–2002.

Fig. 9.16 Vegetation cover depletion from mining use and pollution in Chingola District, Copperbelt Province of Zambia

Spatial Patterns of Deforestation

In the case study sites—Chingola in Zambia and Lubumbashi in the DR Congo—spatial patterns of deforestation were observed relative to roads and key infrastructure. Multiple factors shape land cover change patterns in the Copperbelt, primarily driven by the demand for agricultural land, community livelihoods pushing inhabitants to cut trees for charcoal burning, and urban and industrial infrastructure developments.

Threats to Forests from Mining Pollution

Forests in Zambia and the DR Congo are increasingly threatened by the damaging effects of mining pollution. As the global demand for minerals and metals rises, mining and metallurgical operations are rapidly expanding into previously untouched forested areas, leading to significant environmental degradation. The extraction, processing, and transportation of these resources release various toxic pollutants into surrounding ecosystems, with devastating consequences for the delicate balance of forest habitats.

Importance of Forest Conservation

These forests are bio-reserves of national, regional, and international significance, and their conservation and restoration would have a global environmental impact. Local tribal communities depend on these forests for their daily living, relying on resources such as herbs, fruits, vegetables, leaves, firewood, timber for construction, and animals. Additionally, these forests contain numerous known and potentially useful medicinal and ornamental plants, including orchids. The destruction of biodiversity in these areas, known for their diverse flora and fauna, is concerning, especially as some potential resources remain unexplored.

Forest Resources and Livelihoods

Forests and woodlands harbor the country's rich biological diversity, directly supporting rural communities' livelihoods. Forest food resources, such as fruits, vegetables, mushrooms, roots, and tubers, provide essential micronutrients, especially for children and women. These resources are continuously harvested from indigenous forests for household subsistence and sale. Edible insects, such as caterpillars and termites, and secondary products like honey and beeswax are highly prized forest products in demand both domestically and internationally.

Impact of Mining Operations

Mining operations generate large amounts of airborne particulate matter, such as dust and fumes, which settle on the leaves and soil of surrounding forests. These pollutants interfere with photosynthesis and respiration in trees, leading to decreased growth, reduced resilience, and, in some cases, tree mortality. In localized areas, particularly in the Copperbelt Province, plants have succumbed to air and/or soil pollution from copper mines (Chipundu and Kunda 1994), posing serious environmental risks.

Chemical Contamination

Other sources of pollution include chemical contamination from waste disposal sites and the use of toxic agrochemicals. The deterioration of water quality has been monitored in the Kafue River, originating in the heavily industrialized Copperbelt areas. Metals such as aluminum, iron, nickel, lead, zinc, and cadmium often pollute large areas of wasteland created by mining and other heavy industrial operations. These pollutants affect soil pH and the mobility of essential nutrients like nitrogen, phosphorus, and potassium, discouraging plant recolonization.

Effects on Vegetation

In a subsequent study by Mulenga (1999) in other areas of the Copperbelt Province, pollution, especially from sulfur dioxide emissions from mines, was associated with considerable vegetation and tree death and poor performance. Mulenga reported several deleterious effects on vegetation and trees, including lack of growth, reduction in leaf size and shape, development of chlorotic and necrotic lesions, pronounced stunting, and tree death, along with significant soil pH and fertility decline. He observed a considerable decrease in tree population per unit of land in affected areas compared to areas outside mine locations, with most trees being dwarfed and stunted. The impact of mining pollution on forests is a pressing environmental concern requiring immediate attention and action.

9.9 Most Common Edible Insects

9.9.1 Common Edible Insects in Miombo Forests

In the Miombo forests, caterpillars and termite alates are the most consumed insects. Termite alates, particularly from the *Macrotermes* species, are dispersed during the rainy season and are captured and eaten as a protein source (Chidumayo 1996a). Various types of edible caterpillars, each with different values, are found throughout the Miombo region (Lees 1962; Malaisse 1978).

9.9.2 Economic Potential of Edible Caterpillars

A study conducted in 2000 in the Kopa area of Northern Province (Chidumayo and Mbata 2002) demonstrated that edible caterpillars have significant potential to generate incomes of over \$60 per household, comparable to or even surpassing incomes from agricultural crops. Approximately 98% of households in the area participate in caterpillar harvesting, with 89% collecting them both for consumption and sale. Thus, edible caterpillars serve not only as a food source but also as barter and sale items. In 2000, the price for *G. maja* caterpillars was ZK 15,000 per meda or ZK 11,578 per kg of air-dry weight. Although the price and availability of caterpillars vary annually, income from their sale enables local people to purchase goods and services they otherwise could not afford. Additionally, Chidumayo (1996) noted that people living near the Miombo harvest caterpillars (*Elaphrodes lactea*) between November and December. On a dry basis, these caterpillars contain over 60% protein. Local trade in edible caterpillars is growing, especially in urban areas.

9.9.3 Nutritional Value of Edible Caterpillars

A study from the Congo Basin subregion provides a comprehensive overview of the nutritional values of various caterpillar species and confirms what local people knew empirically (Vantomme et al. 2004). The average percentages of proteins and fats, as well as the energy content of 24 investigated fresh caterpillar species, are 63.5% ± 9.0% for proteins, 15.7% ± 6.3% for fats, and 457 ± 32 Kcal per 100 g (based on dry matter). Due to their high nutritional value, flour made from larvae is traditionally used to prepare a pulp given to children to counter malnutrition (Vantomme et al. 2004). Species particularly rich in calcium (e.g., *Tangoropsis flavinata*), protein (e.g., *Imbrasia epimethea, Imbrasia dione, Antheua insignata*), or iron (e.g., *Cinabra hyperbius*) are given to anemic people and to pregnant and breastfeeding women.

9.9.4 Income Generation from Edible Caterpillars

The study also investigated the income generation aspect of caterpillars. Caterpillars are collected and sold in markets either directly by gatherers or through wholesalers and retailers. N'Gasse (2003) cited by Vantomme (2003) analyzed the revenues of a wholesaler who purchases and sells an average of 24 kg of caterpillars per day during the season (3 months a year). The wholesaler buys caterpillars at 1000 FCFA ($1.70) per kg and sells them at an average price of 1333 FCFA. The estimated profit for the wholesaler was $600 over the season. This income is used for education, food, and healthcare, significantly contributing to improved livelihoods.

9.9.5 Transborder Trade in Edible Insects

There is also significant transborder trade in edible insects within the Central Africa region, as well as with Sudan and Nigeria, and, on a smaller scale, with France and Belgium. Although export figures are rarely available, an inquiry revealed that France and Belgium annually import about 5 tons and 3 tons, respectively, of dried *Imbrasia* species from the Democratic Republic of Congo, valued at about $41,500 in Belgium.

9.9.6 Economic Importance of Edible Caterpillars

A study in Botswana by Zitzmann (1999) revealed that selling *Imbrasia belina* caterpillars contributes 13% of the total household cash income annually, requiring only 5.7% of the labor input for all income-generating activities. This makes *Imbrasia belina* the third largest source of income after the sale of poles and livestock, highlighting its significance to household livelihoods.

9.9.7 Caterpillars in Zimbabwe

In Zimbabwe, for instance, caterpillars like *Gonimbrasia belina* are a major food item. They are collected, transported, and sold industrially, with a price of Z$ 0.60 per 100 g dry weight (mid-1986), comparable to fresh beef (Wilson 1990). Another caterpillar, *Cirina forda*, is commonly found in rural and urban markets (McGregor 1991). According to Gelfand (1971), *Cirina forda*, known as harati, nowha, and madora (*Imbrasia epimethea*), are highly valuable and can be traded for grains such as millet and maize. Chavunduka (1975) also reported that some families can earn a substantial living from selling caterpillars. In regions like Nuanetsi, Belingwe, and parts of Matebeleland, families can collect up to four standard six-tin bags of caterpillars per day, selling them at beer halls for an average price of $24.00 per bag.

9.9.8 Economic Impact of Edible Insects

Edible insects hold significant economic importance in Southern Africa. Quinn (1959) noted that the Pedi people prefer caterpillars to fresh beef, which affects beef sales when caterpillars are available. Processing plants for caterpillars exist in Botswana (Brandon 1987) and South Africa (Dreyer and Wehmeyer 1982). The South African Bureau of Standards estimated annual sales of caterpillars at about 40,000 bags, each containing 40 kg of dried caterpillars. Ferreira highlighted concerns that extensive use of caterpillars as food and income sources may lead to their extinction.

9.9.9 Indirect Contribution to Food Security

Caterpillars also contribute indirectly to food security. A study in Namibia by Chikasa et al. (2002) examined the potential of *Imbrasia belina* worm droppings as feed additives for lactating cattle in north-central Namibia, a region with low rainfall and livestock feed shortages. The worm droppings had high crude protein (8.6%), calcium (3.3%), and phosphorus (0.13%) compared to dry *Colophospermum mopane* leaves. The study found that using these droppings as feed additives increased milk and meat production, improving household economies.

Similarly, Chikasa (2001) found that *Imbrasia belina* worm droppings contained high levels of nitrogen (1.5%) and phosphorus (5.5%), indicating their potential as high-quality manure for crop production. However, their use may be limited to vegetable gardening and potted plants due to the small quantities produced per worm. Households using this practice reduced fertilizer costs, freeing up income for other food purchases.

9.10 Loss of Edible Caterpillars and Mushrooms

Forests are an essential source of wild foods for many urban and rural communities in the Central Africa Copperbelt region. Edible insects, mushrooms, plants, and other foraged items from forests offer crucial nutrition and livelihoods, especially during drought periods. However, the expansion of mining and extractive industries threatens these important wild food sources. Mining operations negatively impact the plants, fungi, and animals thriving in these natural habitats.

In many urban, suburban, and rural areas in Zambia and the DR Congo, forests and trees significantly contribute to food supply and security. They provide essential dietary elements often lacking in agricultural produce (Moore and Vaughan 1994; Nkomesha 1997; Mogaka et al. 2001; Vantomme et al. 2004; Guveya 2006; Ombeni et al. 2022). Moore and Vaughan (1994) observed that gathered foods from forests, such as caterpillars and mushrooms, and their sales or exchange ranked higher than pension money, part-time employment, and vegetable marketing in some parts of the Northern Province in Zambia.

Mogaka et al. (2001) found that forest resources significantly contribute to the total income of poorer households in several parts of Zambia, although they account for only a third of the income of richer households. These findings were supported by a study by Puurstjärvi et al. (2005), which used various published data to estimate the economic value of different forest products as their share of national GDP.

Likewise, a survey by Guveya (2006) in Northwestern and Luapula Provinces highlighted the significant contribution of non-wood forest products (NWFPs) to most household incomes. The survey found that mushrooms, edible caterpillars, chikanda, and thatching grass were the main forest products providing income for the majority of the population.

The study by Nkomesha (1997) revealed that selling mushrooms, fruits, and munkoyo roots generated needed income for several households. Women were the primary collectors of these forest products, while men were mostly involved in marketing. Botolo (2003) emphasized the need for further research to link economic growth and forest development in sub-Saharan Africa, given that most poor people derive their daily livelihoods from forest-related economic activities.

9.10.1 Insect Species Decline

Many edible insect species, such as caterpillars, certain beetle larvae, ants, and crickets, are highly sensitive to environmental changes and pollution. The buildup of heavy metals and other pollutants in the soil and vegetation poisons and kills off insect populations, causing drastic declines in insect biodiversity. While many insects are consumed in Zambia, the insect with the highest commercial value is the edible caterpillar belonging to the family of moths called Saturniidae commonly known as ifishimu (Bemba) and Vinkubala (Nyanga). This is the same family from which Mopani "worms" harvested in Zimbabwe, South Africa, and Botswana come from. Edible caterpillars are found in many areas throughout Zambia, but the epicenter of production is Mpika District in Muchinga Province. At least nine edible caterpillar species have been identified in Mpika. For centuries, the plateau Bemba-Bisa of Northern and Muchinga Provinces has harvested this resource (Fig. 9.17).

Fig. 9.17 Caterpillars from forest trees are used as food and source of income for households in Zambia and the DR Congo

They have developed a unique processing method, which involves roasting the caterpillars in very hot ash called ukusolwela. This gives the Mpika caterpillar a unique smoky flavor that consumers have come to love.

9.10.2 Mushrooms Decline

Vulnerability of Wild Mushrooms to Mining Pollution

Wild mushrooms are highly susceptible to the impacts of mining pollution. The underground mycelia networks essential for fungi growth can be severely damaged by chemical contamination. Many mushroom species cannot survive in polluted soils, leading to their disappearance from affected forest areas. This not only reduces the availability of this nutritious food source but also disrupts broader fungal ecosystems vital for forest health.

Previous Research on Mycorrhizal Fungi in Zambia

Research by Alian and Musenge (1978), Piearce (1978, 1981), Piearce and Francis (1982), and Högberg and Piearce (1986) has documented a wide variety of mycorrhizal fungi associated with various tree species in Zambia, most of which cannot be cultivated artificially. While these species alone may not satisfy the dietary needs of the population, there is an urgent need to further identify and catalog the numerous other mushroom species in the country. At SNR, CBU, in Zambia, the project on mushrooms aimed at studying the prevalence of indigenous mushrooms in various forest ecosystems (Miombo woodlands, closed forests, and some grasslands) around rural villages in six different provinces. It also explored their relationships to biological diversity and nutrient recycling and identified factors influencing their occurrence. Additionally, the project sought to develop and transfer technologies for mushroom domestication to ensure year-round availability as an income source for rural households while empowering women's groups in business and financial management.

Untapped Potential of Mushrooms

The potential of mushrooms as a food source is largely untapped (FAO/GIEWS 2000). Mushrooms grow in various environments, including decaying roots, dead wood, termite mounds, and cultivated land. There is much to be done in sensitizing the public and disseminating information, as well as conducting research to enhance the knowledge and use of mushrooms for food.

Impact of Mushrooms on Food Security and Nutrition

Egal et al. (1996) observed that where local populations use and manage forests for their livelihood, forest products significantly impact their food security and nutrition. In Zambia, most local communities rely on the forest for food and firewood for both home consumption and sale. Promoting indigenous mushroom growing, establishing small agro-based processing units, and developing transportation and marketing systems for these products will improve rural employment opportunities and incomes.

Nutritional Value of Mushrooms

Mushrooms are an excellent source of digestible proteins and fiber, low in fat and energy, and contribute to vitamin and mineral intake (Fig. 9.18). They have high protein content, typically 20–30% crude protein as a percentage of dry matter, and provide all essential amino acids. One of the most cultivated mushrooms, *Agaricus bisporus*, has nutritional values comparable to whole egg protein.

9.11 Global Consumption and Commercial Production

Food of fungal origin is consumed worldwide in large quantities, and commercial production is a rapidly growing industry (Ingram 2002). Edible mushrooms are an excellent source of fiber, vitamins, and some minerals. Mushroom protein is of intermediate nutritional quality between meat and vegetable proteins. Dried fungi and concentrated extracts are also used as medicines and dietary supplements.

In summary, wild mushrooms play a crucial role in food security and nutrition for many communities. However, mining pollution poses a significant threat to their availability and the health of forest ecosystems. Efforts to promote mushroom cultivation and sustainable harvesting can help mitigate these impacts and improve livelihoods.

9.12 Species of Wild Nutraceutical Bulbs, Tubers, and Roots

In recent time, food of fungal origin is consumed all over the world in vast quantities, and commercial production is part of a rapidly growing industry (Ingram 2002). Edible mushrooms have high protein content and are an excellent source of fiber, vitamins, and some minerals. They have fairly high protein content, typically 20–30% crude protein as a percentage of dry matter. Mushroom protein appears to be intermediate in nutritional quality between meat and vegetable proteins. High protein content makes them an ideal food because they contain all the amino acids

Fig. 9.18 Different types of mushrooms are part of the nutritive diets and source of traditional medicines in Zambia and the DR Congo

essential to human nutrition, which are considered a reliable measure of the mushroom nutritive value. Mushrooms are also an important dietary source of amino acids. One of the most cultivated mushrooms worldwide, *Agaricus bisporus*, which was originally imported and now commonly found in Zambia's market, has nutritional values highly comparable to whole egg protein.

In addition to caterpillars and mushrooms which become out of reach to communities as a result of pollution by mining and metallurgic industries, the region harbors a diverse array of wild nutraceutical bulbs, tubers, and roots, whose biological richness is steadily declining in tandem with the loss of forest cover. The wild nutraceutical bulbs, tubers, and roots—including the wild yam locally known as lusala or busala (*Dioscorea* species), the Livingstone potato (*Plectranthus esculentus*), the African potato (*Hypoxis hemerocallidea*), and the bulbs such as orchids and other roots of the *Rhynchosia* species—are regularly collected from forests and wetlands throughout the Central Africa Copperbelt region either for consumption as nutritional food at the household level or sold as essential mainstays of the family sources of income (Fig. 9.19). They have also numerous other health benefits. Hence, the gathering and production of these products for either nutraceutical purposes or domestic consumption are parts of a household's livelihood resilience strategies to prolonged droughts and flooding that threaten food security in the region.

What are the nutraceutical bulbs, tubers, and roots (NBTRs)? The NBTRs include:

- Wild bulbs: These are short modified underground stem—usually globe shaped, sometimes with modified leaves, containing nutrients necessary for the growth of the plant. Bulbs are similar to corms, rhizomes, or tubers. They are primarily bulbs from orchid species such as *Disa engleriana, Disa robusta, Disa zombica, Habenaria clavata, Satyrium amblyosaccos, Satyrium buchananii*, and *Satyrium cursonii* and edible bulbs from other *Satyrium* species. However, some orchid species produce tubers.
- Wild roots: Part of a plant that is found underground, may be swollen or not and may be used as a storage organ and transport medium for nutrients and water, and provides support to the plant. Edible wild roots comprise those of the genera *Eminia*, *Rhynchosia*, and *Vigna*, primarily *Eminia antennulifera* (Baker) Taub., *E. benguellensis* Torre, *E. holubii* (Hemsl.) Taub. (synonym *E. polyadenia* Hauman), *E. harmsiana* De Wild., *Rhynchosia insignis* (O. Hoffm.) R.E.Fr. subsp. Insignis, *R. insignis* (O. Hoffm.) R.E.Fr. subsp. Affinis (De Wild.) Pauwels, *R. heterophylla* Hauman, *R.* 3lavate3 (Hiern) K. Schum, and *Vigna nuda* N.E.Br.
- Wild tubers: Tubers are enlarged underground structures in wild plant species that a plant can use for the storage of nutrients and perennial survival (regrowth in the following growing season) and can also be used for asexual or vegetative reproduction. In terms of shape, underground tubers can be classified as finger-shaped lateral root tubers. There are numerous wild tubers including species such as *Dioscorea odoratissima* (a wild yam), *Sphenostylis stenocarpa* (Chinaka in Zambia), and *Habenaria walleri*. The tubers of the wild yam (*Dioscorea hirti-*

Fig. 9.19 Wild nutraceutical biological diversity species: tubers of wild yams of the *Dioscorea* sp. (above left and right), plants and bulbs of the African potato of the *Hypoxis* species (middle left and right), and leaves and tubers of the of Livingstone potato (*Plectranthus esculentus*)

flora Benth.) also commonly known as lusala or busala and *Plectranthus esculentus* (Umumbu) are important traditional foods for most households in drought-prone areas.

The forest biodiversity including the NRTBs constitutes one of the most important natural resources for households' sustainability especially in rural areas. They provide numerous subsistence goods and needs such as:

Food

The NRTBs are widely used as foods ensuring food security and safety net for the majority of household's livelihood and providing important dictary minerals, vitamins, and proteins.

Beverage

The drinks brewed from some species are increasingly consumed especially during special events such the celebration of a good harvest, a newborn baby, a wedding, etc., becoming the most favorite nonalcoholic beverage found everywhere.

Health

The NRTBs are also a major source of medicines that are widely used to cure numerous illnesses. Treatment with traditional medicines has recently got momentum due to poverty to purchase the drugs at health centers, poor health services in rural settings, and preference over modern medicines.

Income

The NRTBs are traded to earn incomes which contribute to the social security and the economy of the family enabling to meet essential needs.

Employment

Some of the forest dwellers are employed to collect the NRTBs for sale, and the transporters and traders of the products are all employed in the NRTB subsector earning salary that also contributes to the overall family and national economy.

For communities that have depended on foraging wild insects, mushrooms, and other foods from the forest, the loss of these resources can have severe consequences. It threatens food security, cultural practices, and traditional livelihoods. Addressing the environmental devastation caused by irresponsible mining is crucial to preserving these irreplaceable wild edible resources.

Chapter 10
Scaling Up Mitigation Paths for Sustainable Environment

Keywords Remediation · Sustainable practices · Policy · Stakeholder engagement · Management · Awareness

10.1 Environmental Impacts and Mitigation Contexts

Mining activities have profound and multifaceted impacts on the environment, and their consequences were particularly evident in the Central Africa Copperbelt region of Zambia and the DR Congo, which is rich in mineral resources. These impacts include deforestation, soil erosion, water pollution, and biodiversity loss, to name but a few, all of which threaten the sustainability of local ecosystems and communities. The "Scaling up mitigation paths for a sustainable environment" aims to address the pressing need for actionable strategies to combat and reduce the environmental damage caused by mining. It further highlights the challenges and opportunities specific to their unique geological, social, and economic landscapes.

Zambia and the DR Congo share a rich history of mining, especially in copper and cobalt production, which fuels global technological advancements but also poses significant environmental threats. The chapter emphasizes how mining has disrupted natural ecosystems, degraded agricultural land, and compromised water sources that are vital for local communities. It also reflects on the intersection of mining impacts with climate change, suggesting pertinent mitigation efforts undertaken in line with broader environmental sustainability goals. This approach underscores the urgency to implement practical and scalable solutions that not only address immediate mining-induced damage but also foster long-term resilience in the affected areas.

Furthermore, the authors' focus explores innovative mitigation strategies, such as the adoption of cleaner mining technologies, stricter regulatory frameworks, community-driven environmental restoration, and engaging stakeholders—industry representatives, environmental groups, and local communities—in the

A. Muimba-Kankolongo et al., *Mining Impacts on the Environment in the Central African Copperbelt of Zambia and the Democratic Republic of Congo*,
https://doi.org/10.1007/978-3-031-96786-3_10

policymaking process. It also probes into the role of international cooperation and investment in promoting sustainable practices in Zambia and the DR Congo's mining industries. Scaling up these paths, the authors envision a future where mining activities can coexist with thriving ecosystems and sustainable development, ensuring that the benefits of mineral wealth do not come at an irreversible cost to the environment.

10.2 Background

Recurrent droughts in the Copperbelt region have led to food insecurity among several households (Muimba-Kankolongo 2018; Mulenga et al. 2022), but, in addition, food security has been compromised, to a hitherto poorly quantified degree, by the degradation of the environment including farmlands, caused by pollution from mineral extraction and operations of metallurgic industries (Mwitwa et al. 2012b; Kashimbo Kalala et al. 2015). Consequently, it is urgently important for mining and metallurgic industries to implement better environmental and social responsibility activities to prevent environmental pollution and assist communities in the host areas to benefit from their sector activities. Roos et al. (2020) highlighted that besides the economy, the potential benefits of the Environmental Impact Assessment (EIA) include effective compliance with sound environmental standards, sustainable use of natural resources, better decision-making processes by involving stakeholders, improvement in the design of infrastructures, and potential reduction of operating costs. To be effective however, an EIA mechanism needs to be grounded on well-defined legislation and procedural rules where the rights and obligations of all stakeholders are clearly defined and their enforcement ensured through appropriate implementation and compliance monitoring procedures and other instruments. State governments are ultimately responsible for the regulatory and institutional settings within which mining and metallurgical industries operate, and their actions are critical to achieving sustainable benefits from the extractive industries for national economic development.

Nevertheless, according to Larmer (2021), the contrasting political situations of Zambia and the DR Congo in the 1990s led to divergent timings regarding the environmental assessment of mining activity and liberalization. In Haut-Katanga Province, the DR Congo, economic and political collapse during that period meant that effective environmental assessment was delayed until the early 2000s. In Zambia, however, the participatory politics of the Third Republic created a comparatively open context for the discussion of environmental concerns. Media coverage of pollution on the Copperbelt increased dramatically, with the *Times of Zambia* reporting in October 1994 about the environmental damage caused by industrial effluent and toxic gas emissions from mining activities.

While ZCCM (Zambia Consolidated Copper Mines) was reported to be cooperating with the Environmental Council of Zambia (ECZ) to improve its environmental management systems and sought to portray itself as environmentally friendly, the introduction of new environmental laws was not matched by substantially increased monitoring capacity in state bodies such as the ECZ and the Ministry of Mines. On the Copperbelt, the once powerful labor movement was profoundly weakened, limiting its ability to hold mining companies accountable for their environmental impact. This contrast between Zambia and the DR Congo highlights how political and economic factors shaped the timing and nature of environmental assessment and regulation of mining activities in the two countries during this period. Hence, governments in both countries need to improve their governance mechanisms and policies to optimize the benefits from the sector for the social, economic, and environmental development.

10.3 Implications of the Results from Both Countries

While mining and metallurgic exploitations are sources of economic development for Zambia and the DR Congo, they have also caused numerous negative impacts on the surrounding communities and the environment. During our study, it was evident that industrial effluents from waste dumps were released all over the landscape in addition to the spillage of mine waste through broken pipes, causing water flooding and erosion. Both direct and indirect discharges of the wastewater resulting from mining and industrial activities, without purification, in various areas were often polluting water which is important for domestic purposes and destroying the ecosystems comprising the farmlands, the surrounding landscape, and the ambient air. In his book, Larmer (2021) emphasized that some of the most severe mining-related pollution risks stem from ore processing activities aimed at value addition—particularly through the use of local smelters. While the Zambian mineworkers' union often brought up issues related to the community, such as healthcare and education, during discussions with mining companies, concerns about pollution were noticeably absent from the records of their meetings. Similarly, the tripartite forums managing the city of Gécamines company in Lubumbashi in the DR Congo during the 1980s showed a lack of concern regarding pollution. However, the awareness campaign about air pollution remains clearly evident through the thoughtful placement of housing—European and skilled workers' houses are situated farther from pollution sources, whereas those of non-educated employees are nearer. To illustrate this, in Mufulira, for instance, the Kankoyo neighborhood is situated adjacent to the plant, exposing it to persistent air pollution and experiencing damage to homes due to underground mining activities. Meanwhile, European workers and higher-ranking African workers are accommodated farther away from the plant in areas positioned outside the path of the fumes. Furthermore, Larmer (2021) cited Peša (2020b) who had painstakingly found convincing evidence showing how companies handled pollution while sidestepping accountability for its impact on air and water. This occurred in a period when newly independent states focused on boosting mine production rather than addressing environmental issues. Companies' engineers and managers were often confident that future technical advancements would resolve the environmental challenges they were

Table 10.1 Concentration of trace elements (μg/L) in water samples from Zambia and the Democratic Republic of Congo

	Zambia (*n* = 42)		DR Congo (*n* = 35)			
Trace elements	Minimum	Maximum	Minimum	Maximum	WHO standard	*p**
Mn	1.30	1208	2.77	96,140	**50**	<0.001
Co	0.047	96.1	2.52	73,150	**N/A**	<0.001
Ni	0.279	7.13	0.418	42,190	**70**	<0.001
Cu	0.076	3052	5.74	221,900	**2000**	<0.001
Zn	0.837	17,320	0.975	35,670	**5000**	<0.021
As	0.037	1.94	9.238	2313	**10**	<0.001
Cd	0.008	0.208	0.008	552	**3**	<0.001
Pb	0.016	30.2	0.402	37,670	**10**	<0.001
U	0.004	6.49	0.004	200	**30**	<0.001

All values are μg/L, except where indicated. WHO is upper limit for drinking water according to World Health Organization guidelines (not available for Co); Min is minimum value, and Max is maximum value; **p* value for difference between Zambia and DRC (Mann–Whitney test)

creating, including the growing accumulation of waste deposits across the region. Numerous reports and publications highlighted that the worldwide issue of pollution from mining activities, such as effluent discharge and the erosion of old tailing dams, was being tackled through data collection by mining companies using advanced scientific techniques. This approach aimed at preventing the environmental damage already experienced in well-industrialized nations (Larmer 2021).

Regarding pollution of drinking water, the situation appeared to be much worse in the DR Congo than in Zambia (Table 10.1). The findings regarding drinking water from communal taps and boreholes indicate consistently higher metals and metalloid contamination in samples from the DR Congo than in those from Zambia. In Zambia, nearly all samples had concentrations of trace elements well below the WHO limits for drinking water (WHO 2017a, b), whereas a high proportion of samples from the DR Congo considerably exceeded these limits, including for the most toxic elements arsenic, cadmium, and lead. The situation was particularly frightening for lead in tap water in the DR Congo, where most samples exhibited concentrations above the WHO guideline value of 10 μg/L, with some values well above 100 μg/L.

The concentrations of trace metals in food crops were also consistently higher in samples obtained from the DR Congo than in those from Zambia, except for Mn (Table 10.2). Collectively, the findings indicate that crops cultivated close to mining areas were heavily contaminated by trace elements in the DR Congo than in Zambia. Samples of maize, sweet potato, and cassava tubers contained lower concentrations of most trace metals per dry weight than the leafy vegetables, tomato, and common beans. WHO-FAO guidelines for standard limits of trace elements in foods are available for arsenic, cadmium, and lead only (FAO/WHO 2011, 2017). From our results, the standard guidelines were exceeded more in the crop samples obtained from the DR Congo.

Table 10.2 Concentration of trace elements (μg/g dry weight) in samples of food crops from Zambia and the Democratic Republic of Congo

	Zambia (n = 128)		DR Congo (n = 145)		
Trace elements	Min	Max	Min	Max	p*
Mn	3.49	9.06	4.45	880	0.34
Co	0.140	61.9	0.14	186	<0.001
Ni	0.420	75.5	1.04	74.μ5	<0.001
Cu	1.51	21.6	0.04	1605	<0.001
Zn	17.2	6668	5.27	991	<0.001
As	0.880	2.54	0.006	21,06	<0.001
Cd	0.025	9.65	0.028	32.7	<0.001
Pb	0.150	89.5	0.150	104	<0.001
U	0.405	26.6	0.405	27.5	<0.001

All values are μg/g dry weight, except where indicated. *p values for difference between Zambia and the DRC (Mann–Whitney test)

Based on the findings of our study, the excessive concentrations of Pb and Cd found in drinking water and vegetables are arguably the most serious causes of toxicological concern for the people living in the fringe of the mining and metallurgic industries, especially those in the DR Congo. Exposure to Pb, Cd, and As is well known for causing adverse effects on several biological systems in humans, especially neurodevelopment, kidney damage, and cancer disease (Rauh and Margolis 2016; Carver and Gallicchio 2017; Lentini et al. 2018; Satarug 2018; Chen et al. 2019).

Mining-related pollution in Zambia and the DR Congo stems from various factors, including geological and governance differences. Although both countries are located in the Copperbelt region and rely heavily on copper and cobalt mining, the depth and composition of the extracted ores result in distinct industrial processes and environmental pollutants like arsenic (As), cadmium (Cd), lead (Pb), and uranium (U). However, the most significant factor influencing mining pollution levels between Zambia and the DR Congo is their contrasting environmental governance. These disparities trace back to colonial times when Zambia (then Northern Rhodesia) and the DR Congo (then the Belgian Congo) were under British and Belgian rule, respectively. Neither colonial administration prioritized environmental protection (Peša 2020a, b). Following independence, both nations nationalized their mining industries, yet environmental issues remained overlooked until the 1990s.

The apparent mining pollution and differences in mining-related pollution between the DR Congo and Zambia can be attributed largely to environmental governance in the two countries. In the DRC, ASADO/Katanga (2007) and Global Witness (2006) reported that mining companies have hardly complied with the environmental regulatory requirements and the government has not systematically monitored the industries' implementation of the environment assessment protocols. Nationalization of the mining industry in both countries also contributed to environmental issues not being treated seriously. Even though environmental legislation exists in both countries, agencies tasked with implementing such legislation must be well resourced and protected from political and corporate influence. In order to protect the human health and the environmental pollution from mining and metallurgic industries, mitigation measures of the environmental risks related to pollution of soil and groundwater in the areas where minerals are produced must be a national priority.

Zambia has made strides in mitigating mining pollution through initiatives like the Environmental Protection and Pollution Act (ZEMA 2011) and improved environmental management policies and institutions (ZEMA 2017a, b). Zambia's Environmental Council was established to oversee pollution control and human welfare by safeguarding natural resources. These efforts have been bolstered by adequate government funding and support, alongside World Bank-assisted environmental safeguards for extractive industries (World Bank 2015). Such measures have reduced soil and groundwater contamination and enhanced water and sanitation infrastructure in Copperbelt municipalities.

In contrast, the DR Congo has struggled to enforce environmental regulations. Despite revisions to the Mining Code, most recently through the decree n° 18/024 in 2018, compliance among mining companies remains limited, with weak government enforcement of environmental assessments (ASADHO/Katanga 2007; Global Witness 2006). This governance gap is reflected in the World Bank's governance indicators, where Zambia ranks higher (25–50th percentile) than the DRC (0–10th percentile) over the past two decades (World Bank 2020).

10.4 Management of Mining and Metallurgic Industries Operations

Metal extraction and processing operations contribute significantly to the GDP of both the DR Congo and Zambia in terms of foreign currency earnings, employment, and local economic benefits through the implementation of Corporate Social Responsibility programs (Ministry of Mining and Hydrocarbons. 2003; Global Witness 2006; Fraser and Lungu 2007; SADC/FANR 2011; ICMM 2014; Zambia EITI Report 2020). The benefits from mining and metallurgic industries operations are cast in the context of the potential and actual danger they pose to human health and the environment. In both the DR Congo and Zambia, the release of mining effluents and smoke plume containing higher concentrations of heavy metals that contaminate both surface and underground water systems, the landscape, and ambient air often resulting in the removal of large tracts of vegetation have occurred and continue to occur.

10.5 Stakeholders' Awareness Workshops on Impact of Pollution

The workshop gathering was commissioned through an official invitation call from the British Council in Zambia, a statutory body serving as the representative of the Commonwealth funding agency. This call emphasized the importance of collaboration in addressing mining impacts and environmental sustainability challenges in Zambia and the DR Congo. As a Commonwealth partner, the British Council's invitation reinforced its commitment to fostering impactful initiatives that align with the broader goals of sustainable development and environmental restoration in the region. The gathering brought together experts, stakeholders, and community leaders, setting the stage for meaningful discussions and actionable strategies to address pressing issues related to mining and the environment.

BRITISH COUNCIL

21st July, 2011,

Dear All,

RE: INVITATION TO THE DELPHE ANNUAL MEETING

Following a successful networking meeting held last year, it was approved that we could have this meeting on an annual basis. This year the meeting is even more important as we will be looking at changes to the project. Future prospects as well as new reporting systems. The latter is applicable to both running and ending projects so I would like to encourage all of you to attend. Additionally we would like you to attend this meeting with a fellow colleague who is either active or has participated in your project and we feel they shall benefit from the information to be shared.

For those coming from within Lusaka you are welcome to bring 2 other persons with you, if available.

This year our meeting details are as follows.

Venue: Taj Pamodzi Hotel

Date: 5th August, 2011

Time: 09:00 hrs–16:00 hrs

Due to limitations of my budget, I will not be able to cover your transport and accommodation costs. However, I have been given permission that you can use your project monies (under contingency) to cover the above costs. If for any reason this will not be possible, kindly contact me.

I thank you all for your continued commitment to the individual projects you run and I look forward to meeting you all.

Yours Faithfully

Daisy Banda

DELPHE ANNUAL MEETING PROGRAMME

05.08.2011

TAJ PAMODZI HOTEL

09:00	Registration of Participants
09:15	Introductions
09:30	Brief Overview of Delphe
09:45	Delphe Reporting System—Narrative
1045	**Tea Break**
11:00	Delphe Reporting system—Financial
12:00	Question and Ans. Log onto UK contact
1300	**Lunch**
14:00	Project Presentation: Nkrumah University College
14:15	Project Presentation: CBU—Heavy Metal Contamination
14:30	Project Presentation: CBU—Information Literacy Programme
14:45	End of Project Presentation: Mental Health
1500	**Tea Break**
15:15	End of Project Presentation: Multi Grade
15:30	Project Presentation: School of VET
15:45	Q&A
16:00	Close of Meeting

GUIDE FOR PROJECT PRESENTATIONS

- Objectives
- Outputs
- Achievement based on outputs
- Impact on beneficiaries
- Any other Key Highlights/Challenges

Two workshops were organized in Ndola, Copperbelt Province of Zambia, and in Lubumbashi, Katanga Province of DR Congo, assembling local decision-makers and several stakeholders from representatives of mining companies, civil society, environmental agencies, and scientists from government research agencies and high learning institutions and international and local nongovernmental organizations (NGOs). Members of the local communities mainly from areas where samples were obtained were also included. During the meeting, our present findings and those from some of previous studies on mining pollution and related population health issues were presented to create the public awareness of the challenges and influence the dialogue for policies on sustainable management of mining operations and feedback from participants. A special session was then organized to shape the future of mineral exploitation and metallurgic industry operations as a way forward to anchor consolidated policy governance to mitigate the undergoing environmental degradation for sustainable socioeconomic development and community livelihood in the Copperbelt region.

Workshop participants freely discussed the various negative mining issues (Table 10.3) and expressed their views and with a united voice suggested to recommend good governance of mining exploitation and metallurgic industry operations

Table 10.3 Reaction of stakeholders in the Central Africa Copperbelt region of Zambia and the DR Congo and suggestions for mining impacts mitigation

Identified mining operation issues	Impacts on environment and the population	Participants' statement	
		DRC	Zambia
Damage of natural habitat at mining and other sites away.	Death of forest trees and the vegetation cover. Considerable reduction of wild foods such as fruits and insects from forests in the process of changing rainfall patterns, thus undermining the local livelihoods.	✓	✓
Land degradation due to spillage of toxic chemicals and waste dumping.	Poor growth of forests and natural vegetation near mining arcas and at dumping sites considerable loss of biodiversity. Soil pollution from contaminated air and irrigation with polluted water. Decline in agricultural production and quality of products. Increased soil erosion and sedimentation load in rivers.	✓	✓
Water table pollution due to acid drainage and leakage.	Deterioration in water quality resulting in drinking unsafe ground and surface water. High incidence of waterborne diseases among the population such as diarrhea. Drying of streams.	✓	✓
River pollution from operations, broken pipes, and illegal processing.	Browning of rivers from spilling chemicals such as artisanal miners by sieving and cleaning in the rivers the earthen products they collect to extract minerals. Death of fish and other aquatic animals and plants.	✓	✓
Pollution through effluent from mineral processing.	Long-term pollution of water. Destruction of natural resources via erosion leading to increased sedimentation of toxic chemical loads.	✓	✓
Population health hazard through air pollution due to smoke emission in the atmosphere.	Increased levels of toxic chemical buildup in peoples' bodies. Widespread respiratory problems from wind-blown and inhalation of smoke particularly in children and old people such as asthma, bronchitis, and pulmonic coughing. Birth defects from lead and cadmium poisoning mostly affecting both breast-feeding mothers and babies. Other sicknesses such as goiters, skin disease, and higher blood concentrations with trace metals.	✓	✓

(continued)

Table 10.3 (continued)

Identified mining operation issues	Impacts on environment and the population	Participants' statement	
		DRC	Zambia
Health problems from aerial inhalation of contaminated dust.	High incidence of respiratory complications including communicable diseases like tuberculosis and death particularly among artisanal miners. Increased incidence of coughing mainly in children and old people due to diseases like asthma, bronchitis, and pulmonary dysfunction.	✓	✓
Health hazard due to exposure to extreme noise and vibration.	Intolerable stress from continuous loud operation equipment. Frequent blasting from mining exploitation and increased traffic of heavy trucks carrying minerals and other mining equipment disturbing the population and wildlife.	✓	✓
Violation of human rights.	Opposition to mines is suppressed by the use of police force and military deployment using violence, torture, arrests, and harassment. Mining areas often associated with militarization and abuses by inhuman treatment. Communities rights violated through displacement resettlement in different locations without compensation for loss of land. Logistical support provided to police, army, and independent security which sometimes causes death of the people. Many underage children work in the mines.	✓	✓
Manual labor risk mainly in artisanal mining.	Long-term disability due to accidents working underground and exposure to contaminated dust. Loss of life for many miners during times when they are trapped under collapsing mineshafts. Neither government officials nor representatives of mining companies accept responsibility for the lives and welfare of miners.	✓	✓
Town residents and rural communities have complained to courts.	Settlement of matters always taking long and often rulings are not in favor of residents despite having misused and their rights violated.	✓	✓

to mitigate the pollution of natural resources. All participants recognized that respective governments were not doing enough to discourage unsustainable mining exploitation. Local authorities in Lubumbashi highlighted the fears they face when dealing with the presence of the general public that has just come out of war, and by knowing such pollution information, it could erupt into a riot and general disobedience. Therefore, members of the community could not participate in the workshop in the DR Congo. However, it was acknowledged that while there are some laws regulating mining exploitation in both the DR Congo and Zambia, these seem to be incomplete or simply not enforced to their full extent. In Lubumbashi in the DR Congo, for instance, mining waste is dumped anywhere including in rivers and in the city center even where several thousands of inhabitants live, polluting the environment, the land for agricultural production, and water and destroying the entire food chain. Though the mines were important to national economy by providing financial resources to governments, basic infrastructures for community services, and employment for the population, the level of pollution they cause brings more negative than positive effects such as human health problems and food insecurity. Participants expressed strong views for the inevitable needs to find trade-offs between social equity, ecological integrity, and economic development in the mining sector through recommendations to submit to the government authorities dealing with mineral and metallurgic industries.

10.6 Policy and Legislation in the Copperbelt of Zambia and the DR Congo

The DR Congo environmental legislations are based on the "Receuil des Textes Juridiques en Matières d'Environnement en RD Congo" that includes the "Décret n° 038/2003 du 26 mars 2003" (Code Minier 2018), while the Environmental Management (Licensing) Regulations SI. No. 112 of 2013 regulate all activities relating to air and water pollution, waste management, hazardous waste, pesticides and toxic substances, and ozone depleting substances in Zambia (ZEMA 2013). Pieces of legislation within the two countries are significantly adequate on paper. However, their implementation is shrouded in political interference, corruption in some of the implementing agencies, poorly funded and staffed implementing agencies, lack of monitoring systems of implementing agencies, and lack of accountability of national governments to their citizens regarding environmental problems arising from operations of mining and metallurgic industries.

These potential negative effects of mining and metallurgic industries and related operations require stringent implementation of national and international regulations, sound monitoring systems, and removal corruptions in institutions that are tasked with the implementation of environmental legislation. Moreover, the government environmental management agencies operate as policymaking and self-monitoring agencies, creating inefficiency and inertia and resulting in some mining companies' failure to adhere to national and international environmental safeguards. In the event that pollution has occurred, it takes from 5 to 24 h even more for inspectors to arrive at the scene of disaster. The time lapse is adequate for culprits to clean and minimize the perceptive impact of the pollution. This is worsened by the inadequacies of modern monitoring equipment in national agencies despite the huge revenues that national governments accrue from mining and metallurgic industries operations.

To reinforce regulations for environmental protection, governments of the two countries should take several measures such as conducting regular reviews and updates of existing environmental laws to ensure they are up to date and effective in addressing present environmental challenges making mining and metallurgic industries accountable for responsible environmental stewardship. Otherwise, they should further strengthen enforcement mechanisms, such as increasing fines and penalties for noncompliance, which can also deter individuals and businesses from violating environmental regulations. Additionally, implementing transparent monitoring and reporting systems to track environmental performance and compliance with regulations would also help hold polluters accountable for their actions.

Numerous ecological, economic, and social issues about the impacts of mining and metallurgical industries arose out of this study. Generally, they are complex in nature and necessitate multiple perspectives and immediate attention to tackle. From our observations in the field, it is recommended that:

- The Ministry of Mines and Minerals Development and that of Green Economy and Environment in Zambia and the Ministry of Mines and that of Environment and Sustainable Development in the DR Congo together with local related departments in the Copperbelt Province in Zambia and the former Haut-Katanga Province in the DR Congo in partnership with several stakeholders including nongovernmental organizations, community-based organizations like churches, bilateral agencies, and the private sector need to take mining and metallurgic companies to task and engage them in discussion on the mitigation strategies for the environmental pollution issues.
- It was worth noting that the legislation, which should regulate industrial corporate social responsibility (CSR) calling for the improvement of the local community's livelihood, though existent, is not enforced in both the DR Congo and Zambia. Concerned government agencies need particularly to exert a very-high-profile regulatory control over the misleading claims for good practices by numerous companies and to put in place rigorous regulations and policy prescriptions taking into account the implementation of appropriate CRS to benefit the local communities where mining activities take place.
- KCM in Zambia and Gécamines in the DR Congo as well as several other mining companies operating in the Central Africa Copperbelt should always endeavor to take further steps to ensure that reliable data on SO2, dust, and other emissions as well as water analysis are collected and publicly disseminated.
- Governments should respect their own laws related to mining and not sign development agreements that supersede popularly constituted legislation. The fact that governments sign separate agreements with industrial firms without disclosing the contents of such agreements places the industries above national laws.
- It was clear that in both countries, the ministries responsible for the environment, public health, social welfare, labor, finance, mining, immigration, and national planning are the key institutions that are supposed to have a role in governing mineral extraction and processing. Even though an institutional framework exists for monitoring the impact of mining and metallurgic industries operations on the environment and local communities, local government institutional structures

cannot exercise their authority over the industries without the approval from the sectorial ministries and sometimes from the head of state. Coordinated efforts, at all levels, are necessary to sustainably manage the environment in the Central Africa Copperbelt.

- Adequate resources and support need to be provided to environmental protection agencies to enhance their capacity to enforce regulations effectively.
- Engaging with stakeholders, including industry representatives, environmental groups, and local communities, in the policymaking process can help ensure that regulations are balanced and address the concerns of all parties involved. Local communities in all mining areas are aware of both the socioeconomic and environmental impacts resulting from mineral extraction and processing, the operations which significantly outweigh short-term economic gains. Communities require constant awareness on the impacts of industrial operations through messages such as radio, television, home delivered leaflets, community meetings, and school drama programs. This will certainly help them avoid contaminated areas that have the potential to affect their health and overall livelihood.
- Finally, the safety of residents and miners remains of paramount concern especially in view of series of contamination and fatal accidents occurring regularly at different mine sites. Public disclosure of the health and safety reports is primordial to restoring public confidence in the management of the mines and metallurgic industries in the DR Congo and Zambia.

From the foregoing, it is clear that the perpetual discharge of wastewater and fume plume from the operations of mining and metallurgic industries, respectively, in the streams and rivers and the atmosphere considerably threatens the environment and communities' livelihood. The induced pollution affects the scarce water, agricultural farmlands, biodiversity and forest natural resources, and the population health of the two countries (Riediker et al. 2019). Most community inhabitants claimed they suffer from numerous sicknesses such as cancer, respiratory difficulties, and effects on their auditory organs, blood pressure (cardiovascular system), and physiological well-being due to noise pollution from mining operations. Overall, extractive industries are not depraved developmental organizations per se. They can significantly contribute to national economic growth and development, but they must be appropriately and effectively regulated. This would surely strengthen environmental regulations of mining and metallurgic activities and the rights of affected communities. It is not only communities who suffer but also the governments of both countries, as they are unable to obtain profitable benefits from mineral exploitations while being burdened with the environmental and social damage industries cause.

The full responsibilities lie on the hands of state governmental agencies like the Zambia Environmental Management Agency (ZEMA) and the Labour Administration and the Labour Inspectorate of the DR Congo's Ministry of Labor that regulate and enforce laws which will ensure monitoring of obligatory compliance regulations of mining and metallurgic industries to implement fully and correctly the submitted environmental management and monitoring plans and empower the local communities.

Chapter 11
Conclusion and Recommendations

Keywords Findings · Community involvement · Policy · Environmental stewardship · Long-term strategies

At the end, the book wraps up with the conclusion and recommendations in which the perspectives of mining sustainability to protect the environment and ensure community health and food and nutrition security in the region are presented. They serve as a critical culmination of the analysis presented in the book, synthesizing key insights on the environmental impacts of mining in Zambia and the DR Congo. They aim to provide a reflective summary of the challenges identified, drawing connections between the socio-environmental consequences of mining and the pressing need for sustainable practices, at the same time acknowledging the broader implications of mining on the ecosystems, communities, and economies of these resource-rich nations while emphasizing the urgency of charting a path toward environmental stewardship and resilience.

It is essential to stress that Zambia and the DR Congo, as major players in global copper and cobalt production, face unique environmental challenges that demand tailored solutions. These include soil degradation, air and water pollution, food contamination, and deforestation and biodiversity loss, all of which pose significant threats to local livelihoods and well-being. This chapter not only consolidates the evidence of these impacts but also underscores the interdependence of human well-being and environmental health, arguing that sustainable mining practices are indispensable for the long-term prosperity of the Central Africa Copperbelt region.

Finally, the "Conclusion and Recommendations" chapter also envisions a future shaped by collaborative efforts among governments, industry stakeholders, local communities, and international organizations. It emphasizes practical recommendations, such as strengthening regulatory frameworks, promoting cleaner technologies, and investing in community-driven environmental initiatives. Highlighting these strategies, the book calls for a collective commitment to mitigate environmental damage while ensuring that mining continues to contribute to the economic

A. Muimba-Kankolongo et al., *Mining Impacts on the Environment in the Central African Copperbelt of Zambia and the Democratic Republic of Congo*, https://doi.org/10.1007/978-3-031-96786-3_11

development of Zambia and the DR Congo. Ultimately, the chapter serves as both a reflection and a roadmap, guiding stakeholders toward a more sustainable and equitable mining sector.

11.1 Conclusion

Throughout the book paragraphs, we argue that our findings contribute to the knowledge of environmental degradation with persuasive attestation of mining and metallurgical operations pouring huge quantities of noxious metal elements into the ecosystems in the Central Africa Copperbelt region of Zambia and the DR Congo, causing severe environmental contamination of water, food crops, and ambient air. This pollution negatively impacts the well-being of surrounding communities, with children being more vulnerable than adults to pollutants. Besides focusing on financial gain and tax contributions, governments in the two countries have failed instituting consistent checking and administrative execution of safeguards and environmental protection regulations. There is an urgent need for stringent legislative and policy reinforcement of industrial activities to mitigate negative impacts on environmental protection, food and nutrition safety, and population health.

The devastating impacts of mining pollution on the environment, including water, food, air, and biodiversity, cannot be understated. The long-term consequences of irresponsible mining practices affect natural ecosystems, wildlife, human health, and livelihoods. Urgent action is required to regulate and mitigate the effects of mining pollution to protect our planet for future generations. By promoting sustainable mining practices and investing in innovative technologies, we can work toward a cleaner and healthier environment for all living beings.

It is evident that mineral extraction activities have significant negative impacts on biodiversity, including habitat destruction, water pollution, soil degradation, and disruption of ecosystems. These impacts threaten the survival of many species and lead to biodiversity loss. It is crucial for mining companies, governments, and stakeholders to recognize these impacts and take effective measures to minimize and mitigate them through sustainable practices, reclamation efforts, and biodiversity conservation initiatives. Ultimately, safeguarding biodiversity is essential for the health of ecosystems and the well-being of future generations.

The impacts of mining pollution on communities' livelihoods are significant and far-reaching. It is crucial to recognize the devastating effects that mining pollution can have on the environment, health, and socioeconomic well-being of communities. In conclusion, urgent action is needed to address these issues and mitigate the negative consequences of mining pollution on communities' livelihoods. Collaboration between stakeholders, implementation of sustainable practices, and stringent regulations are essential steps toward ensuring the well-being of affected communities and the preservation of their livelihoods.

Mining and metallurgical industries can be major contributors to environmental pollution due to activities such as extraction, processing, and disposal of materials.

There are several solutions that can help mitigate and reduce the environmental impact of these industries:

1. Implementing cleaner and more efficient technologies in mining and metallurgical processes can help reduce emissions and waste generation. This includes using advanced equipment, processes, and materials that are more environmentally friendly.
2. Encouraging the recycling and reuse of materials within the mining and metallurgical industries can help reduce the need for new extraction and processing, thereby reducing environmental impact.
3. Proper management of waste generated from mining and metallurgical activities is crucial. Implementing effective waste management practices, such as recycling, treatment, and safe disposal of waste, can help minimize environmental pollution.
4. Mining and metallurgical industries often require large quantities of water for processing operations. Implementing water conservation and recycling measures can help reduce water usage and minimize impacts on water quality and availability.
5. Rehabilitating mined sites after extraction activities can help restore ecosystems and reduce the environmental impact of mining operations. This can involve recontouring the land, replanting vegetation, and restoring natural habitats.
6. Enforcing strict environmental regulations and standards for mining and metallurgical industries can help ensure that they operate in an environmentally responsible manner. Regular monitoring and enforcement of these regulations are essential to prevent pollution.
7. Engaging with local communities and stakeholders to understand their concerns and involve them in decision-making processes can help promote sustainable practices and minimize environmental impacts of mining and metallurgical activities.

Overall, improving water and the ambient air quality for communities requires a holistic approach that involves various measures aimed at reducing pollution and promoting sustainable practices. To enhance water quality, it is crucial to implement proper wastewater treatment systems to remove pollutants before discharging water into rivers and streams. Additionally, enforcing regulations on industrial and agricultural activities to prevent the contamination of water sources with harmful chemicals and nutrients is essential. Investing in infrastructure for safe drinking water supply and promoting water conservation practices can also help improve water quality for communities.

In terms of ambient air quality, reducing emissions from vehicles, industries, and power plants is key to decreasing air pollution levels. Implementing stricter emission standards and promoting the use of clean energy sources such as solar and wind can help reduce air pollutants. Creating green spaces and promoting tree planting in urban areas can also improve air quality by absorbing pollutants and providing fresh oxygen. Community engagement and awareness-raising initiatives on the importance of air quality can encourage individuals to adopt environmentally friendly practices such as carpooling, using public transportation, and reducing energy consumption.

Moreover, curbing biodiversity loss due to mining pollution requires a multifaceted approach that involves stringent regulatory frameworks, sustainable practices, and community engagement. Governments must enforce strict environmental regulations that limit the impact of mining activities on sensitive ecosystems and biodiversity-rich areas. Implementing comprehensive monitoring and reporting systems can help track the environmental impact of mining operations and ensure compliance with regulations. Mining companies should adopt sustainable practices such as reclamation, restoration, and biodiversity offsetting to mitigate the negative effects of mining pollution on biodiversity. Engaging with local communities and indigenous groups to incorporate their knowledge and concerns into decision-making processes can help ensure that biodiversity conservation is prioritized in mining operations. Additionally, investing in research and technology for cleaner mining practices, such as using green technologies and recycling waste materials, can help minimize the environmental footprint of mining activities and protect biodiversity for future generations.

Mitigating mining pollution is crucial for maintaining a sustainable environment. One way to achieve this is by implementing stricter regulations and monitoring practices to ensure that mining activities adhere to environmental standards. This can include monitoring water quality, air emissions, and soil contamination to prevent harmful pollutants from entering the environment. Additionally, utilizing technologies such as tailings dams and liners can help contain and properly store mining waste to prevent leaching into surrounding areas. Promoting responsible mining practices, such as reclamation and restoration of mined areas, can also help mitigate the environmental impact of mining activities. Collaboration between governments, mining companies, and local communities is vital to ensure sustainable mining practices that protect the environment for future generations.

By implementing these solutions, mining and metallurgical industries can reduce their environmental footprint and work toward sustainable and responsible operations.

11.2 Recommendations

Addressing mining environmental pollution issues which is crucial, concerted efforts from the private and public sectors, international organizations, and both bilateral and multilateral donor communities are needed to reinforce and implement guidelines regulating mining and metallurgic operations for protecting the environment, ensure food and nutrition safety, and preserve population health in the Central Africa Copperbelt.

(a) There exist several potentially effective techniques for heavy metal management—such as cultivation of heavy metal-tolerant crop species and land remediation methods, advanced filtration systems, state-of-the-art filtration technologies to reduce water contamination from runoff, soil remediation technologies, and use of bioremediation and phytoremediation techniques to restore contaminated soil using plants and microorganisms that can detoxify pollutants—that are readily available for use. These should be adopted and integrated into the overall management strategies for heavy metals in the region to protect the local communities and ensure their sustainable livelihood.

(b) In addition, governments should enforce regulations that require mining companies to adhere to specific pollution control measures. These regulations should include guidelines for waste disposal, water treatment, and air quality standards. For this, they need to implement stringent environmental regulations that mandate regular monitoring and maintenance, hold mining companies financially responsible for environmental restoration and pollution caused, and encourage the adoption of sustainable mining practices, such as using eco-friendly technologies, reducing water and energy consumption, and implementing reclamation plans to restore affected areas.

(c) Regular monitoring of mining activities and pollution levels can also help detect potential issues early and ensure compliance with environmental regulations. Utilizing advanced technologies such as satellite imagery and drones to enable real-time monitoring of mining activities and their environmental impacts, while conducting regular audits and assessments to ensure compliance with established environmental standards. Inspections should be mostly carried out by independent bodies to maintain transparency.

(d) Support research and development efforts to discover new technologies and methods for minimizing environmental impact in the mining industry. Efforts should be devoted to funding studies aimed at developing more sustainable mining practices and pollution mitigation techniques. These can lead to the development of cleaner extraction techniques and innovative pollution control measures.

(e) Engage local communities and stakeholders in decision-making processes related to mining activities to ensure their needs are met, their traditional knowledge is integrated, and their concerns are actively heard and addressed. This will help build trust between mining companies and local residents.

(f) Ensure the promotion of reclamation and remediation efforts on the mining and metallurgic companies to create and implement comprehensive reclamation and remediation plans to restore land affected by their activities and minimize long-term environmental damage and restore ecosystems. Additionally, initiate educational campaigns that inform about the impacts of mining pollution and promote sustainable practices.

(g) Raise the level of transparency and accountability in reporting mining extraction activities and operations of metallurgic companies on pollution data to the public. Holding these companies accountable for their environmental impact can incentivize them to improve their operation practices.

(h) Foster collaboration and partnership between governments, mining companies, environmental organizations, and local communities to work together toward finding sustainable solutions to mining pollution issues. Partnerships can bring together different expertise and resources to address these complex challenges.

By implementing these recommendations, a conducive footprint for reducing the environmental impact of mining and metallurgical companies' activities would be achieved to safeguard the health of Zambia and the DR Congo ecosystems and communities' livelihood for future generations.

Earth Organization (2022) reported that biodiversity loss can come in the form of habitat loss through land converted to agriculture, combined with land degradation through intensive farming practices—which is the principal contributor to the decline and extinction of species—and it can also come in the form of unsustainable food production, where our meat and dairy consumption habits are to blame, as livestock farming requires disproportionate areas of land to be used.

Species of plants and animals that are vulnerable to extinction require everyone's utmost attention to ensure that significant changes at a policy and a collective individual level are made. Thus, the following implementations are necessary:

11.2.1 Biodiversity Conservation

This is the one of most important solutions to biodiversity loss: conserving biodiversity that is at risk of extinction by protecting them with adequate conservation strategies. Biodiversity conservation covers a wide range of activities that can be done. Protecting habitats is an extremely important biodiversity conservation activity, done by identifying the habitats facing threats and eliminating these threats in order to maintain the natural area. This also comes in the form of leaving wildlife undisturbed, especially nesting and denning areas, and wildlife habitats can be promoted by setting up man-made bird and bat houses.

Limiting and modifying agricultural activities also falls into the category of biodiversity conservation. This can be done by conserving water in wetlands and reducing irrigation and by managing livestock grazing through maintaining good-quality range conditions and leaving areas ungrazed. Biodiversity conservation can also come in the form of domestic conservation. This involves an individual taking responsibility for any wildlife they have direct control over. Maintaining your garden by eradicating and controlling weed growth can be particularly beneficial to conserving wildlife. Reducing disturbance to wildlife and monitoring pets and their behavior with wild animals are also domestic biodiversity conservation solutions.

11.2.2 Pressure on Governments

Pressure needs to be put on governments to draft, pass, and enforce legislation to protect biodiversity. All governments should strive to create an environment that welcomes intergovernmental bodies and international policymakers to collaborate in advocacy in regard to issues of biodiversity. Unsustainable food production can also be a cause of biodiversity loss. Therefore, governments can ensure that company policies are put in place to encourage more sustainable methods by ensuring sustainable fishing methods through certification of seafood products, for example. They can also protect national parks and other areas with flora and fauna through laws, and they can provide incentives and subsidies to farmers to encourage production methods toward sustainable land stewardship.

11.2.3 Funding Directed to Innovative Solutions

Technology is advancing at such a rapid pace, so directing funding and research into technological biodiversity loss prevention methods can be extremely beneficial. One of the main causes of biodiversity loss is habitat loss and fragmentation caused by deforestation, so a simple solution would be to plant more trees; however, tree planting by hand can be slow and labor intensive. Biocarbon engineering came up with this solution to tackle this issue by first mapping the area of regeneration via the drone. Next, the drone will fly around 2–3 meters above the surface and then shoot biodegradable seed pods into the soil which contain all the nutrients the tree needs to start growing. Biocarbon engineering estimates this speeds up the planting of trees by ten times and at 15% of the cost.

11.2.4 Substitute Products

Obtaining the resources to create the products we consume is severely damaging to biodiversity. Examples include meat consumption, baked goods containing palm oil, mass-produced cheap clothing, and the use of plastic straws. One of the simplest solutions to biodiversity loss is by substitute products with sustainable and environmentally friendly replacements. Land use does not exclusively limit to raising cattle but is often dedicated to producing food for cows and other livestock.

11.2.5 Vertical Ocean Farming

Human action has affected the oceans too. Sea life has been impacted through pollution and climate change, but the changes impacting the ecosystem the most are those caused by overfishing. Overfishing occurs when fish are caught from the ocean at a faster rate than the populations can replenish. Many other sea creatures, such as oysters, have been severely overfished, and the practices to catch them now involve scraping the bottom of the ocean, catching all sorts of unintended fish, and destroying the habitat at the same time.

GreenWave, a charity which promotes the use of restorative vertical ocean farming, was created to address this issue. Co-founder Bren Smith describes vertical ocean farming as an "underwater garden" where they grow kelp, mussels, scallops, and oysters, by mimicking the habitats that would have previously existed in the ocean.

Having these gardens removes the need to trawl the ocean floor with large nets, and they can even provide food for other wildlife, such as fish and seals. An additional benefit about vertical ocean farming methods is that the seaweeds and shellfish require no fresh water, feed, or fertilizer, which dramatically reduces the overall environmental impact and keeps costs down.

11.2.6 Change Your Individual Choices

Biodiversity loss is directly caused by anthropogenic actions, making it crucial to adopt deliberate and sustainable choices that support it. Additionally, pollution from mining and metallurgical industries in the Central Africa Copperbelt region degrades critical habitats including forests, grasslands (both dry and wetlands), and aquatic ecosystems (lakes, rivers, and oceans). This destruction often accelerates the downfall, displacement, or extinction of local species that rely on these habitats for subsistence. Pollution disrupts essential ecosystem interactions, such as parasite–host interactions and mutualistic interactions, and may favor certain species over others, leading to non-native species taking hold and outcompeting with native ones, further deteriorating biodiversity loss.

Biodiversity sustains indispensable ecosystem services such as species genetic hybridization, water cleansing, climate change legislation, and soil nutrient quality. The loss of biodiversity reduces these services, negatively affecting community maintenances and welfare. It also hampers ecosystems' aptitude to acclimate, increasing species susceptibility and precariousness. Many communities in the Central Africa Copperbelt region of Zambia and the DR Congo depend on biodiversity for wild foods, ethnic distinctiveness, and lucrative activities like tourism, wild item collection, and trawling. Therefore, pollution-induced biodiversity loss can have severe socioeconomic consequences for both local and global communities.

Funding Agency This study was funded by the British government through its program on Development Partnerships in Higher Education (DelPHE) and overseen by the Association of Commonwealth Universities (ACU) in the United Kingdom under the project Data for the DFID's 25 priority (Public Service Agreement) countries under the grant IATI Identifier: GB-1-111543.

Conflicts of Interest The authors declare that they have no financial ties or affiliations with mining companies and metallurgical industries or any NGOs that may influence the perspectives or conclusions drawn within the pages of this book.

References

Adejumobi II, Agre P, Anautshu DO et al (2022) Diversity, trait preferences, management and utilization of yams landraces (Dioscorea species): an orphan crop in DR Congo. Research Square, pp 1–34. https://www.researchsquare.com/. Accessed 29 Nov 2024

Agence Nationale pour la Promotion des Investissements (ANAPI) (2021) Investir dans le sous-secteur de l'eau en RDC. Cahier Sectoriel, Kinshasa/Gombe, p 80. https://anapi.cd/

Aguilar-Manjarrez J, Bannayan-Avval M (2001) A global mapping system for bambara groundnut production. Food Agriculture Org. Technology & Engineering. 47 p. https://books.google.ca/books/about/A_Global_Mapping_System_for_Bambara_Grou.html?id=I7rWhGvy4O4C&redir_esc=y

Ambika Asati, Mohnish Pichhode, & Kumar Nikhil, 2016. Effect of Heavy Metals on Plants: An Overview. IJAIEM 5(3), 56–66. https://paper.researchbib.com/view/paper/74028

Amnesty International (2018) "This is what we die for." Human rights abuses in the Democratic Republic of the Congo power the global trade in cobalt. Amnesty International Ltd, London. https://www.amnesty.org/en/documents/afr62/3183/2016/en/. Accessed 24 Oct 2024

Angon PB, Islam S, Shreejana KC et al (2024) Sources, effects and present perspectives of heavy metals contamination: soil, plants and human food chain. Heliyon 10(2024):e28357. https://www.cell.com/heliyon. Accessed 2 Apr 2025

ASADHO/Katanga (2007) The impact of extractive industry activities on the rights of local communities. In: NIZA (ed) NIZA briefing paper. Groundwork. http://old.groundwork.org/special-reports/ASHADHO. Accessed 2 Mar 2016

ASADO/Katanga (2007) The impact of extractive industry activities on the rights of local communities. In: NIZA (ed) NIZA Briefing Paper. Groundwork, 2007. 8pp. https://www.groundwork.org.za/specialreports/ASHADHO%20English.pdf

Asati A, Pichhode M, Nikhil K (2016) Effect of heavy metals on plants: an overview. IJAIEM 5(3):56–66. https://paper.researchbib.com/view/paper/74028

Ashley Steel E, Bwembelo L, Mulani A et al (2022) Wild foods from forests: quantities collected across Zambia. Br Ecol Soc People Nat 4(5):1159–1175. https://doi.org/10.1002/pan3.10367

Ashton PJ, Love D, Mahachi H et al (2001) An overview of the impact of mining and mineral processing operations on water resources and water quality in the Zambezi, Limpopo and Olifants Catchments in Southern Africa. Contract report to the mining, minerals and sustainable development (SOUTHERN AFRICA) project, by CSIR-Envirotek, Pretoria, South Africa

A. Muimba-Kankolongo et al., *Mining Impacts on the Environment in the Central African Copperbelt of Zambia and the Democratic Republic of Congo*, https://doi.org/10.1007/978-3-031-96786-3

and Geology Department, University of Zimbabwe, Harare, Zimbabwe. Report no. ENV-P-C 2001-042, p xvi + https://www.iied.org/mining-minerals-sustainable-development-mmsd

ATSDR (2012) Toxicological profile for manganese. Agency for Toxic Substances and Disease Registry. U.S. Department of Health and Human Services, Public Health Service. 556pp. Giorgia, USA. https://stacks.cdc.gov/view/cdc/12399. Accessed 1 Jul 2025

AVRDC (2003) Vegetables for life: confronting the crisis in Africa. AVRDC publication no. 03-564. AVRDC—the World Vegetable Center, Shanhua. https://www.Worldveg.tind.io/record/31448/files/eb0018.pdf. Accessed 18 Aug 2009

Banque Centrale du Congo (2013) Rapport annuel 2013 de la Banque Centrale du Congo. 328pp. Kinshasa, République Démocratique du Congo. https://www.bcc.cd/publications/rapports-annuels/rapport-annuel-2013

Banza Lubaba Nkulu C, Casas L, Haufroid V, De Putter T, Saenen ND, Kayembe-Kitenge T, Musa Obadia P, Kyanika Wa Mukoma D, Lunda Ilunga JM, Nawrot TS, Luboya Numbi O, Smolders E, Nemery B (2018) Sustainability of artisanal mining of cobalt in DR Congo. Nat Sustain 1(9):495–504. https://doi.org/10.1038/s41893-018-0139-4

Banza Lubaba Nkulu C, Casas L, Haufroid V et al (2009a) Sustainability of artisanal mining of cobalt in DR Congo. Nat Sustain 1:495–504. https://www.nature.com/articles/s41893-018-0139-4. Accessed 18 Aug 2009

Banza Lubaba Nkulu C, Nawrot TS, Haufroid V et al (2009b) High human exposure to cobalt and other metals in Katanga, a mining area of the Democratic Republic of Congo. Environ Res 109:745–752. https://doi.org/10.1016/j.envres.2009.04.012

Banza Lubaba Nkulu C, Casas L, Haufroid V et al (2018) Sustainability of artisanal mining of cobalt in DR Congo. Nat Sustain 1, 495–504. https://doi.org/10.1038/s41893-018-0139-4

Bernard A, Schadeck C, Cardenas A (1991) Potentiation of diabetic glomerulopathy in uninephrectomized rats subchronically exposed to cadmium. Toxicol Lett 58:51–57. https://doi.org/10.1016/0378-4274(91)90190-H

Bhat SA, Hassan T, Majid S (2019) Heavy metal toxicity and their harmful effects on living organisms—a review. Int J Med Sci Diagn Res 3(1):106–122. https://www.ijmsdr.com. Accessed 2 Apr 2025

Blacksmith Institute report (2007) The world's worst polluted places: the top ten (of the dirty thirty), New York, USA, p 72. https://www.Blacksmithinstitute.org

Bokanga M (1995) Cassava; opportunities for the food, feed and other industries in Africa. In: Agbor Agber T, Brauman A, Griffon D, Trecher S (eds) Transformation Alimentaire du Manioc. ORSTOM Editions, Paris, pp 557–569. https://www.semanticscholar.org/paper/Cassava-%3A-Opportunities-for-the-food-%2C-feed-%2C-and-Egbe/1d31968907eb6d587079f32cc9985403361b8cd3

Boocock CN (2002) Environmental impacts of foreign direct investments in mining sector of Sub-Saharan Africa. In: OECD and GFII. Foreign direct investment and the environment: lessons from the mining sector. OECD, Paris. https://www.econbiz.de/Record/environmental-impacts-of-foreign-direct-investment-in-the-mining-sector-in-sub-saharan-africa-boocock-colin-noy/10001724159

Bradbury JH, Holloway WD (1988) Chemistry of tropical root crops: significance for nutrition and agriculture in the Pacific. ACIAR monograph no. 6. Australian Centre for International Agricultural Research, Canberra. https://www.aciar.gov.au/publication/books-and-manuals/chemistry-tropical-root-crops-significance-nutrition-and-agriculture-pacific. Accessed 12 Feb 2007

Broadman HG (2007) Africa's Silk Road: China and India's new economic frontier. The International Bank for Reconstruction and Development/World Bank, Washington, DC. http://www.worldbank.org

Brummett BH, Boyle SH, Siegler IC, Kuhn CM, Ashley-Koch A, Jonassaint CR et al (2008) Effects of environmental stress and gender on associations among symptoms of depression and the serotonin transporter gene linked polymorphic region (5-HTTLPR). Behav Genet 38:34–43. https://doi.org/10.1007/s10519-007-9172-1

Buchet JP, Lauwerys R, Roels H (1990) Renal effects of cadmium burden of the general population. Lancet 336:699–702. https://doi.org/10.1016/0140-6736(90)92201-R

Butler RA (2006) Deforestation in the Congo Rainforest. Mongabay.com. A place out of time: tropical rainforests and the perils they face, Los Angeles, California. http://rainforests.mongabay.com. Accessed 21 May 2009

Buzea C, Pacheco I (2020) Heavy metals: definition, toxicity, and uptake in plants. In: Faisal M, Saquib Q, Alatar AA, Al-Khedhairy AA (eds) Cellular and molecular phytotoxicity of heavy metals. Nanotechnology in the life sciences. Springer, Cham. https://doi.org/10.1007/978-3-030-45975-8_1

CAADP (2013) 2013 Report [CAADP Implementation Support]. Agriculture, Food Security and Nutrition. 12p. Midrand, South Africa. https://nepad.org/

Campbell D, Fiebig M, Mailloux M et al (2011) Zambia environmental threats and opportunities assessment (ETOA). The United States Agency for International Development (USAID)/Zambia mission, p 99. https://www.usaid.gov/

Carver A, Gallicchio VS (2017) Heavy metals and cancer. IntechOpen, Open Access. https://www.intechopen.com/search?query=Austin%20and%20Gallicchio%202017

Central Statistical Office (2000) Census of population and housing. Housing and household characteristics. Analytical report. Government Printers, Lusaka. https://www.zamstats.gov.zm/

Chalwe A, Malambo C, Zimba K et al (1998) Cassava root rots and associated micro-organisms in Zambia. In: Akoroda MO, Teri JM (eds) Proceedings of the conference on food security and diversification in SADC countries: the role of cassava and sweetpotato on improving nutritional status of east and southern Africans: role of root crops, Lusaka, Zambia, pp 317–324. https://istrc-ab.org/. Accessed 15 Oct 2005

Chapoto A, Chisanga B, Kabisa M (2018) Zambia agriculture status report

Chavunduka DM (1975) Insects as a source of protein to the African. Rhodesia Sci News 9:217–220. https://www.amazon.ca/Rhodesia-Science-News-Vol-1975/dp/B00D8ASEUM. Accessed 15 Oct 2005

Chen C, Park T, Wang X et al (2019) China and India lead in greening of the world through land-use management. Nat Sustain 2:122–129. https://doi.org/10.1038/s41893-019-0220-7

Cheyns K, Banza Lubaba Nkulu C, Kabamba Ngombe L et al (2014) Pathways of human exposure to cobalt in Katanga: a mining area of the DR Congo. Sci Total Environ 490:313–321. https://doi.org/10.1016/j.scitotenv.2014.05.014

Chidumayo EN (1996a) Land use planning. The Zambia Forestry Action Program (ZFAP) Secretariat, Lusaka. https://zm.chm-cbd.net/zambia-forestry-action-programme-zfap. Accessed 15 Nov 2010

Chidumayo EN (1996b) Urbanisation, charcoal production and deforestation. In: Campbell B (ed) The Miombo in transition: woodlands and welfare in Africa. CFIOR, Bogor, p 95. https://doi.org/10.17528/cifor/000465

Chidumayo EN, Mbata KJ (2002) Shifting cultivation, edible caterpillars and livelihoods in the Kopa area of Northern Zambia. For Trees Livelihoods 12:175–193. https://doi.org/10.1080/14728028.2002.9752423

Chikasa P, Hamid OY, Horn-Nghishikungu LN (2002) Potential of Imbrasia belina to be used as manure for crop production. Dissertation, Namibia Ogongo Agricultural College

Chipundu PM, Kunda DM (eds) (1994) State of the environment in Zambia. The Environmental Council of Zambia, Lusaka, p 115. https://www.zema.org.zm/docs/state-of-environment-report-1994/

CIAT (2004) Regional bean variety testing in Southern Africa. Highlights, CIAT in Africa no. 20, pp 1–2. https://alliancebioversityciat.org/publications-data/regional-bean-variety-testing-southern-africa

Coalition for Rainforest Nations (2008) Mining and deforestation. Non-wood news no. 16. EcoNook, New York, USA. https://www.rainforestcoalition.org/deforestation-and-degradation/

Census of Business Establishments, 2010. In: Infrastructure and Cities for Economic Development ("ICED") (2018). Kitwe: Inclusive Urban Economic Development Diagnostic. Volume: URB001. 8pp. https://ukgreencitiesandinfrastructure.org/. Accessed 17 Sept 2024

Central Statistics Office (CSO), 2008. Labor force survey report, LFS – 2008. Labor Statistics Branch. Lusaka, Zambia. 125pp. https://www.zamstats.gov.zm/census-and-statistics/

Central Statistics Office (CSO), 1990. Census of Population, Housing and Agriculture: Vol. 10. Analytical Report. Lusaka, Zambia. 115pp. https://www.zamstats.gov.zm/census-and-statistics/

Central Statistics Office (CSO), 2003. An edition of 2000 census of population and housing (2003). Lusaka, Zambia. 10pp. https://www.zamstats.gov.zm/census-and-statistics/

Code Minier (2018) Décret no 038/2003 du 26 mars 2003 portant règlement minier tel que modifié et complété par le décret no 18/024 du 08 juin 2018. 59eme Année. No. Spécial du 12 Juin 2018. Imprimerie CEDI, Centre Protestant d'Éditions et de Diffusion. Kinshasa, RDC. http://congomines.org/system/attachments/assets/000/001/467/original/J.O._no_speìcial_du_28_mars_2018_CODE_MINIER.PDF.pdf?1523182711. Accessed 14 May 2020

Concas A, Ardau C, Cristini A et al (2006) Mobility of heavy metals from tailings to stream waters in a mining activity contaminated site, Chemosphere 63(2), 244–253, https://doi.org/10.1016/j.chemosphere.2005.08.024

Coursey DG (1967) Yams. An account of the nature, origins, cultivation and utilization of the useful members of the Dioscoreaceae. Longmans, Green and Co. Ltd., London, p 230. https://archive.org/details/yams0000unse/page/n6/mode/1up

Cuvelier J (2014) Work and masculinity in Katanga's artisanal mines. Afr Spectr 49:3–26. https://doi.org/10.1177/000203971404900201

Davey C (2023) The environmental impacts of cobalt mining in Congo. Earth Organization, Pine Mountain Club. https://earth.org/cobalt-mining-in-congo/

Deena R (2024) 15 biggest environmental problems of 2024. Earth Organization, Pine Mountain Club. https://earth.org/cobalt-mining-in-congo/

de Milliano WAJ (1992) Pearl millet diseases in Southern Africa. In: de Milliano WAJ, Frederiksen RA, Bengston GD (eds) Sorghum and millets diseases: a second world review, Patancheru, India, p 378. https://1lib.sk/book/2976165/73e0f7/sorghum-and-millets-diseases-a-second-world-review.html. Accessed 30 May 2008

De Putter T, Decrée S, Banza Lubaba Nkulu C (2011) Mining the Katanga (DRC) Copperbelt: geological aspects and impacts on public health and the environment—towards a holistic approach. In: Kribek B (ed) IGCP/SIDA Project 594, Inaugural Workshop; Kitwe, Zambia. Czech Geological Survey, pp 14–17. https://www.geology.cz/igcp594/kitwe/PROCEEDINGS-OF-THE%20WORKSHOP.pdf

De Putter T, Decrée S, Banza Lubaba Nkulu C (2011) Mining the Katanga (DRC) Copperbelt: geological aspects and impacts on public health and the environment—towards a holistic approach. In: Kribek B (ed) IGCP/SIDA project 594, inaugural workshop; Kitwe, Zambia. Czech Geological Survey, pp 14–17. https://www.publication/272510675_Mining_the_Katanga_DRC_Copperbelt_geological_aspects_and_impacts_on_public_health_and_the_environment_-_towards_a_holistic_approach. Accessed 17 Mar 2024

Earth Organization (2022) The remarkable benefits of biodiversity, California, USA. https://earth.org/cobalt-mining-in-congo/

Earth Organization (2024) 15 biggest environmental problems of 2024, California, USA. https://earth.org/cobalt-mining-in-congo/

Egal F, Ngoma A, Ndione PD (1996) Integration of food security and nutrition in forestry planning: the role of participatory approaches (A:\Unasylva - No_ 202 - Forests, food security and sustainable livelihoods.htm). https://www.fao.org/forestry/unasylva/en/

Elenge MM, De Brouwer C (2011) Identification of hazards in the workplaces of artisanal mining in Katanga. Int J Occup Med Environ Health 24:57–66. https://doi.org/10.2478/s13382-011-0012-4

Etzel RA, Landrigan PJ (eds) (2024) Textbook of children's environmental health, 2nd edn. Oxford Academic, New York. https://doi.org/10.1093/oso/9780197662526.001.0001

Extractive Industries Transparency Initiative (EITI) (2014) Democratic Republic of Congo. Executive Committee of the Extractive Industries Transparency Initiative report, Kinshasa, Gombe, p 195. https://eiti.org/countries/democratic-republic-congo. Accessed 12 Mar 2016

Faber M, Laurie S, Ball AM et al (2010) A crop-based approach to address vitamin A deficiency in South Africa—a regional manual for SADC. ARC-Roodeplaat, Pretoria. https://www.arc.agric.za/arc-vopi/Pages/Crop%20Science/Improved-Nutrition-A-Food-Based-Approach.aspx. Accessed 12 Sept 2011

Faber, M., Laurie, S.M., van Jaarsveld, P.J., 2016. Total β-carotene content of orange sweetpotato cultivated under optimal conditions and at a rural village. African Journal of Biotechnology, 12(25): 3947–3951. https://doi.org/10.4314/ajb.v12i25

Fanshawe, D.B., 1973. Check list of the woody plants of Zambia showing their distribution. Forest Research Bulletin No. 22. Ministry of Lands and Natural Resources, Division of Forest Research. 48pp. Government Printers. Lusaka, Zambia. https://www.mlnr.gov.zm/

FAO (1986) African agriculture: the next 25 years (main report, Rome, Italy). https://agris.fao.org/search/en/providers/122621/records/6477516e5eb437ddff747d8c

FAO (1988) Sorghum and millets in human nutrition. Food and Nutrition Division, Rome, Italy. https://www.fao.org/4/T0818e/T0818E01.htm#Preface

FAO/WHO, 1993. Evaluation of certain food additives and contaminants: Forty-first report of the Joint FAO/WHO Expert Committee on Food Additives, WHO technical report series; 837. Geneva, Switzerland. 64pp. https://www.who.int/publications/i/item/9241208376

FAO (2002) State of the food insecurity in the world, p 4. https://www.fao.org/4/y7352e/y7352e00.htm. Accessed 27 Jun 2008

FAO (2003) A perspective on water control in Southern Africa: support to regional investment initiatives. Land and water discussion paper 1, Rome, Italy. https://www.fao.org/4/y5096e/y5096e00.htm. Accessed 30 Dec 2023

FAO (2020) Biodiversity for food and agriculture and ecosystem services—Thematic Study for The State of the World's Biodiversity for Food and Agriculture. Rome, Italy. https://doi.org/10.4060/cb0649en

FAO/GIEWS (2000) Special feature: forests and food security in Africa. FAO Africa report no.1, April 2000-6, Rome, Italy. https://www.fao.org/4/x4868e/x4868e11.htm

FAO/UNEP (2021) Global assessment of soil pollution—summary for policy makers. FAO, Rome, p 84. https://doi.org/10.4060/cb4827en

FAO/WFP (1998) Crop and food supply assessment mission to Zambia. Special report of Food and Agriculture Organization and World Food Program of the United Nations, Rome, Italy, p 12

FAO/WFP (2005) Special report by the Food and Agriculture of the United Nations (FAO) and World Food Program (WFP) on crop and food supply assessment mission to Zambia, Rome, Italy

FAO/WHO (2011) Joint FAO/WHO Food Standards Programme Codex Committee on Contaminants in Foods, Food CF/5 INF/1. https://www.fao.org/fao-who-codexalimentarius/committees/committee/en/?committee=CCCF. Accessed 23 Jul 2025

FAO/WHO (2017) Evaluation of certain contaminants in food. Eighty-third report of the Joint FAO/WHO Expert Committee on Food Additives. In: WHO technical report series: no. 1002, Geneva, Switzerland. WHO, Geneva. https://apps.who.int/iris/handle/10665/254893. Accessed 2 Sept 2020

Feeney P (2001) The limitations of corporate social responsibility on Zambia's Copperbelt. Konkola Copper Mines (KCM): environmental management plan (May 2001). Report. http://www.minesandcommunities.org

Feng T, Shi Y, Wang X et al (2025) Synergies of air pollution control policies: a review. J Environ Manag 377:124655. https://doi.org/10.1016/j.jenvman.2025.124655

Fraser A, Lungu J (2007) For whom the windfalls? Winners and losers in the privatization of Zambia's Copper Mines, Lusaka Civil Society Trade Network of Zambia (CSTNZ): Catholic Centre for Justice, Development and Peace (CCJDP). http://woek.de/web/cms/upload/pdf/kasa/publikationen/fraser_200

Galea KS, MacCalman L, Jones K (2015) Urinary biomarker concentrations of captan, chlormequat, chlorpyrifos and cypermethrin in UK adults and children living near agricultural land, J. Expo. Sci. Environ. Epidemiol. 25(6): 623–631. https://doi.org/10.1038/jes.2015.54

Galhena DH, Freed R, Maredia KM (2013) Home gardens: a promising approach to enhance household food security and wellbeing. Agric Food Secur 2:8. https://doi.org/10.1186/2048-7010-2-8

Geenen S, Marysse S (2016) Democratic Republic of the Congo: mining sector. In: Encyclopedia of mineral and energy policy. Springer, Berlin, pp 1–6. https://link.springer.com/referenceworkentry/10.1007/978-3-642-40871-7_112-1

Gelfand M (1971) Diet and traditional in an African culture. E and S Livingstone, Edinburgh. https://archive.org/details/diettraditionina0000gelf/page/n6/mode/1up. Accessed 31 Jan 2009

Global Witness (2006) Digging in corruption: fraud, abuse and exploitation in Katanga's copper and cobalt mines. Global Witness Publishing, Washington, DC. https://cdn.globalwitness.org/archive/files/import/kat-doc-engl-lowres

Global Business Reports (2014) Mining in Africa's Copperbelt Zambia and Democratic Republic of Congo. Eng Mining J:44. https://www.gbreports.com/publication/copperbelt-mining-2014-emj-release. Accessed 12 Jan 2015

Gordon DK (2021) Kingdoms and associations: copper's changing political economy during the nineteenth century, p 155. In: Larmer M, Guene E, Henriet B et al (eds) Across the Copperbelt, urban & social change in Central Africa's borderland communities. Boydell & Brewer, Rochester, p 437. https://boydellandbrewer.com/9781847012661/across-the-copperbelt/. Accessed 11 Apr 2025

Guveya E (2006) Final draft report of the mid-term baseline survey for forest resource management project (FRMP). In: Ministry of Tourism, Environment and Natural Resources (MTENR), Forestry Department, Lusaka. http://www.mtenr.gov.zm/. Accessed 23 May 2009

Haggblade S, Nyembe M (2007) Commercial dynamics in Zambia's cassava value chain. Working paper no. 32. Food security research project, Lusaka. https://doi.org/10.22004/ag.econ.54491

Haller H, Ito S, Izzo JL et al (2011) Olmesartan for the delay or prevention of microalbuminuria in type 2 diabetes. N Engl J Med 364(10):907–917. https://doi.org/10.1056/nejmoa1007994

Harkonen J, Haapasalo H, Hanninen K (2015) Productisation: a review and research agenda. Int J Prod Econ 164:65–82. https://doi.org/10.1016/j.ijpe.2015.02.024

Health Effects Institute (2022) The state of air quality and health impacts in Africa. A report from the state of global air initiative, Boston, p 23. https://www.stateofglobalair.org/resources/report/state-air-quality-and-health-impacts-africa. Accessed 24 Apr 2024

Hilson G (2020) The 'Zambia Model': a blueprint for formalizing artisanal and small-scale mining in Sub-Saharan Africa? Resour Policy 68:101765. https://doi.org/10.1016/j.resourpol.2020.101765

Hoet P, Jacquerye C, Deumer G et al (2013) Reference values and upper reference limits for 26 trace elements in the urine of adults living in Belgium. Clin Chem Lab Med 51:839–849. https://doi.org/10.1515/cclm-2012-0688

Holland PM, Abramson RD, Watson R, Gelfand DH (1991) Detection of specific polymerase chain reaction product by utilizing the 5′--3′ exonuclease activity of Thermus aquaticus DNA polymerase. Proc Natl Acad Sci U S A 88(16):7276–7280. https://doi.org/10.1073/pnas.88.16.7276

Human Rights Watch (2019) "We have to be worried": the impact of lead contamination on children's rights in Kabwe, Zambia. Human Rights Watch, New York. https://www.hrw.org/report/2019/08/23/we-have-be-worried/impact-lead-contamination-childrens-rights-kabwe-zambia. Accessed 20 Nov 2020

ICRISAT (2006) Spreading the word on fertilizer in Zimbabwe. Global theme on agroecosystems, p 24. https://oar.icrisat.org/2383/. Accessed 11 Feb 2007

Infrastructure and Cities for Economic Development ("ICED") (2018) Kitwe: inclusive urban economic development diagnostic. Volume: URB001, 8pp. http://icedfacility.org/?s=Kitwe. Accessed 17 Sept 2024

Ingram S (2002) The real nutritional value of fungi. http://www.world-of-fungi.org/mostly_medical/stephanie_ingram/nutritional_value.htm. Accessed 4 Nov 2008

Institut National de la Statistique (INS) (1984) Résultats provisoires du Recensement Scientifique de la Population du 1er juillet 1984, Kinshasa, Gombe, p 71. https://ins-rdc.org/recensements/recensements/. Accessed 2 Sept 2008

International Bank for Reconstruction and Development/The World Bank (2016) Zambia mining investment and governance review, Washington, DC, p 82. https://documents1.worldbank.org/curated/en/305921468198529463/text/105820-REVISED-PUBLIC-Zambia-Report-ONLINE.txt

International Council on Mining and Metals (ICMM) (2014) Enhancing mining's contribution to the Zambian Economy and Society: mining, partnerships for development, London, UK, p 149. https://www.icmm.com. Accessed 22 Oct 2024

International Development Research Centre (IDRC) (1996) The costs of mining in Latin America. https://www.idrc.ca/reports/1996. Accessed 11 Feb 2007

International Fund for Agricultural Development (FAD)/Food and Agriculture Organization of the United Nations (2000) The world cassava economy: facts, trends and outlook. Food & Agriculture Organisation, Rome, p 46. https://www.fao.org/4/x4007e/X4007E00.htm

International Institute for Applied Systems Analysis (IIASA) (2020) Bending the curve of biodiversity loss, Laxenburg, Austria. https://iiasa.ac.at/news/sep-2020/bending-curve-of-biodiversity-loss. Accessed 19 Mar 2024

International Institute of Tropical Agriculture (IITA) (1990) Cassava in tropical Africa: a reference manual. Balding & Mansell International, Wisbech, p 176. https://cgspace.cgiar.org/items/1f13ff7f-bd3d-4fd1-bfce-601aa9c40a18

International Labor Organization (2020) Mapping interventions addressing child labour and working conditions in artisanal mineral supply chains. 60pp. Geneva, Switzerland. https://www.ilo.org/publications/mapping-interventions-addressing-child-labour-andworking-conditions

IPEC, 2008. Action against child labour: IPEC highlights 2008 / International Labour Office, ILO International Programme on the Elimination of Child Labour. – Geneva: ILO, 2009 – 1 v. – 76 p. https://docslib.org/doc/8475789/international-programme-on-the-elimination-of-child-labour-ipec. Accessed 28 March 2024.

Järup L, Berglund M, Elinder CG et al (1998) Health effects of cadmium exposure—a review of the literature and a risk estimate. Scand J Work Environ Health 24: 1-51. https://pubmed.ncbi.nlm.nih.gov/9569444/

Kabata-Pendias A (2011) Trace elements in soils and plants, 4th edn. Taylor and Francis Group, LLC, Boca Raton, p 534. http://www.taylorandfrancis.com. Accessed 1 Apr 2025

Kambani SM (2003) Small-scale mining and cleaner production issues in Zambia. J Clean Prod 11:141–146. https://doi.org/10.1016/S0959-6526(02)00033-1

Kambewa E (2010) Cassava commercialization in Malawi. MSU International Development working paper 109. Department of Agricultural, Food, and Resource Economics, Department of Economics, Michigan State University, Michigan, 29pp. https://www.canr.msu.edu/resources/cassava-commercialization-in-malawi. Accessed 14 Aug 2013

Kanninen M, Murdiyarso D, Seymour F et al (2007) Do trees grow on money?: the implications of deforestation research for policies to promote REDD. For Perspect 4:71. https://doi.org/10.17528/cifor/002347

Karikari SK, Wigglesworth DJ, Kwerepe BC et al (1997) Bambara groundnut: Botswana country report, pp 11–18. In: Heller J, Begemann F, Mushonga J (eds) Bambara groundnut Vigna subterranea (L) Verdc. promoting the conservation and use of underutilized and neglected crops. Proceedings of the workshop on conservation and improvement of bambara groundnut, Harare, Zimbabwe. IPGRI, Rome. https://www.researchgate.net/publication/244993439_Bambara_groundnut_Vigna_subterranea_L_Verdc

Kashimbo Kalala S, Lukens L, Mbikayi E et al (2015) Intoxication alimentaire aux éléments traces métalliques (ETM) de trois espèces maraichères cultivées sur le sol du bord de la rivière Lubumbashi (Lubumbashi-Katanga/R.D. Congo). Int J Innov Sci Res 15:102–112. http://www.ijisr.issr-journals.org/abstract.php?article=IJISR-15-065-11. Accessed 25 Oct 2020

Khan A, Khan S, Khan MA et al (2015) The uptake and bioaccumulation of heavy metals by food plants, their effects on plants nutrients, and associated health risk: a review. Environ Sci Pollut Res 22:13772–13799. https://doi.org/10.1007/s11356-015-4881-0

Kimiywe J, Waudo J, Mbithe D, Maundu P (2008) Utilization and medicinal value of indigenous leafy vegetables consumed in urban and peri-urban Nairobi. AJFAND 7(4):1–15. https://www.researchgate.net/publication/43554730_Utilization_and_Medicinal_Value_of_Indigenous_Leafy_Vegetables_Consumed_in_Urban_and_Peri-Urban_Nairobi

Konkola Copper Mines (KCM) plc (2007) Policy on safety, health, environment and quality. Contributing to sustainable development. August 2007 revision 8. KCM Plc, Chingola. http://kcm.co.zm/corporate-profile/company-policies/. Accessed 1 Feb 2013

KPMG Global Mining Institute (2014) Democratic Republic of Congo: country mining guide-strategic series, p 40. http://kpmg.com/mining. Accessed 21 Dec 2021

KPMG International Cooperative (2014) Democratic Republic of Congo: Country mining guide. Strategy Series ©. 40pp. Amstelven, Nederlands. https://assets.kpmg/content/dam/kpmg/pdf/2014/09/democratic-republic-congo-mining-guide.pdf

Kříbek B, Majer V, Knésl I et al (2011) Impact assessment of mining and processing of copper and cobalt ores on the environment in Copperbelt, Zambia: an overview. In: Kříbek B (ed) Mining and the environment in Africa. Proceedings of the inaugural workshop IGCP/SIDA no. 594, Kitwe/Zambia, October 17th-18th, 2011. Czech Geological Survey, Prague. ISBN 978-80-7075-119-0. https://www.publication/272510675_Mining_the_Katanga_DRC_Copperbelt_geological_aspects_and_impacts_on_public_health_and_the_environment_-_towards_a_holistic_approach. Accessed 17 Mar 2024

Lai O (2022) Explainer: what is biodiversity loss? Earth Organization, Pine Mountain Club. https://earth.org/biodiversity-loss-definition-and-examples/

Lancaster PA, Brooks JE (1983) Cassava leaves as human food. Econ Bot 37:331–348. https://doi.org/10.1007/BF02858890

Larison JR, Likens GE, Fitzpatrick JW, Crock JG (2000) Cadmium toxicity among wildlife in the Colorado Rocky Mountains. Nature 406(6792):181–183. https://doi.org/10.1038/3501

Larmer M (2021) Living for the city: social change and knowledge production in the Central African Copperbelt. Cambridge University Press, Cambridge, p 398. https://doi.org/10.1017/9781108973120

Larmer M, Guene E, Henriet B et al (2021) Across the copperbelt: urban & social change in Central Africa's borderland communities. James Currey Publisher, Imprint of Boydell & Brewer Ltd., Rochester, p 437. https://www.boydellandbrewer.com. Accessed 20 Aug 2024

Latham MC (1979) Human nutrition in tropical Africa. FAO, Rome

Laurie SM (2010) Agronomic performance, consumer acceptability and nutrient content of new sweet potato varieties in South Africa. Dissertation, thesis, University of the Free State

Lees HMN (1962) Working plan for forests supplying the Copperbelt and Western Provinces. Forest Department. Government Printers, Lusaka

Lentini P, Zanoli L, Granata A et al (2018) Kidney and heavy metals—the role of environmental exposure (Review). Mol Med Rep 15:3413–3419. https://doi.org/10.3892/mmr.2017.6389

Leung WTW, Butrum RR, Cheng FH (1972) Proximate composition, mineral and vitamin contens of East Asian foods. Food Composition Tables for Use in East Asia. Washington DC, USA. https://www.fao.org/4/X6878E/X6878E00.htm

Liu C, Bayer A, Cosgrove SE et al (2011). Clinical practice guidelines by the infectious diseases society of America for the treatment of methicillin-resistant Staphylococcus aureus infections in adults and children: executive summary. Clin. Infect. Dis. 52(3):285–92. https://doi.org/10.1093/cid/cir034

Lora-Wainwright A (2017) Resigned activism: living with pollution in rural China. MIT Press Direct, Cambridge, p 229. https://doi.org/10.7551/mitpress/9780262036320.001.0001

Lubaba Nkulu Banza C, Nawrot TS, Haufroid S et al (2009b) High human exposure to cobalt and other metals in Katanga, a mining area of the Democratic Republic of Congo, Environ. Res. 109(6), 745–752. https://doi.org/10.1016/j.envres.2009.04.012

Luboya N, Kabila B, Manengo T et al (2004) Le saturnisme du couple mère-nouveau-né á Lubumbashi (RDC). Revue d'information et d'actualités médicales 4:3–6. Université de Lubumbashi, Faculté de médecine. Presses universitaires de Lubumbashi, RDC

Lukombo S, Nluta S, Bongenda M, Enene J, Kimfuema V, Nsuanda A, Mahungu NM (2022) Transformation du manioc en RD Congo. In: Mahungu NM, Ndonda A, Kendenga T, Bidiaka S (eds) Le Manioc en RD Congo. IITA, INERA, Ibadan, Nigeria, pp 241–289

Lydall MI, Auchterlonie A (2011) The Democratic Republic of Congo and Zambia: a growing global "Hotspot" for copper-cobalt mineral investment and exploitation. In: Proceedings of the 6th Southern Africa base metals conference, Phalaborwa, South Africa, p 14. https://www.saimm.co.za/Conferences/BM2011/. Accessed 17 May 2020

Mabeta J, Mweemba B, Mwitwa J (2018) Key drivers of biodiversity loss in Zambia. Policy brief # 3. Biodiversity finance initiative (BIOFIN)-Zambia, 8pp

Madisa ME, Assefa Y, Obopile M (2010) Assessment of production constraints, crop and pest management practices in peri-urban vegetable farms of Botswana. Egypt Acad J Biolog Sci 1(1):1–11. https://www.eajbs.eg.net

Mahungu NM, Yamagushi J, Almazan AM et al (1987) Reduction of cyanide during processing of cassava into some traditional African foods. J Food Agric 1:11–15. https://www.researchgate.net/publication/290162018_Reduction_of_cyanide_during_processing_of_cassava_into_some_traditional_African_foods

Malaisse F (1978) The miombo ecosystem. In: Tropical forest ecosystem. UNESCO/UNEP/FAO, Paris, pp 589–600

Martin OV (2023) Synergistic effects of chemical mixtures: how frequent is rare? Curr Opin Toxicol 36:100424. https://doi.org/10.1016/j.cotox.2023.100424

Martin OV, Scholze M, Ermler S et al (2021) Ten years of research on synergisms and antagonisms in chemical mixtures: a systematic review and quantitative reappraisal of mixture studies. Environ Int 146:106206. https://doi.org/10.1016/J.ENVINT.2020.106206

Massomo SMS, Mabagala RB, Mortensen CN, Hockenhull J, Swai IS (2005) Cabbage production in Tanzania: challenges faced by smallholder farmers in the management of black rot disease. J Sustain Agric 26(4):119–141. https://doi.org/10.1300/J064v26n04_08

Matakala PW, Kokwe M, Statz J (2015) Zambia national strategy to reduce emissions from deforestation and forest degradation (REDD+). Ministry of Lands, Natural Resources and Environmental Protection. 74pp. Lusaka, Zambia. https://www.mlnr.gov.zm/

Mauyo LW, Okalebo JK, Kirkby RA et al (2007) Legal and institutional constraints to Kenya-Uganda crossborder bean marketing. Afr J Agric Res 2:578–582. https://www.academia.edu/15486693/Legal_and_institutional_constraints_to_Kenya_Uganda_cross_border_bean_marketing

Mbewe DN, Mwala MS, Mwila GP (1995) Country report for Zambia. In: Heller J, Begemann F, Mushonga J (eds) Promoting the conservation and use of underutilized and neglected crops, pp 64–67. 9. Proceedings of the workshop on conservation and improvement of bambara groundnut (*Vigna subterranean* (L.) Verdc.) 14–16 November, 1995, Harare, Zimbabwe. International Plant Genetic Resources Institute, Rome. https://cgspace.cgiar.org/items/aa92175f-c947-4346-8569-a3c24ec5867d

McGrath JM, Spargo J, Penn CJ (2014) Soil fertility and plant nutrition. In: Van Alfen N (ed) Encyclopedia of agriculture and food systems, vol 5. Elsevier, San Diego, pp 166–184. http://www.elsevier.com/locate/permissionusematerial. Accessed 1 Apr 2025

Midtgård U, Binderup ML (1994) The Nordic Expert Group for criteria documentation of health risks from chemicals. 114. Cobalt and cobalt compounds. Arbete och Hälsa 39:1–66. National Institute of Occupational Health, Solna, Sweden. https://api.semanticscholar.org/CorpusID:106534242

Miller RL, Tegen I, Perlwitz JP (2004). Surface radiative forcing by soil dust aerosols and the hydrologic cycle. J. Geophys. Res., 109, D04203, https://doi.org/10.1016/j.chemosphere.2005.08.024" https://doi.org/10.1029/2003JD004085

Minde IJ, Jumbe CBL (1997) Situation analysis and outlook of cassava and sweetpotato in SADC countries, Lilongwe, Malawi. https://api.semanticscholar.org/CorpusID:169223438. Accessed 12 Mar 1998

Ministry of Finance and National Planning (MFNP) GRZ (2002) Zambia poverty reduction strategic paper: 2002-2004. MFNP, Planning and Economic Management Department, Lusaka, p 402. https://www.imf.org/External/NP/prsp/2002/zmb/01/. Accessed 10 Nov 2008

Ministry of Lands and Natural Resources (2018) Artisanal and small-scale mining handbook for Zambia. Geological Survey of Denmark and Greenland, Lusaka, p 217. https://www.mlnr.gov.zm/. Accessed 24 Jun 2024

Ministry of Lands Natural Resources and Environmental Protection (MNLR) (2015) United Nations convention on biological diversity, fifth national report, Lusaka, Zambia, p 66. https://chm.cbd.int/database/record?documentID=204863. Accessed 24 Jun 2024

Ministry of Mines and Minerals Development (2022) A glance into the 2022 national mineral resources development policy. Unlocking Zambia's potential. Policy monitoring and research centre, p 4. https://policycommons.net/artifacts/3183513/oped/3982120/

Ministry of Mining and Hydrocarbons (2003) Guide for mining investors. Ministry of Mines, Kinshasa. http://congomines.org/reports/47-guide-de-l-investisseur-du-secteur-des-mines-et-hydrocarbures. Assessed 1 Feb 2013

Ministère de l'Environnement, Conservation de la Nature et Développement Durable (2016) Environnement et Développement Durable RDC: Stratégie et Plan d'Action Nationaux sur la Biodiversité (2016–2020). 91pp. Kinshasa, République Démocratique du Congo. https://medd.gouv.cd/environnement-et-developpement-durablerdc-strategie-et-plan-daction-nationaux-sur-la-biodiversite/

Ministère de l'économie nationale (2022) Le Répertoire Des Entreprises Mines Et Carrières RD Congo. 10pp. Kinshasa, République Démocratique du Congo. https://docslib.org/doc/1281240/le-r%C3%A9pertoire-des-entreprises-mines-et-carri%C3%A8res-rdcrepertoire-des-entrepises-mines-et-carrieres-de-la-rd-congo

Ministère des Mines (2013) Statistiques minières de l'exercice 2013. République Démocratique du Congo, p 35. https://congomines.org/reports/653-ministere-des-mines-statistiques-minieres-de-l-exercice-2013. Accessed 26 Jan 2024

Missao de Inquerito Agricola-MIA (1969) A agricultura tradicional de Mocambique, vol 1. Missão de Inquerito Agricola de Mocambique Lourenco Marques. https://books.google.ca/books/about/A_agricultura_tradicional_de_Mocambique.html?id=9JdNAAAAMAAJ&redir_esc=y

Mtaita TA, Msuya D (1994) Current status of Allium production in Tanzania. Acta Hortic 358:71–74. https://doi.org/10.17660/ACTAHORTIC

Mufungizi I (2025) Mineral potential facing socio-economic development challenges: case study of the Democratic Republic of Congo, a 'geological scandal'. Int Geol Rev 67(6):754–776. https://doi.org/10.1080/00206814.2024.2401575

Mugisa IO, Karungi J, Akello B, Ochwo-Ssemakula MKN, Biruma M, Okello DK, Otim G (2016) Determinants of groundnut rosette virus disease occurrence in Uganda. Crop Protect 79:117–123. https://doi.org/10.1016/j.cropro.2015.10.019

Muimba-Kankolongo A (2018) Food crops production by smallholder farmers in Southern Africa: challenges and opportunities for improvement, 1st edn. Academic Press/Elsevier, Cambridge. https://www.elsevier.com/books/food-crop-production-by-smallholder-farmers-in-southern-africa/muimba-kankolongo/978-0-12-814383-4. Accessed 28 Mar 2018

Muimba-Kankolongo A, Banza Lubaba Nkulu C, Mwitwa J et al (2021) Contamination of water and food crops by trace elements in the African Copperbelt: a collaborative cross-border study in Zambia and the Democratic Republic of Congo. Environ Adv 6:100103. https://doi.org/10.1016/j.envadv.2021.100103

Muimba-Kankolongo A, Banza Lubaba Nkulu C, Mwitwa J et al (2022) Impacts of trace metals pollution of water, food crops, and ambient air on population health in Zambia and the DR Congo. J Environ Public Health 2022:4515115. https://doi.org/10.1155/2022/4515115

Mukelabai N, Rweyemamu M, Meadows K (2006) Strengthening livelihoods through food and nutrition security in vulnerable SADC countries. In: Mid-term review of OSRO/RAF/510-511/SAF. Regional Interagency Coordination Support Office (RIACSO), Johannesburg. 58pp

Mulenga C, Clarke C, Meincken M (2022) Bioaccumulation of Cu, Fe, Mn and Zn in native Brachystegia longifolia naturally growing in a copper mining environment of Mufulira, Zambia. Environ Monit Assess 194:8. https://doi.org/10.1007/s10661-021-09656-0

Mulenga M (1999) Impact of sulphur dioxide on vegetation on the Copperbelt: a case study for Kitwe and Mufulira. BSc degree thesis, Copperbelt University

Muliokela SW (ed) (1997) Zambia seed technology handbook. Ministry of Agriculture, Food and Fisheries, Arlov, p 394. https://www.agriculture.gov.zm/. Accessed 4 Mar 2005

Musa Obadia P, Kayembe-Kitenge T, Haufroid V et al (2018) Preeclampsia and blood lead (and other metals) in Lubumbashi, DR Congo. Environ Res 167:468–471. https://doi.org/10.1016/j.envres.2018.07.032

Mushagalusa Balasha A, Murhula BB, Munahua DM (2019) Yard farming in the city of Lubumbashi: resident perceptions of home gardens in their community. J City Dev 1(1):46–53. https://doi.org/10.12691/jcd-1-1-8

Mwaanga P, Silondwa M, Kasali G et al (2019) Preliminary review of mine air pollution in Zambia. Heliyon 5(9):e02485. https://doi.org/10.1016/j.heliyon.2019.e02485

Mwanza N (2006) Biodiversity overview. https://zm.chm-cbd.net/biodiversity-overview

Mwitwa C, Kabemba C (2007) Copper boom in Zambia: boom for whom? Southern Africa Resource Watch (SARW), Resource insight no. 3, Johannesburg, South Africa, p 20. http://www.sarwatch.org/. Accessed 19 May 2020

Mwitwa J, German L, Muimba-Kankolongo A et al (2012a) Governance and sustainability challenges in landscapes shaped by mining: mining-forestry linkages and impacts in the Copperbelt of Zambia and the DR Congo. For Policy Econ 25:19–30. https://resourcegovernance.org/sites/default/files/ZamCopperBoom.pdf. Accessed 29 Jan 2012

Mwitwa J, Muimba-Kankolongo A, German L (2012b) Copper mining, forests and livelihoods in DR Congo and Zambia: forest governance, forest-based livelihoods and copper mining linkages. Lambert Academic Publishing, Saarbrücken, p 65. https://lap-publishing.com/#

Mwitwa J, Muimba-Kankolongo A, German L, Puntodewo A (2009) A report of the scoping study in Zambia and the Democratic Republic of Congo. Copper belt University, School of Natural Resources and Center for International Forestry Research. 92pp. Kitwe, Zambia

Nachinanga Gubula Siaciti (2022) The role of artisanal and small-scale mining in enhancing sustainable livelihoods in Zambia. Dissertation, University of Zambia

Nair GM, Ravindran CS, Moorthy SN et al (1989) Sweetpotato indigenous technologies and recent advances of production, processing and utilization in India. In: Proceedings of the international symposium on sweetpotato, Baybay, Philippines, pp 301–312

National Research Council (1991) Managing global forest trees genetic resources. Board on Agriculture National Research Council/The National Academy of Science, Washington, DC. https://doi.org/10.1016/0378-1127(93)90091-Z

National Water and Sanitation Council (NWASCO) (2021) Urban and peri-urban water supply and sanitation sector report, Lusaka, Zambia, p 116. https://www.nwasco.org.zm/. Accessed 18 Mar 2024

Ncube E, Banda C, Mundike J (2012) Air pollution on the Copperbelt Province of Zambia: effects of sulphur dioxide on vegetation and humans. J Nat Environ Sci 3:34–41. https://www.academyjournals.net

Nemery B, Banza Lubaba Nkulu C (2018) Assessing exposure to metals using biomonitoring: achievements and challenges experienced through surveys in low- and middle-income countries. Toxicol Lett 298:13–18. https://doi.org/10.1016/J.TOXLET.2018.06.004

Nenguwo N (2004) Review of vegetable production and marketing (supply chain analysis) increasing the value and quality assurance for the fresh vegetables and herbs supply chain to Sun International Hotels in Zambia. Technical Report Submitted to: Regional Center for Southern Africa, US Agency for International Development Gaborone, Botswana May. https://scholar.google.com/scholar_lookup?hl=en&publication_year=2004&author=N.+Nenguwo&title=Review+of+vegetable+production+and+marketing+%28supply+chain+analysis%29+increasing+the+value+and+quality+assurance+for+the+fresh+Ne

Nestares T, Barrionuevo M, Urbano G, López-Frías M (2001) Nutritional assessment of protein from beans (Phaseolus vulgaris L) processed at different pH values, in growing rats. J Sci Food Agric 81(15):1522–1529. https://doi.org/10.1002/jsfa.965

Nguyen L (2024) 5 biggest environmental issues in Africa in 2024. Earth Organization, Pine Mountain Club. https://earth.org/environmental-issues-in-africa/. Accessed 30 May 2024

Nhamo L, Matchaya G, Mabhaudhi T, Nhlengethwa S, Nhemachena C, Mpandeli S (2019) Cereal production trends under climate change: impacts and adaptation strategies in southern Africa. Agriculture 9(2):30. https://doi.org/10.3390/agriculture9020030

Nigam SN, Ramanatha RV, Gibbons RW (1983) Utilization of natural hybrids in the improvement of groundnuts (Arachis hypogaea). Exp Agric 19:355–359

Ningabo A, Ousmane S, Pumulo Pitiya R (2022) Zambia's copperbelt area and copper mining: a review. J Geosci Environ Prot 10:67–75. https://doi.org/10.4236/gep.2022.103005

Njovu FC, Chuzu P, Nguvulu CZ (2004) Key threats and opportunities for conservation of biodiversity and natural resources in the Copperbelt areas of biological significance: an analytical interpretive report. Draft copy-unpublished

Nsiami MC, Katanga KJ, Bwenda KC et al (2006) Vers une gestion communautaire de la forêt claire (miombo) au Katanga. Rapport, Ministère de l'environnement et des forêts, République Démocratique du Congo. https://medd.gouv.cd/. Accessed 15 Sept 2007

Nutrition Smart Agriculture Country Profile (2020) Nutrition smart agriculture in the Democratic Republic of Congo. Country profile report, p 19. https://thedocs.worldbank.org/en/doc/449071598888572537-0090022020/. Accessed 25 Nov 2024

NWASCO (2004) Water sector reform in Zambia. Lusaka, Zambia, p 44. https://www.nwasco.org.zm/. Accessed 18 Mar 2024

Nweke FI (2005) The cassava transformation in Africa. In: Proceedings of the validation forum on global cassava development strategies, A review of cassava in Africa with country case studies on Nigeria, Ghana, the United Republic of Tanzania, Uganda and Benin, vol 2. IFAD/FAO, Rome. https://www.fao.org/4/a0154e/A0154E02.htm#ch1. Accessed 4 Dec 2017

Nyasulu P, Cohen C, De Gouveia L, Feldman C, Klugman KP and von Gottberg A (2011) Increased risk of death in human immunodeficiency virus-infected children with pneumococcal meningitis in South Africa, 2003-2005. Pediatr Infect Dis J 30(12):1075–80. https://doi.org/10.1097/INF.0b013e31822cca05

Nyirenda SP, Sileshi GW, Belmain SR et al (2011) Farmers' ethnoecological knowledge of vegetable pests and pesticidal plant use in Malawi and Zambia. Afr J Agric Res 6:1525–1537. http://www.academicjournals.org/AJAR

Obilana AB (2003) Overview: importance of millets in Africa. In: Belto PS, Taylor JPN (eds) Afripo, workshops on the proteins of sorghum and millets: enhancing nutritional and functional properties for Africa, 2-4 April 2003, Pretoria, South Africa. http://www.afripro.org.uk/papers/Paper02Obilana.pdf. Accessed 18 Mar 2014

Okigbo BN (1990) Vegetables in tropical Africa. In: Opeña RT, Kyomo ML (eds) Proceedings of a workshop on vegetable research and development in SADCC countries, 9–13 July, 1990. Asian Vegetable Research and Development Center, Arusha, Tanzania, p. 178. AVRDC Publication no. 90–328. https://searchworkslb.stanford.edu/view/1839343

Ominawo IA, Momoh AH, Usman A (1999) Proximate composition of functional properties of 4 cultivars of Bambara groundnut (*Voandezia subterranean*). Plant Foods Hum Nutr 53:153–158. https://doi.org/10.1023/a:1008027030095

Otoo JA (1996) Rapid multiplication of cassava. In: IITA research guide 51. IITA, Ibadan. http://www.iita.org/cms/details/trn_mat/IRG51/irg51.htm. Accessed 27 Jan 2008

Peters D (2008) Assessment of the potential of sweetpotato as livestock feed in East Africa: Rwanda, Uganda, and Kenya A report presented to The International Potato Center (CIP). Nairobi, Kenya. 64 p. https://scholar.google.com/scholar_lookup?&title=Assessment%20of%20the%20potential%20of%20sweetpotato%20as%20livestock%20feed%20in%20East%20Africa%3A%20Rwanda%2C%20Uganda%2Cand%20Kenya&publication_year=2008&author=Peters%2CD.#d=gs_cit&t=1761671736332&u=%2Fscholar%3Fq%3Dinfo%3Ayr2iP4Y7XuAJ%3Ascholar.google.com%2F%26output%3Dcite%26scirp%3D0%26hl%3Den

Peša I (2020a) Crops and copper: agriculture and urbanism on the central African copperbelt, 1950–2000. J South Afr Stud 46:527–545. https://doi.org/10.1080/03057070.2020.1750872

Peša I (2020b) Mining, waste and environmental thought on the Central African Copperbelt, 1950–2000. Environ Hist Camb 28(2). https://doi.org/10.3197/096734019X15755402985703

Peša I (2021) Between waste and profit: environmental values on the Central African Copperbelt. Extr Ind Soc 8:100793. https://doi.org/10.1016/j.exis.2020.08.004

Purseglove JW (1976) Origins and migrations of crops in tropical Africa, pp 291–309. In: Harla JR, de Wet JMJ, Stember ABL (eds) Origins of African plant domestication. Aldine, Chicago, p 498. https://doi.org/10.1515/9783110806373.291

Quinn PJ (1959) Foods and feeding habits of the Pedi. Witwatersrand University, Johannesburg, p 278. https://doi.org/10.2307/1158120

Raemaekers R (ed) (2001) Crop production in tropical Africa. Directorate General for International Co-operation (DGIC), p 1450

Ramírez-Bahena MH, García-Fraile P, Peix A et al (2008) Revision of the taxonomic status of the species *Rhizobium leguminosarum* (Frank 1879) Frank 1889, R. phaseoli Dangeard 1926AL and R. trifolii Dangeard 1926AL. R. trifolii is a later synonym of R. leguminosarum. Reclassification of the strain Rhizobium leguminosarum DSM 30132T (NCIMB 11478T) into the new species *Rhizobium pisi* sp. nov. IJSEM 58:2484–2490. https://doi.org/10.1099/ijs.0.65621-0

Rauh VA, Margolis AE (2016) Research review: environmental exposures, neurodevelopment, and child mental health—new paradigms for the study of brain and behavioral effects. J Child Psychol Psychiatry 57(7):775–793. https://doi.org/10.1111/jcpp.12537

Reed D (ed) (1996) Structural adjustment, the environment, and sustainable development. Earthscan Publications Ltd, London. 386pp

Riediker M, Zink D, Kreyling W et al (2019) Particle toxicology and health—where are we? Part Fibre Toxicol 16:1–33. https://doi.org/10.1186/s12989-019-0302-8

Roos C, Cilliers DP, Retief FP, Alberts RC, Bond AJ (2020) Regulators' perceptions of environmental impact assessment (EIA) benefits in a sustainable development context. Environ Impact Assess Rev 81:106360. https://doi.org/10.1016/j.eiar.2019.106360

SADC-FANR (2004) Enhancing agriculture and food security for poverty reduction in the SADC region: key issues paper for extra-ordinary summit. Dar-es-Salaam, Tanzania, p 37. https://www.sadc.int/document/declaration-agriculture-food-security-2004. Accessed 13 Sept 2008

SADC/FANR (2011) Regional agricultural policies (RAP): country summary agricultural policy review reports. Directorate of Food. In: Agriculture and Natural Resources, SADC/Food, Agriculture and Natural Resources, Gaborone, Botswana, p 393. https://www.sadc.int/files/7113/5293/3509/Regional_Agricultural_Policy_Review_Reports_2011.pdf. Accessed 1 Feb 2013

Satarug S (2018) Dietary cadmium intake and its effects on kidneys. Toxics 6(1):15. https://doi.org/10.3390/toxics6010015

Siachoono M (2018) Biodiversity in Zambia (Chapter 13). In: Pullaiah T (ed) Global biodiversity, vol 3. Selected countries in Africa. Apple Academic Press Inc, Waretown, p 384. http://www.crcpress.com/

Smyth KJF (1968) Bambara groundnut or 'Njugu Mawe' (Voandzeia subterranea). Ukiriguru Res. Notes 12. Ministry of Agriculture, Ukiriguru

Santopinto JJ, Fox KA, Goldberg RJ et al (2003) Creatinine clearance and adverse hospital outcomes in patients with acute coronary syndromes: findings from the global registry of acute coronary events (GRACE), Heart 89(9): 1003–8. https://doi.org/10.1136/heart.89.9.1003

Soroko D (1998) Cassava and sweetpotato: can they resolve Zambia's food security problems? Opening speech by USAID mission to Zambia. SARRNET scientific workshop, Lusaka, Zambia. https://cgspace.cgiar.org/items/58080da5-4cdd-47aa-ac81-20b7f65972a1. Accessed 12 Jan 2009

Southern Africa Root Crops Research Network (SARRNET) (2008) Improving rural livelihoods in Southern Africa through root and tuber crops. IITA, SADC NARS and Stakeholders' proposal to SADC-FANR, Lilongwe, Malawi. https://cgspace.cgiar.org/items/58080da5-4cdd-47aa-ac81-20b7f65972a1. Accessed 12 Jan 2009

Statista (2025) Mine production of cobalt in the Democratic Republic of Congo from 2010 to 2024. https://www.statista.com/statistics/264930/global-cobalt-reserves/. Accessed 30 Mar 2025

Swanevelder CJ (1998) Bambara-food for Africa (Vigna subterranean-bambara groundnut). National Department of Agriculture, ARC-Grain Crops Institute, Pretoria. 16pp

Swift MJ, Izac A-MN, van Noordwijk M (2004) Biodiversity and ecosystem services in agricultural landscapes—are we asking the right questions? Agric Ecosyst Environ 104(1):113–134. https://doi.org/10.1016/j.agee.2004.01.013

Tchatchou B, Sonwa DJ, Ifo S, Tiani AM (2015) Deforestation and forest degradation in the Congo Basin: state of knowledge, current causes and perspectives. CIFOR, Bogor, Indonesia. https://www.ciforicraf.org/knowledge/publication/5894/

Thevenon F, Adatte T, Poté J et al (2012) Recent human-induced trophic change in the large and deep perialpine Lake Lucerne (Switzerland) compared to historical geochemical variations. Palaeogeogr Palaeoclimatol Palaeoecol 363–364:37–47. https://doi.org/10.1016/j.palaeo.2012.08.010

Thevenon F, Guédron S, Chiaradia M et al (2011) (Pre-) historic changes in natural and anthropogenic heavy metals deposition inferred from two contrasting Swiss Alpine lakes. Quat Sci Rev 30(1–2):224–233. https://www.sciencedirect.com/science/article/abs/pii/S0277379110003793. Accessed 12 Mar 2025

Thevenon F, Poté J (2012) Water pollution history of Switzerland recorded by sediments of the large and deep perialpine Lakes Lucerne and Geneva. Water Air Soil Pollut 223:6157–6169. https://doi.org/10.1007/s11270-012-1347-6

Tittonell P, van Wijk MT, Rufino MC, Vrugt JA, Giller KE (2007) Analysing trade-offs in resource and labour allocation by smallholder farmers using inverse modelling techniques: a case-study from Kakamega district, western Kenya. Agric Syst 95(1–3):76–95. https://doi.org/10.1016/j.agsy.2007.04.002

Tshibaka T, Lumpungu K (1989) Trends and prospects for cassava in Zaire. Working paper no. 4 on cassava. IFPRI, Washington, DC. Insect Sci Appl 21:23–32. https://ebrary.ifpri.org/digital/collection/p15738coll2/id/126077. Accessed 7 Sept 2007

Tshonda O, Leonard G, Laghmouch M, Mpundu M (2018) Lorsque richesses économiques et pouvoirs politiques forcent une identité régionale. Tome 2: bassin du cuivre-matrice et horizon. Musée royal de l'Afrique centrale. 296pp. Leuvensesteenweg, Belgique. https://www.africamuseum.be

Tsurukawa N, Prakash S, Manhart A (2011) Social impacts of artisanal cobalt mining in Katanga, Democratic Republic of Congo. Öko-Institut, Freiburg, p 75. https://www.oeko.de/en/publications/social-impacts-of-artisanal-cobalt-mining-in-katanga-democratic-republic-of-congo/. Accessed 14 Mar 2024

Umpula E, Van Woudenberg A (2024) Toxic pollution from industrial mines challenges clean energy claims. UK corporate watchdog RAID and DRC-based African Resources Watch (AFREWATCH). https://www.afrewatch.com. Accessed 19 Feb 2024

UN-Water (2011) Policy brief on water quality. Water quality challenges and impacts, Geneva, Switzerland, p 22. https://www.unwater.org/publications/un-water-policy-brief-water-quality. Accessed 17 Nov 2024

UN-Water Decade Program (2015) Water and health: information brief. Action on water, sanitation and hygiene (WASH), Zaragoza, Spain, p 6. https://www.unwater.org/water-facts/wash-water-sanitation-and-hygiene. Accessed 17 Nov 2024

UNICEF (2019) Silent suffocation in Africa: air pollution is a growing menace, affecting the poorest children the most, New York, US, p 14. https://www.unicef.org/reports/silent-suffocation-in-africa-air-pollution-2019. Accessed 22 Oct 2024

UNICEF Zambia (2024) Annual Report 2024-UNICEF Zambia supported children and families through drought, disease, and difficult conditions, improving health, education and protection. Lusaka, Zambia. 18pp. https://www.unicef.org/zambia/reports/unicef-zambia-country-office-annual-report-2024. Accessed 20 Nov 2018

United Nations Development Program (UNDP) (2009) Province du Katanga: Pauvreté et Conditions de Vie des Ménages, Kinshasa, RD Congo, p 20. https://docslib.org/doc/1595711/province-du-katanga-profil-resume-pauvrete-et-conditions-de-vie-des-menages. Accessed 17 Nov 2024

United Nations Environment Program Global Environment Monitoring System (GEMS)/Water Program (2008) Water quality for ecosystem and human health, 2nd edn, Ontario, Canada, p 130. https://www.unep.org/resources/report/water-quality-ecosystem-and-human-health. Accessed 18 Mar 2024

US Agency for International Development (USAID)/Sustainable Water Partnership (2021) Democratic Republic of the Congo water resources: profile overview, Washington, DC, USA, p 11. https://www.globalwaters.org/resources/assets/democratic-republic-congo-water-resources-profile. Accessed 17 Nov 2024

USDA (2016) Food composition databases of nutrient database for standard reference. https://ndb.nal.usda.gov/ndb/search/list. Accessed 18 Dec 2018

USDA Agricultural Research Service (ARS) (2008) Nutrient data laboratory. https://www.ars.usda.gov/northeast-area/beltsville-md/beltsville-human-nutrition-research-center/nutrient-data-laboratory/. Accessed 17 Nov 2024

USDA National Nutrient Database for Standard Reference, Legacy (2018). https://ndb.nal.usda.gov/ndb/. Accessed 18 Dec 2018

Vantomme P, Göhler D, N'deckere-Ziangba F (2004) Contribution of forest insects to food security and forest conservation: the example of caterpillars in the sub-region of the Congo Basin (Cameroon, Central African Republic, Congo Brazzaville and Democratic Republic of Congo). Wildlife Policy Briefing 3:1–4. https://odi.org/en/publications/contribution-of-forest-insects-to-food-security-and-forest-conservation-the-example-of-caterpillars-in-central-africa/. Accessed 12 Jan 2007

Vörösmarty CJ, McIntyre PB, Gessner MO et al (2010) Global threats to human water security and river biodiversity. Nature 467:555–561. https://doi.org/10.1038/nature09440

Waddington SR, Sakale WD, Mekuria M (2004) Progress in lifting soil fertility in Southern Africa. In: Proceedings of the 4th international crop science congress, 26 September-1 October 2004, Brisbane, Australia. https://www.cropscience.org.au. Accessed 12 Jan 2007

Wang, D., Fu, X., Zhang, J., Xu, C., Hu, Q. & W. Lin, 1992. Association between blood lead level during pregnancy and birth weight: A meta-analysis. Am. J. Ind. Med. 63(12): 1085–1094. https://doi.org/10.1002/ajim.23187

Wang D, Fu X, Zhang J, Xu C, Hu Q, Lin W (2020) Association between blood lead level during pregnancy and birth weight: a meta-analysis. Am J Ind Med 63(12):1085–1094. https://doi.org/10.1002/ajim.23187

WHO (1992) IPCS, cadmium. Environmental health criteria, p 134. World Health Organization, International Program on Chemical Safety, Geneva. http://www.inchem.org/documents/ehc/ehc/ehc134.htm. Accessed 4 Sept 2007

WHO (1995) IPCS, inorganic lead. Environmental health criteria, p 165. World Health Organization, International Program on Chemical Safety, Geneva. https://www.who.int/teams/environment-climate-change-and-health/chemical-safety-and-health/health-impacts/chemicals/lead. Accessed 22 Nov 2024

WHO (2003) The right to water. WHO Library Cataloguing-in-Publication Data, Geneva, p 44. https://iris.who.int/handle/10665/42661. Accessed 22 Nov 2024

WHO (2017a) Guidelines for drinking-water quality: fourth edition incorporating the first addendum. Geneva: Licence: CC BY-NC-SA 3.0 IGO 631. https://www.who.int/publications/i/item/9789241549950. Accessed 2 Sept 2020

WHO (2017b) Guidelines for drinking-water quality: fourth edition incorporating the first addendum. World Health Organization, Geneva. Licence: CC BY-NC-SA 3.0 IGO 631. https://www.who.int/publications/i/item/9789241549950. Accessed 22 Nov 2024

WHO (2021) Global air quality guidelines. Particulate matter (PM2.5 and PM10), ozone, nitrogen dioxide, sulfur dioxide and carbon monoxide, Geneva, Switzerland, p 300. https://www.who.int/publications/i/item/9789240034228. Accessed 22 Nov 2024

Wilburn DR (2011) Cobalt mineral exploration and supply from 1995 through 2013: U.S. Geological survey scientific investigations report 2011–5084, p 16. https://pubs.usgs.gov/sir/2011/5084/. Accessed 12 Jan 2007

Wilkin P (2001) Dioscoreaceae of South-Central Africa. Kew Bull 56:361–404. https://doi.org/10.2307/4110963

Wilson KB (1990) Biological dynamics and human welfare: a case study of population, health and nutrition in Southern Zimbabwe. Dissertation, University College London

Wolfire D, Brunner J, Sizer N et al (1998) Forests and the Democratic Republic of Congo: opportunity in a time of crisis. World Resources Institute, Washington, DC, p 30. https://www.wri.org/research/forests-and-democratic-republic-congo. Accessed 12 Jan 2007

World Bank (2015) Zambia Copperbelt Environment Project (CEP). World Bank, Washington, DC, p 97. http://documents.worldbank.org/curated/en/615451468170065644/pdf/ICR19040P070960sclosed0Dec027020110.pdf. Accessed 17 Mar 2016

World Bank (2015a) Zambia Copperbelt Environment Project (CEP). Washington, DC, USA. 143pp. https://projects.worldbank.org/en/projects-operations/project-detail/P070962

World Bank (2015b) Zambia Copperbelt Environment Project (CEP). Washington, DC, USA, p 97. http://documents.worldbank.org/curated/en/615451468170065644/pdf/ICR19040P070960sclosed0Dec027020110.pdf. Accessed 17 Mar 2016

World Bank (2018) Zambia—systematic country diagnostic (SCD). Report no. 124032-ZM. World Bank, Washington, DC, p 131. http://documents.worldbank.org/curated/en/290011522954283481/Zambia-Systematic-Country-Diagnostic. Accessed 19 Jun 2024

World Bank (2020) Worldwide governance indicators. http://info.worldbank.org/governance/wgi/Home/Reports. Accessed 5 Feb 2021

World Wide Fund for Nature (WWF) (2016) Water in the Zambian economy: exploring shared risks and opportunities in the Kafue Flats. 2nd ed. 64pp. Gland, Switzerland. https://www.panda.org. Accessed 26 Oct 2025

World Wildlife Fund (WWF) (2020) In: Almond REA, Grooten M, Petersen T (eds) Living planet report 2020—bending the curve of biodiversity loss. WWF, Gland, p 83. https://www.worldwildlife.org/publications/living-planet-report-2020. Accessed 12 Jan 2024

Yabe J, Nakayama SMM, Ikenaka Y et al (2015) Lead poisoning in children from townships in the vicinity of a lead-zinc mine in Kabwe, Zambia. Chemosphere 119:941–947. https://doi.org/10.1016/j.chemosphere.2014.09.028

Yager TR (2012) The Mineral Industry of Congo (Kinshasa). 2012 Minerals Yearbook. U.S. Department of the Interior, U.S. Geological Survey for Congo (Kinshasa). 11pp. https://www.usgs.gov/centers/national-minerals-information-center/minerals-yearbookmetals-and-minerals

Yager TR (2014) The mineral industry of Congo (Kinshasa). Minerals. In: v. III. Area reports: international—Africa and the Middle East. U.S. geological survey yearbook, pp 11.1–11.10. http://minerals.usgs.gov/minerals/pubs/country/africa.html#cg. Accessed 22 Feb 2018

Yearbook rapport économique (2023) République Démocratique du Congo 2023-2024 - Vers l'émergence à l'horizon 2040. En partenariat avec l'Agence Nationale pour la Promotion des Investissements (ANAPI), p 114. http://www.dev.investindrc.cd/fr/ANAPI. Accessed 22 Feb 2018

Zambia Consolidated Copper Mines-IH (ZCCM-IH) (2005) Preparation of Phase 2 of a Consolidated Environmental Management Plan-Project, Zambia Consolidated Copper Mines-IH (ZCCM-IH), Kitwe, Zambia. https://www.zccm-ih.com.zm/. Accessed 7 Jul 2011

Zambia Consolidated Copper Mines-IH (ZCCM-IH) (2005) Preparation of phase 2 of a consolidated environmental management plan-project. Summary report, Kitwe, Zambia, p 110

Zambia EITI Report (2020) Zambia Extractive Industries Transparency Initiative. Centre Urbain Nord, Tunis, p 168. https://eiti.org/documents/zambia-2020-eiti-report. Accessed 12 Mar 2016

Zambia Environmental Management Agency (ZEMA) (2013) Limits for effluent and waste water. Regulatory 7 (2). Statutory Instruments, Third Schedule. https://www.zema.org.zm/environmental_management_act/

ZAWA (2015) Country Report for the Tenth Session of Parties (COP-10) to Ramsar Convention on Wetlands of International Importance, Uruguay.

ZEMA (2011) The Environmental Management Act (EMA) No. 12 of 2011. Ministry of Legal Affaires, Government of Zambia 2011. Government Printer. Lusaka, Zambia, p 177. http://www.zema.org.zm/index.php/environmentallegislation

ZEMA (2017a) Zambia environment outlook report 4 (ZEO-4). Zambia Environmental Management Agency, Lusaka, p 142. http://www.zema.org.zm/index.php/download/zambia-environmental-outlook-report-4-zeo-4/. Accessed 17 May 2020

ZEMA (2017b) Zambia environment outlook report 4 (ZEO-4). Zambia Environmental Management Agency, Lusaka, p 142. https://www.zema.org.zm/docs/zambia-environment-outlook-report-4/. Accessed 24 Jun 2024

Index

A. Muimba-Kankolongo et al., *Mining Impacts on the Environment in the Central African Copperbelt of Zambia and the Democratic Republic of Congo*, https://doi.org/10.1007/978-3-031-96786-3

MIX
Papier aus verantwortungsvollen Quellen
Paper from responsible sources
FSC® C105338

If you have any concerns about our products,
you can contact us on
ProductSafety@springernature.com

In case Publisher is established outside the EU,
the EU authorized representative is:
Springer Nature Customer Service Center GmbH
Europaplatz 3, 69115 Heidelberg, Germany

Printed by Libri Plureos GmbH
in Hamburg, Germany